ISBN 978-1-334-41363-6
PIBN 10700811

This book is a reproduction of an important historical work. Forgotten Books uses
state-of-the-art technology to digitally reconstruct the work, preserving the original format
whilst repairing imperfections present in the aged copy. In rare cases, an imperfection in
the original, such as a blemish or missing page, may be replicated in our edition. We do,
however, repair the vast majority of imperfections successfully; any imperfections that
remain are intentionally left to preserve the state of such historical works.

THE

JOURNAL OF MICROSCOPY

AND

NATURAL SCIENCE:

THE JOURNAL OF

THE POSTAL MICROSCOPICAL SOCIETY.

Editor:

ALFRED ALLEN,

Hon. Sec. P.M.S.

Associate Editors:

PROF. **V.** A. LATHAM, M.D., D.D.S., F.R.M.S., ETC.,
Chicago University, U.S.A.;

J. STEVENSON BROWN, *President Montreal Micro. Soc.,*
Montreal, Canada;

FILANDRO VICENTINI, M.D., *Chieti, Italy.*

VOL. V. THIRD SERIES.
VOL. XIV. OLD SERIES.

London:

BAILLIERE, TINDALL, & COX, 20 KING WILLIAM ST., STRAND.
SIMPKIN, MARSHALL, HAMILTON, KENT, & Co.; LIMITED.

Bristol:

Preface.

N many ways the completion of the present volume of the *International Journal of Microscopy and Natural Science* compels us to say a few words respecting it.

On glancing through the volume, we cannot but feel a just pride in the variety and general excellence of the many papers and articles it contains. Amongst the important contributions, the continuation of the translation of Dr. F. Vicentini's memoir on the "Bacteria of the Sputa and Cryptogamic Flora of the Mouth," which is completed in the present part (October), deserves especial mention. Indeed, so well has this memoir been received, that we intend to offer it shortly in book form ; a preface to the work has been written by Prof. W. D. Miller of Berlin, and Dr. Vicentini has appended a bibliography and other important additions. The continuation of Mr. H. C. Vine's valuable papers on the "Predacious and Parasitic Enemies of Aphides" has been of great interest, and our best thanks are due to Mr. Vine for his contributions. Our thanks are also due to Mr. C. D. Soar for his papers on the "British Hydrachnidæ," to Mr. W. Thomson for his papers on "The Influence of Light on Life," and "On the Study of the Micro Fungi," to our past-president, Mr. G. H. Bryan, for his lucid article on "Benham's 'Artificial Spectrum Top,'" to Mr. C. J. Watkins for his interesting contribution on "The Denizens of an Old Cherry Tree," to Prof. V. A. Latham for the valuable account of the "Methods and Formulæ used in Examining Blood." The Presidential Addresses of Dr. H. L. Browne, Dr. J. Hall, the Rev. E. T. Stubbs, and Dr. J. S. Turner also deserve mention. Lastly, we must offer our best thanks for the kind assistance we have received from so many of our Subscribers.

THE INTERNATIONAL
JOURNAL OF MICROSCOPY & NATURAL SCIENCE

THE JOURNAL OF THE POSTAL MICROSCOPICAL SOCIETY.

"Knowledge is not given us to keep, but to impart; its worth is lost in concealment."

[The Editor does not hold himself responsible for the views of the authors of the papers published.]

The Denizens of an Old Cherry Tree, With Notes of its Surroundings.

By C. J. WATKINS.   Plates I. & II.

MORE than twenty years ago this cherry tree—then approaching its prime—was a convenient place to hang the saccharine snares that allured the sweet-loving Lepidoptera haunting our garden, which is on a strip of Liassic clay that forms one bank of the little river which winds its course along the Cotteswold Valley where we have long resided.

Standing in our garden and looking up stream we observe to the right a hilly ridge, about 700 feet above sea level, and crowned with a wood, famous for the botanical rarities found there and on the surrounding hills by Oade Roberts* and his young disciple, the late Edward Newman (to whom we owe so much for his valued labours in British Entomology), who searched these hills and dales for floral and insect treasures, and who, in his later years, in one of his letters to the writer, wished to know whether the scenes of his boyhood were changed.   Alas, they are !   The noble old beech woods of our hill-sides are fast

* See Withering's *British Plants;* 7th Edition, 1830.

disappearing, and the woodman's axe ceases not in its yearly slaughter. Few, very few, of the old forest giants spread their sturdy limbs, where for centuries they stood the rude shocks of countless gales, and sheltered from the heat and storm our forefathers and their herds. The owners of the soil care not about replacing the famous fallen with young stock of the same race. They are eager to reap in their time, so the quicker-growing spruce and larch now stand where the grand old trees once lifted their heads to the sky.

If our valley has been rich in floral treasures, we can point to far more ancient stores of Nature in the wonderful Oolitic coral reef on which the old wood we have mentioned stands. This reef extends for miles, and we have traced it on the hills which form the other boundary of our valley, telling us of pre-historic times, when the old ocean beat its restless waves against the coral shores of this inland sea, and we can imagine our hill tops bearing those curious trees and plants peculiar to that different climate and epoch. Even then a few kinds of insects (chiefly Neuroptera) sported over the waters, while the Ammonites floated on the surface of the waves, and the terrible Plesiosaurus, like a gigantic swan, searched for its prey on the incoming tide.

During this changing period the rocks forming our hills were deposited in seas rich with microscopic life, whose tiny chambered shells (Foraminifera) form a considerable portion of these vast Oolitic limestone strata which yields to-day the famous building freestone of this district, and the coral rag stone used for metalling our roads and for building walls, in which we have admired many a choice example of those wonderful organisms that once spread their beautiful tentacles in the vast seas and lagoons on which the eye of man never gazed !

> "Frail were their frames, ephemeral their lives,
> Their masonry imperishable."

To return to the present appearance of the landscape, as seen from the bottom of our valley and looking towards the sources of our river, we observe a hill like a promontory standing out in relief, at whose base, the two streams forming the sources of the river, join in one where three valleys meet, their sides ascending and forming a series of hills, over which we have, in bygone days, spent many a pleasant hour, hunting for insects in their native haunts, and, in the proper season, giving chase to that rarity of our native butterflies—now, we fear, nearly extinct—the Large Blue (*Lycæna arion*), king among its brethren, born amid the bee-haunted wild thyme, whose opening flowers told us the time to look for this coveted prize in its perfect state. To capture a specimen sailing along before the stiff breeze (which often prevails on these hills), one needed to excel in running to keep the insect in view, especially when our chase extended over such rough places and disused quarries as usually form the haunt of this species.

But our limbs are not so strong as they were in those days, and the roads to these lovely spots are long, so we are now content to observe those insects which are to be found nearer home ; and, if the reader will bear with us, we will attempt to describe some of the marvels of insect life observed in the stump of our old cherry tree during three months of 1893.

Some years ago this cherry tree was struck by lightning, and soon after showed signs of decay. The limbs broke off during the strong gales which sometimes sweep up the valley from the Bristol Channel with terrific force ; but the stump, now only some seven feet high and about ten inches in diameter at its base, was left standing, as it formed a convenient post on which a metal clothes line had been attached for years. For the last three years signs of its being inhabited were freely shown by the numerous particles of wood scattered from time to time on a bed of marjoram growing round its base, on the leaves of which the ejected light coloured " jaw dust " from above, showed in striking contrast to the cólour of the herb, and which induced our old gardener to remark that the old stump was " wivvil e'ten and dakadey." We had many times watched to see whether any insect passed in or out of the holes from which the dust had fallen, but without success until October 23rd, 1892, when at mid-day, in the sunlight, we saw a black wood wasp (*Pemphredon*) actively examining the mouths of the larger burrows. Wishing to secure it for identification, we managed to get the agile creature into a glass tube, but, accidentally dropping the cork, our prisoner was gone in an instant.

Now, knowing what to look for, we waited and watched, and within ten minutes we were rewarded with another visit from this determined little creature, evidently bent on some maternal cares in the interior of the stump. Selecting a hole, it entered, but when it emerged it was to enter again the glass prison from whence it was transferred to the killing bottle, and now graces our collection as one of the first specimens of *Pemphredon lugubris* we had taken, although specimens of this genus had been observed, in previous seasons, during the hottest sunshine, on the leaves of our fruit trees, moving with that wonderful activity of wing so characteristic of the Fossorial Hymenoptera, to which these highly specialised insects belong.

We now desired to remove the stump for careful examination indoors during the winter, but business matters and ill-health prevented the carrying out of our intention. However, during March, 1893, a high wind broke off the upper portion of the trunk, which we removed indoors, and its examination brought to light such a numerous assemblage of burrows, galleries, and tunnels, occupied by different insects in various stages of their life histories, that, during April, we had the stump cut down, and, after sawing it into suitable blocks, we carefully divided each block into portions with a strong pocket knife, and these portions, with their living inmates, were put into a series of large glass-topped boxes, very special pieces being stored in smaller boxes. The careful cutting up of this rotten stump occupied the writer, for several hours each day, for more than a week, but the fascinating pleasure of unveiling these insect marvels was ample reward for any pains that may have been taken in doing so.

The first inhabitants to attract our attention were several Coleopterous larvæ, in unstained tortuous burrows of from one-eighth to a quarter of an inch in diameter. Fig. 2, Plate I., shows a portion of the rotten cherry-wood with a burrow containing a full-grown larva, and near it an active nymph or pupal form of the same insect, while above the wood is seen the perfect insect, a blue-black beetle (*Melandrya caraboides*). The three stages are drawn from nature and of the natural size, although (as in the case of most wood feeders) the species is very variable in size, specimens of the imago differing in length from seven-sixteenths to over three-quarters of an inch. When first emerged from its pupal form, this beetle is of a very light brown colour, and several hours elapse ere the rich, deep, normal shade of blue-black finally appears. Its first appearance forms such a strange contrast to that of the more matured insect, that we killed and set a newly-emerged specimen for our collection. The perfect insects made their appearance in our boxes at various intervals from April 27th to May 8th, on which latter date we found several half-grown larvæ in the borings. We frequently observed the imago with its head projecting from the mouth of the burrow made by the larvæ, into which it rapidly backed on our attempting to capture it. These beetles are quick runners, and appear to emerge from the burrows only after dark. The general hue of the larva is cream colour, with pale-brown head. The jaws are darker and of a horny nature, admirably adapted for eating its way into the wood, and at the same time forming the burrows needed for its protection then and also in its future stages. Fig. 1, Pl. II., shows magnified dorsal view of the jaws of the larva.

The three pairs of legs, which are situated on the three segments next the head of the larva, enable it to move along its burrow, and, combined with the sudden contraction of the segments, to perform the curious feat of repeatedly turning quite over sideways, which feat we witnessed by placing several larvæ on a level surface. When put into a glass tube, they could in this manner move along it, from which we infer that the larva not only propels itself corkscrew fashion, but removes by the same means the accumulating wood refuse of its labours. The larval stage appears to continue for some considerable time, probably for more than a year, but the pupal form seems to end in a few weeks. In this stage it can move along the burrow by means of its angular spine-studded segments and the two spine-tipped processes on the lower side of the last segment. These burrows were more or less filled with the wood particles evacuated by the larvæ, whose sole pabulum appears to be the rotten wood. From this strange food Nature provides nutriment and builds up the insect through its wonderful transformations.

We observed no insect parasites in the burrows of this beetle, but on May 31st, on raising a block of the cherry wood containing borings occupied by the young larvæ of this beetle, we found a female specimen of *Pompilus spissus* at rest on the wood, as if only lately emerged from the borings near it. This exceedingly active and predacious insect belongs to a genus whose members chiefly burrow in sand and generally prey on spiders. Mr. Edward Saunders, who kindly identified the Aculeates bred from our cherry stump, referring to this species, says :—

"*Pompilus spissus* out of the stump is very interesting. I wonder if it always inhabits such localities. I can find no actual account of its habits, but always imagined that it nested in banks, sands, etc., as its allies. It, however, lacks the comb of spines on the front tarsi of its female, which are probably for digging purposes (as most of the sand species possess them), so possibly it may be a wood frequenter; but it is quite unlike the other *Pompili* if it is. *P. niger* and *P. minutulus* are also combless; possibly their habits are similar."

Of other beetles we found dead specimens of *Calathus cisteloides* and *Sinodendron cylindricum*, while the living imagines of the following species were taken in April from the cherry bark when cutting open the blocks, viz. :—*Cis boleti, C. hispidus, C. nitidus, Anaspis frontalis,* and *A. fasciata.*

The next inmates to claim our attention were the nymphs and perfect forms of the solitary black wasps of the genus *Pemphredon,* of which two species, *P. lugubris* (previously mentioned) and *P. Shuckardi,* appeared in our boxes in the perfect state from April 29th to the middle of May, *P. lugubris* being the most numerous of any of the denizens of the stump. Its crooked and stained burrows running in all directions, we sometimes met with a long winding boring ending in several short branches, as in Fig. 3, Pl. I. In one of these short branch borings will be seen the legless larva; in another gallery will be found the nymph or pupal form of *P. lugubris* (this genus makes no cocoon), while the imago, a female, is represented resting on the wood near. These are all drawn natural size.

In our examination of the burrows of this species in April, the larval stage was nearly over, as only two of the pale yellow, helpless larvæ were found, and these in a few days passed into the nymph stage, which, compared to the larval life, is of short duration. The first imagines to appear were chiefly males, and out of eighty specimens bred the proportion of males to females was ten to one. A male emerged with one pair of wings only and no abdomen; the fly, however, was very active and lived for several hours, and it was the only cripple observed.

The numerous borings of this industrious insect prove a subject of great interest. They are not formed by the grub for the sake of its food, as in the case of the beetle previously described; but the mother *Pemphredon*, with her cutting and toothed mandibles (Fig. 2, Pl. II.), gnaws away the wood, and removes it with great dexterity with her legs, which are armed with spines that act like rakes (Fig. 5, Pl. II.), thus toiling on alone until the required burrow is ready to receive the fruits of her labours, and form a home for her progeny which she is destined never to see ! She now sallies forth on a hunting expedition among the Aphides, which are captured, stung, and paralysed, and let us hope rendered insensible. In this condition the prey is brought home and closely packed at the end of one of the burrows in sufficient quantity to provide food for one of her family while in its larval state. An egg is now laid in this larder,* and over all sufficient of the wood *debris* is packed to form a partition, against which another larder is stocked and another egg deposited, the process being repeated until her egg-laying powers are exhausted, and now, her mission being accomplished, she dies as the summer fades ; while in their separate retreats her tiny progeny emerge from their egg-cradles to find themselves surrounded with food, which, owing to the prescient action of the parent, is preserved from decay until the larvæ are full grown, and no longer require a carnivorous pabulum. In the perfect state these Fossores feed on the nectar of flowers and the sap of plants—a most interesting fact, for here we have a herbivorous insect hunting and providing food for its carnivorous young ! In the larval stage *P. lugubris* is subject to the attack of the larvæ of an Ichneumon fly (*Perithous varius*, Fig. 14, Pl. II.), of which three specimens emerged from these borings ; and we can only infer that the parent Ichneumon gained access to the burrow while its occupants were in the larval stage, when it deposited its egg near or in its future prey.

* The remains of these larders consist chiefly of legs, heads, cornicles, wings, and the extremities of antennæ of aphides, some of which were sent to Mr. H. C. A. Vine for identification, who kindly reported the genus to be *Siphonophora*, and the species probably *S. rosæ* and *S. pisi*.

Another parasite of *P. lugubris*, nine specimens of which appeared in our boxes between April 27th and June 1st, was a brilliant member of the Golden Wasp tribe (*Chrysididæ*), viz., *Omalus auratus* (Fig. 10, Plate II.), resembling, in splendour of colours, the well-known ruby-tail fly (*Chrysis ignita*), parasitic in nests of mud-wasps (*Odynerus*), etc., but only about half the size of that species. The *Chrysididæ* are, in every sense, Cuckoo flies, for they make no nests, but lay their eggs in those of other Hymenoptera, especially in those where provision has been stored for the larvæ. The little parasitic larva is hatched, and, it is said, feeds in the larder with the larva of the original tenant, and when the store of food is consumed it finishes up, not by ejecting its foster-brother from the nest as the Cuckoo is said to do, but turns on its companion, and sucks its juices. In the life-history of our *Omalus* we had proof that this was the case, as, in each instance, we found the pupal case of *Omalus* (Fig. 13, Plate II.) in the burrow of *Pemphredon*, and close to the remains of a larder, showing the Aphides to have been eaten as well as the Pemphredon larva.

In speaking of a species of Cuckoo-fly (*Hedychrum*) allied to *Omalus*, a French naturalist said he observed it enter the dwelling of a Sand-bee in order to lay its egg. The intruder was discovered by the bee, who at once proceeded to turn out its enemy by laying hold of her with its mandibles. But the Hedychrum rolled herself up like a ball, and was invulnerable. The bee carried her out, gave her a good shaking, bit her wings off, and left her. She, however, had her way for all that, and crawled back again into the nest, and laid her egg. From the structure of the tarsi and serrated claws of *Omalus* (Fig. 11, Plate II.), we imagine our Cuckoo fly quite equal to the task, even without wings, of returning to the borings of *Pemphredon*, after having been summarily ejected by the enraged fossore. From the burrows of *Pemphredon* appeared several species of Diptera, and taking them in the order of Mr. Verrall's British list, we have first three species of the family *Mycetophilidæ*, or "Fungus gnats" of the genus *Sciaria*, the imagines of which emerged from April to June, about forty specimens being observed. The larvæ of these species appear to feed on the mouldy contents of the larders of the fossores, especially where the store food had not, from some cause, been eaten by the larva for whom it was intended.

On April 15th, in an untouched larder of *Crabro* (a genus of fossores allied to *Pemphredon*), we found two larvæ of *Sciaria* feeding on mouldy imagines of *Rhingia rostrata*—the snout fly. These small thread-like worms turned to curious horned pupæ by May 7th, and the resulting flies appeared in about three weeks, showing this small dipteron to be parasitic on other and much larger diptera, and possibly on Aphides, etc., in other fossorial larders.

Continuing the list of Diptera from borings of *Pemphredon*, we find the family *Chironomidæ* represented by a species of *Ceratopogon*, whose pupæ we found on June 2nd, and flies of both sexes emerged a week later. We did not find the larva of this tiny fly, but Mr. Theobald, in his *British Flies*, says of the larvæ of this genus, " that they inhabit manures, and others are found under the bark of decaying trees." Some of the females of this genus are blood suckers, and in early summer are a troublesome pest to persons having tender skin, alighting on one's brow, neck, or wrists, causing the most intense itching, and often severe inflammation in the parts attacked. A gnat fly in comparison of size is quite large to this minute, but terrible insect.

The next species of Diptera are two female specimens of *Trichomyia*, bred about the end of May, following which is a dipteron new to the British list, *Brachycoma erratica*, Mgn., of the family *Tachinidæ*, kindly identified, with the other Diptera of our cherry stump, by Dr. Meade, of Bradford, who has fully described this new species on page 110 in the May number of *The Entomologist's Monthly Magazine*, 1894. During May, 1893, we bred two specimens (male and female) of this species from pupæ found in the burrows of *Pemphredon*—the male appearing on May 10th. Roughly speaking, this fly is much darker, but similar in general appearance to a small specimen of the house fly (*Musca domestica*).

Of the large family *Anthomyidæ* our stump produced two examples, viz., one female *Hyetodesia errans* on May 13th, and several specimens of the well marked, striking species *Hylemyia festiva*, Zett. This last species is of rare occurrence in England. It is described by Dr. Meade at page 222 *Entomologist's Monthly Magazine*, October, 1893, and on page 285 of the same magazine for December, 1893, he records the bred specimens above, the pupæ of which were found in April in burrows of *Pemphredon lugubris*, the flies, of both sexes, emerging during April and May. In each case, the pupa of *H. festiva* was found close to the aphidal remains of the *Pemphredon* larder, and in two instances a pupa was adjacent to a pupal case of *Omalus*, showing the parasitical dipteron connected in its economy with the parasitic hymenopteron. From the same burrows of *Pemphredon*, during April and May, appeared three specimens of *Lonchæa vaginalis*, the last bred dipteron of our list.

We now come to the four remaining fossorial hymenoptera of our cherry stump, all species of the extensive genus *Crabro*, whose predaceous members provision their nests with Diptera. This genus may always be determined by the constant neuration of the wing (see Fig. 8, Pl. II.). The species have been tabulated by Mr. Edward Saunders in *The Transactions of the Entomological Society*, 1880, pp. 280–281; also in his chief work on *The Hymenoptera Aculeata of the British Isles*, now being issued. Fig. 1, Pl. I., shows a male of *Crabro leucostomus*, a black shining insect, and also four of the tough, sepia-coloured cocoons spun by its larvæ, from one of which the *Crabro* has emerged; these are all drawn natural size. The female is larger than the male, and, from more than forty cocoons observed in our boxes, only seven perfect imagines appeared during May, all of which were males. In the burrow, each cocoon is in the position previously occupied by the stores of the larder, some of the remains of which are usually seen attached to one end of it, consisting chiefly of the harder portions of the pretty green, metallic-looking dipteron, *Microchrysa polita*, sometimes mixed with fragments of a brown and yellow species of *Melanostoma*, probably *M. mellinum*, a genus of the family *Syrphidæ*. As this species was in the pupal stage when the cherry wood was first examined in April, we did not find the larva. In the larval stage this *Crabro* is subject to the attacks of the larva of a small hymenopteron, *Pteromalus apum*, shown magnified (Fig. 9, Pl. II.). It is parasitic on many bees and wasps, and is a member of the large family *Chalcididæ*, of which over one thousand British species have been recorded. Each larva of *Crabro leucostomus* attacked by this little fly is destined to support several specimens of its larval enemy, which do not actually kill it until its cocoon is spun, in which, after making a last meal of their host, they turn to pupæ. Those under observation appeared in the perfect state in June, from six to nine flies emerging from one cocoon of their host, while from one cocoon of a larger *Crabro—C. cephalotes*— we took twenty-seven naked pupæ of this parasite.

Another parasite of *Crabro leucostoma* is an Ichneumon, *Phygadeuon gravipes*, of which two females emerged from the burrows in June. The wings of this dark brown fly almost equal in expanse those of its host, and its size indicates that the *Crabro* cocoon is occupied by only one of its enemy.* Our remaining species of *Crabro* have the abdomen banded more or less with black and yellow, and are easily mistaken by the non-entomologist for small wasps, which they resemble, not only in shape and colour, but in the possession of a powerful sting, so that care should be exercised in the examination of living specimens. Like other members of the genus they store their nests with living Diptera, packed together in a state of torpor, and in their general economy they appear to be similar to each other. The first of these (*Crabro vagus*) is shown enlarged at Fig. 5, Pl. I., of which thirteen specimens—chiefly males—appeared in June. It may be distinguished from the thirty British species of the genus by the entirely black third segment of the abdomen. Its cocoon is the same shape, but a trifle smaller, than that of *Crabro chrysostomus* (Fig. 4, Pl. I.), and its brown colour is a shade darker than the cocoon of that species. So far as we have examined the larder remains of this *Crabro*, the stored Diptera appear to belong to the genera *Musca* and *Pollenia*, but it may bring home prey of other allied genera, as in an untouched larder of a *Crabro*, which we inferred was made by *C. vagus*, but which did not produce a *Crabro*, owing to the failure or non-deposition of the egg, we found two whole male specimens of *Musca corvina*, and one whole female *Stomoxys calcitrans*.

From the burrows of this *Crabro* a female specimen of *Anthophora furcata* (Fig. 6, Pl. I.) emerged in June—the only bee met with in the stump. Mr. Saunders, in speaking of this species,

---

* We are indebted to Mr. T. R. Billups for his kind identification of the Parasitic Hymenoptera of our stump.

says :—" How *A. furcata* came to exist as a single female in the stump is curious. One would have thought that more examples would have been in the same burrow; possibly, however, these died." After all the specimens of *C. vagus* had appeared, we carefully examined the pieces of cherry wood in which it had burrowed, but no other example of the bee in any stage could be found. The gallery of the burrow from which the bee had emerged had been much enlarged, and contained some wood *debris*, of much finer particles than that in the other portions of the *Crabro* burrow, from which we infer that the parent bee had cleaned out, enlarged, and utilised for its nidus, a portion of the *Crabro* home.

We now come to the most numerous, but one of the varied, inhabitants of the stump, viz., *Crabro chrysostomus* (Fig. 4, Pl. I.), where a female is drawn natural size with three of the vandyke-brown cocoons *in situ* in the burrows, with their anal ends resting on and partially. attached to the larder remains, which appear to be chiefly specimens of the genus *Platychirus*. An untouched larder in a portion of wood, consisting of several borings, with cocoons of this *Crabro*, contained eight specimens of *Platychirus fulviventris*. Of this interesting fossore we bred sixty-four males and twelve females during May and June. The males, when handled or touched, exhale a rich odour. The female can emit the scent, but in a much less degree. This agreeable perfume we have noticed when examining certain species of bees (*Halicti*), but have not met with it in connection with other species of *Crabro*. *C. chrysostomus* is subject to the attacks of the parasitic hymenopteron *Pteromalus apum*, mentioned previously in connection with *C. leucostomus*.

The last inmate of our stump to notice is *Crabro cephalotes* (Fig. 6, Plate II.), the finest fossore of our series, and of which ten males and four females made their exit from the cocoons. The cocoon is shaped like *C. chrysostomus*, and of the same colour, but rather larger. Fig. 4, Pl. II., shows a pupa removed from its cocoon. The burrows of this species contain the remains, and sometimes whole specimens of the largest dipterous prey found in the stump. In one nidus we found three whole specimens of *Rhingia rostrata*, and in another burrow were examples of *Syrphus ribesii*, insects fully half as large as their captor, and of strong flight. The dexterity and power of the *Crabro* in securing and bringing home such living prey is obvious, for, alone and unassisted, the burdened insect unerringly flies home through the brilliant sunlight, and, aided by a sense to us unknown, deposits her burden in a chamber where the light scarcely, or never, enters. This fine *Crabro* is also subject to the attack of the little parasite *Pteromalus apum*, and, as noted elsewhere, we found in one of its cocoons twenty-seven pupæ of its enemy, while another cocoon examined contained seventeen living specimens.

The marvellous industry displayed by the female fossores in forming their varied burrows and galleries of each nidus is a subject of the greatest interest, and when we examine the organs and limbs of these active creatures, our wonder is increased at their beautiful adaptation to the functions performed. We cannot but admire the suitability of such an instrument as the mandible of a *Crabro* (Fig. 3, Pl. II.) for reducing the wood into fragments suitable for removal by the rakes and brushes of the leg (Fig. 7, Pl. II), and, did time and space permit, we should like to point out many other interesting features in the economy of these creatures. The appended list of insects found in this rotten stump is not exhaustive, and its Fauna could have been much extended with notices of the Crustaceans, Myriapods, and Arachnoidea, met with in the mined wood; while, to a student of Fungi, several, and perhaps some uncommon forms, especially in connection with the fossorial stores, could have been observed. One species, at least, appears to form part of the pabulum of the larvæ of the dipterous genus *Sciara*.

If by this brief and imperfect notice of the denizens of our cherry stump we may arouse the interest of any reader to undertake the investigation of some of the hidden marvels of the insect world, our labour of love—performed under many difficulties—will not have been in vain.

> "There are deep thoughts of tranquil joy,
> For those who *thus* their minds employ;
> And trace the wise design that lurks
> In holy Nature's meanest works."

o a Cherry Tree Stump
During April, May, & June, 1893,

By CHARLES J. WATKINS, PAINSWICK, GLOUCESTERSHIRE.

Family or Group.	Genus and Species.	Date, etc.
COLEOPTERA.		
CARABIDÆ ...	Calathus cisteloides, *Illiger* ...	One specimen found dead in burrows.
LUCANIDÆ ...	Sinodendron cylindricum, *Linne* ...	One ,, ,, ,, ,,
CISSIDÆ... ...	Cis boleti, *Scopoli* ...	Twelve specimens found alive in cherry bark.
,, ...	Cis hispidus, *Paykull* ...	Two ,, ,, ,, ,,
,, ...	Cis nitidus, *Herbst* ...	One ,, ,, ,, ,,
MELANDRYDÆ ...	Melandrya caraboides, *Linné* ...	Thirty-three bred from larvæ, April and May.
MORDELLIDÆ ...	Anaspis fasciata, *Forster* ...	One found alive in cherry bark.
,, ...	Anaspis frontalis, *Linné* ...	One ,, ,, ,, ,,
ORTHOPTERA.		
FORFICULIDÆ .	Forficula auricularia, *Linné* ...	One found dead in a cell, in which it just fitted, and remains of others in other burrows.
HYMENOPTERA.		
TENTHREDINIDÆ ...	Emphytus perla, *Klug* ...	One female, April 26th, emerged from borings.
CYNIPIDÆ ...	Synergus facialis, *Dalman* ...	One female, May 15th, ,, ,, ,,
CHRYSIDIDÆ ...	Omalus Auratus, *Dahlbom* ...	Nine, May & June, from burrows of *Pemphredon.*
ICHNEUMONIDÆ ...	Phygadeuon gravipes, *Gravenhorst* ...	Two females, June, burrows of *Crabro leucostomus.*
,, ...	Perithous varius, *Gravenhorst* ...	One male and two females, May, from *Pemphredon* burrows.
CHALCIDIDÆ ...	Pteromalus apum, *Westwood* ...	Fifty specimens, May & June, fr. *Crabro* burrows.

FOSSORES—POMPILIDÆ	Pompilus spissus, *Schiödte* ...	One female, May 31.
" PEMPHREDONIDÆ	Pemphredon lugubris, *Latreille* ...	Eighty, both sexes, April and May.
" "	Pemphredon Shuckardi, *Moraw* ...	Four specimens, April and May.
" CRABRONIDÆ	Crabro leucostomus, *Linné* ...	Seven males only from forty pupæ, May.
" "	" cephalotes, *Panzer* ...	Ten males, four females, May and June.
" "	" chrysostomus, *Lep.* ...	Sixty-four males & twelve females, May & June.
" "	" vagus, *Linné* ...	Thirteen, chiefly males, June.
ANTHOPHILA—APIDÆ	Anthophora furcata, *Panzer* ...	One female, May, in burrow of *Crabro vagus.*

LEPIDOPTERA.

TINEÆ .. .		A micro-moth emerged the end of May from pupa found on the bark, & not yet identified.

HEMIPTERA.

HOMOPTERA—APHIDIDÆ	Siphhonophora (rosæ ?)	Numerous in "larders" of *Pemphredon.*
" "	" (pisi ?) ...	" "

DIPTERA.

MYCETOPHILIDÆ ...	Sciara nervosa, *Meigen*	These flies began to apppear on April 27, and continued coming out up to June, when about forty specimens had emerged. Two larvæ of *S. præcox* were found, April 15, in an untouched larder of *Crabro* feeding on a mouldy specimen of *Rhingia rostrata*; these were kept separate, and turned to pupæ end of April ; one imago emerged end of May. All these species probably feed in *Crabro* larders.
"	" nitidicollis, *Meigen* ...	
"	" (præcox, *Meigen* ?) ...	
"	" (pulcaria, *Meigen* ?) ...	

Family or Group.	Genus and Species.	Date, etc.
PSYCHODIDÆ	Trichomyia (urbica, *Haliday* ?)	Two specimens appeared, end of May.
TACHINIDÆ	Brachycoma erratica, *Meigen*	New British species, two specimens, male and female, May, from borings *Pemphredon*. Described by Dr. Meade, in May, *Entomological Monthly Magazine*, 1894, p. 110.
ANTHOMYIDÆ	Hyetodesia errans, *Fallén*	One female, May 13, from fossorial borings.
„	Hylemyia festiva, *Zetterstedt*	Rare, bred both sexes ; sixteen specimens from borings of *Pemphredon*, emerged from April 26 to May 13. Described by Dr. Meade in *Entomological Monthly Magazine*, Oct., '93, & page 285. Do., Dec., '93, records specimens above.
SAPROMYZIDÆ	Lonchea vaginalis, *Fallén*	Three females, April and May, from borings of *Pemphredon*.

The following Diptera we identified among the stored prey in partially consumed and in untouched "larders" of the four species of *Crabro*. Numerous stores still remain to be examined.

Family or Group.	Genus and Species.	Date, etc.
STRATIOMYIDÆ	Microchrysa polita, *Linné*	Both sexes, in nidus of *Crabro leucostomus*.
SYRPHIDÆ	Melanostoma (mellinum ?)	„ „ „ „
„	Pyrophaena ocymi, *F.*	Females „ „ (Sp. ?)
„	Platychirus albimanus. *F.*	„ „ „ „
„	„ fulviventris, *Mcq.*	Eight in one larder of *Crabro chrysostomus*.
„	„ clypeatus, *Mg.*	In nidus of *Crabro* (Sp. ?)
„	Syrphus balteatus, *Deg.*	One female in one larder of *Crabro cephalotes*.
„	„ ribesii, *Linné*	Two females in same larder „ „
„	Rhingia rostrata, *Linné*	Three specimens in one larder „ „
MUSCIDÆ	Pollenia (Sp. ?)	In nidus of *Crabro* (sp. ?).
„	Musca corvina, *F.*	Two males in one larder of *Crabro* (*vagus* ?).

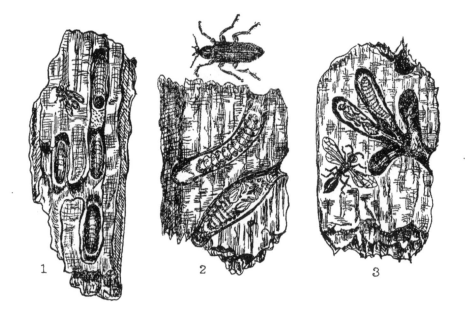

1   2   3

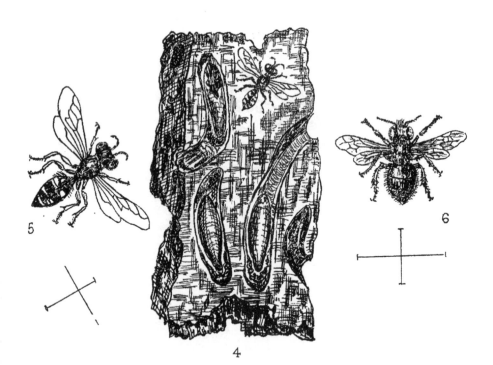

5   4   6

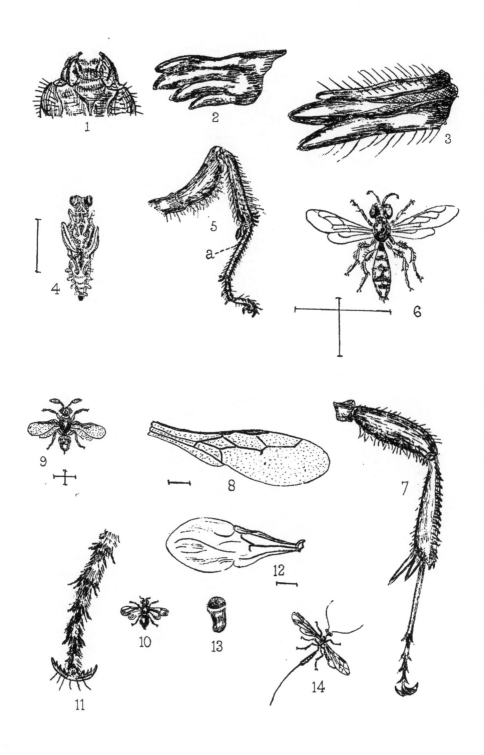

SUMMARY OF INSECTS OBSERVED IN THE CHERRY STUMP.

Order.	Groups or Families.	Genera.	Species.	No. of Specimens.
COLEOPTERA	... 5	5	8	52
ORTHOPTERA	... 1	1	1	1
HYMENOPTERA	... 9	10	14	258
LEPIDOPTERA	... 1	1	1	1
HEMIPTERA	... 1	1	2	12
DIPTERA ...	... 9	17	23	90
TOTAL ...	26	35	49	414

EXPLANATION OF PLATES I. AND II.

PLATE I.

Fig. 1.—Portion of rotten Cherry wood, showing nidus of *Crabro leucostomus* (Linn.), with four cocoons, from one of which a male imago has emerged. All drawn natural size.

,, 2.—Portion of rotten Cherry wood burrowed by the larva of *Melandrya caraboides* (Linn.) shown in its burrow, with the active pupa in an adjacent burrow. The imago is seen near the wood at the top of this figure. All natural size.

,, 3.—Portion of rotten Cherry wood showing nidus of *Pemphredon lugubris* (Latr) ; two of the galleries are occupied by the larval and pupal forms respectively, and near the latter is a female imago. All natural size.

,, 4.—Portion of rotten Cherry wood showing borings of a female *Crabro chrysostomus* (Lep.) containing three cocoons, with their lower ends resting on the dipterous remains of the larval larder. From the upper cocoon the female imago shown has emerged. All natural size.

,, 5.—*Crabro vagus* (Linn), female. Enlarged.

,, 6.—*Anthophora furcata* (Panz.), female, enlarged. Bred from boring of *Crabro vagus*.

PLATE II.

Fig. 1.—Upper side of head of larva of *Melandrya caraboides* showing toothed mandibles adapted for cutting and tearing wood (2-inch objective).

,, 2.—Toothed mandible or jaw of female *Pemphredon lugubris* (1-inch objective).

Fig. 3.—Toothed mandible or jaw of female *Crabro cephalotes*, Panz. (1-inch objective).

,,   4.—Pupa of *Crabro cephalotes*, taken from a cocoon in the Cherry wood, enlarged.

,,   5.—Front leg of female *Pemphredon lugubris*, showing brushes and rakes for clearing out the wood *debris* from the burrows (2-inch objective).

,,   5a.–Spur or calcar used for cleaning the Antennæ, etc. (2-inch objective).

,,   6.—*Crabro cephalotes*, Panz., female, enlarged.

,,   7.—Posterior leg of *Crabro chrysostomus*, Lep., showing spined tibia adapted for clearing its burrows, etc. (2-inch objective).

,,   8.—Anterior wing of *Crabro*, enlarged, typical of the genus.

,,   9.—*Pteromalus apum*, Westw., Hymenopteron, parasitic on *Crabro leucostomus*, Linn. ; enlarged.

,, 10.—*Omalus auratus*, Dahlb , Hymenopteron, parasitic in borings of *Pemphredon lugubris*, Latr. (natural size).

,, 11.—Portion of middle leg of *Omalus auratus*, Dahlb., showing spined tarsi and serrated claws (½-inch objective).

,, 12.—Anterior wing of *Omalus auratus*, Dahlb., enlarged.

,, 13.—Cocoon of *Omalus auratus*, Dahlb., taken from boring of *Pemphredon lugubris*, Latr., in the rotten Cherry wood (natural size).

,, 14.—Parasitic Ichneumon, *Perithous varius*, Gr., female, bred from borings of *Pemphredon* (natural size).

Drawn by Charles J. Watkins.

Concerning famines in India, which were formerly often terrible, Mr. C. E. D. Black, in his third decennial report of progress, does not deny the existence of "habitually starving millions," but maintains that, taking the country as a whole, it can always furnish food enough for all its inhabitants. The difficulty has hitherto been in moving the surplus of one or other locality to the spots where deficiency exists. This has now been mainly overcome, and the days when grain was selling at famine prices in one district and rotting on the ground in another are gone. Registered meteorological observations indicate that, as a rule, two-thirds of India are affected each year, either favourably or prejudicially, differently from the other third. There is no record of a universal failure of crops, any more than of a general harvest above the average.

The Development of the Germ Theory.

BY H. LANGLEY BROWNE, F.R.C.S.E.*

A S this is one of the occasions upon which custom decrees the breaking of the rule that silence is golden, and having nothing in the way of original research or discovery to bring before you, I think it may be not uninteresting if I draw your attention for a few moments to the " Modern Development of the Germ Theory of Disease."

The theory itself is by no means a new one, though the proofs of it have only been forthcoming recently and even yet are not complete. Linnæus published an article in 1730, attributing smallpox, measles, plague, syphilis, dysentery, and whooping-cough to the agency of minute animals. Ehrenberg in 1828 found numerous organisms in water and dust, and designated them " infusion animals," and in remarking upon the highly organised structure, astonishing minuteness, and fecundity of these animals, argued that they might cause many of the diseases affecting man ; and Schwann in 1837 declared that the atmosphere was laden with germs of fermentation and putrefaction.

In 1839 Sir Henry Holland published in his *Medical Notes and Reflections* an article on the " Hypothesis of Animalcule Life as a Cause of Disease " ; an article so clear and logical that I propose to review it at some length. He thought that certain diseases, and particularly some of the epidemic and contagious ones, were derived from some species of animalcule life, existing in the atmosphere under certain circumstances, and capable, by application to the lining membrane, of acting as a virus on the human body. He maintained that there are conditions of animal life in the atmosphere as minute, as numerous, as variously diffused as those visible by the aid of the microscope in water, and that we are carried so far by our actual knowledge of these minute forms of existence, and by such uniform gradations of change, that we must not suppose the series to stop because we can no longer draw evidence from our own senses. This would

---

* President's Address, read at the Annual Meeting of the Birmingham and Midland Branch of the British Medical Association June 14, 1894.

imply a sudden breach of continuity such as we find in no other part of the scale of animal life, and be contrary to the opinion of Locke, " that in all the corporeal world we see no chasms or gaps."

Admitting, however, that even if we never reach actual proofs of the existence of these minute forms of life, be they insect or of other kind, he argues that the probability of their existence is little lessened by the failure, considering the obstacles which have to be surmounted. And, if existing, we may presume that they have many points of affinity with well-known insect genera, such as sudden generation at irregular and often distant periods under certain conditions of season or locality, and the diffusion of the swarms so generated over wide tracts of country often following particular lines of movement. Whatever is true or peculiar to the habits of insects, or the forms of animalcule life, obvious to our senses, is likely to be equally applicable to those whose minuteness removes them further from our observation. Their generation may be presumed to be even more dependent on casualties of season or place, their movements determined by causes of which we are less cognizant, and their power of morbidly affecting the body, to be in some proportion to their multitude and minuteness.

Claiming that these considerations are of great interest to the general theory of disease, he says they have close reference to contagious exanthemata in particular ; and though not sanctioned by any direct proofs in relation to these diseases, yet fully justify the prosecution of the research through every possible channel and render plausible, at least, the arguments of those who have ventured to support the opinion that they depend upon the action and phenomena of parasitic life.

Having regarded the question as to whether the epidemic tendency of carbuncular boils has not depended on causes from without, giving hydrophobia and glanders as instances of disease conveyed from animals to man ; noting that epidemic influenzas are in no way produced by atmospheric states or changes, he then takes up the subject of cholera, and proceeds in the most masterly manner to elucidate his theory by the remarkable phenomena shown in this disease.

After tracing the history of its diffusion for twenty years under every climate, in every place, the disease being always absolutely identical, he states that we have not the smallest reason, from knowledge or analogy, to assume that any gaseous, mineral, or vegetable matter diffused from the atmosphere or exhaling from the earth, could create a disorder or spread it in a manner so remarkable over the globe ; and equally inapplicable, for the same reasons, is every theory founded on the temperature, habits, food, or other conditions of particular communities.

Further, he says, it must be kept in mind that we have to deal here with immigrating malaria, a wandering cause of disease, capable not merely of being diffused through the atmosphere and conveyed along vast tracts or lines over the globe, affecting different places with a varying intensity which no known condition of earth, atmosphere, or human habits can explain, but also possessing the power of reproducing itself so as to spread the disorder by fresh creation of the virus which first evolved it. This faculty of reproduction stands foremost among the conditions essential to a right theory of cholera ; all our reasoning stops short unless under recognition of the fact so stated. Without it there would seem a physical impossibility in explaining the phenomena of the disease, and particularly its distribution and succession in different places and seasons. A thorough study of these singular details of its history, keeping this principle constantly in view, will not only confirm and illustrate the latter, but will lead us to organic life as the only conceivable source and subject of such reproduction. It is against all the analogy of nature to suppose this power to belong to inorganic matter. Either animal or vegetable life, in their simpler forms, must furnish the material cause we seek for, since to them alone can belong the faculty of renewing indefinitely the active cause of the disease.

These enlightened views were held by Sir Henry Holland, and give the fullest early account of the germ theory of disease which I am able to bring before you. They form a brilliant example of the scientific use of the imagination, and although again very forcibly brought forward by Henle in 1840, it was a theory hardly accepted by the profession, and very many years elapsed before any proofs were forthcoming in support of it.

Indeed, if we pass over a period of thirty-six years and take von Ziemssen's *Cyclopædia of Medicine*, published in 1875, we find Liebermeister, in his introduction to *Infectious Diseases*, recording nearly the same views without being able to bring forward much more actual proof of them.   He refers to the hypotheses produced up to our time by both the learned and unlearned, most of them not very poetic, but about as vague as the descriptions of the old poets who wrote of the "death-dealing shafts of Apollo;" or as the views advanced in a well-known novel, where cholera is associated with the footsteps of the wandering Jew.   He mentions that in all great epidemics, since the time of the Athenian plague, there has been a revival of the popular notion that the wells were poisoned, and says that this idea had certainly the advantage over most of the hypotheses which were clothed in scientific guise, inasmuch as it supposed a real cause.

Referring to the theory of contagium vivum, and alluding to its antiquity, he shows how, in later times, it had been maintained with decided ill luck.   The statements of the first observers who believed that they had found the organisms that were at the basis of all epidemic diseases, were soon recognised as too hasty or overdrawn, the animalculæ of smallpox, of cholera, and of choleraic fungi, proved to be quite common infusorial organisms, such as can be found in all decomposing substances. About the middle of the present century, the judgment of condemnation of this theory was almost unanimous; it has been regarded pretty generally as an unreal, unscientific play of fancy.   Among the medical authorities, Henle was probably the last who elaborated the theory of a contagium vivum.   This he did in 1853, with as much modesty as thoroughness, though even as early as 1840 he had maintained it with convincing logic.

Liebermeister gives very fully his own reasons for maintaining the correctness of the theory :—New investigations on the appearance, mode of propagation, and significance of low organisms; the fact that the poisons of infectious diseases can reproduce themselves to an unlimited extent; that with a minimum quantity of vaccine virus we can vaccinate a child, from this child ten, and so on *ad infinitum ;* as with vaccine virus so with variola, measles, scarlet fever, typhus, etc., the poison can be multiplied to an endless extent.

Among chemical actions, it is chiefly the processes of fermentation and decomposition which, by their capacity for extension by means of the smallest possible quantity of matter, show the most striking analogy to contagious diseases; and we know now that these ferment processes are all associated with the multiplication of low organisms, so that the theory of fermentation becomes virtually identical with the theory of a contagium vivum.

Liebermeister insists on the fact that contagious diseases never originate spontaneously, but are dependent on a transmission, a continued propagation, of a disease-poison. He notes also that the contagious diseases of the silkworm have been proved to be parasitic, and the history of the development of the parasite has been followed pretty thoroughly; but he omits altogether any reference to Pasteur, whose work this was, and in his search for proofs for his theory is apparently ignorant of the work of Lister in this country.

Immense advances have been made since 1875 in the discovery of the special micro-organisms of the diseases then recognised as infectious, and many additions have been made to the list of these latter. The denizens of this invisible world which have so long escaped observation are now being dragged into the light of day and subjected to the closest investigation. The study of micro-organisms has become a necessary part of medical education, and attempts at classification have been made, and many terrible and prodigiously long names coined to express withal the groupings, functions, and forms of these little beings.

To enable students to pursue these studies with greater ease, the German Government spent £12,000 in bringing the lens to a higher state of perfection, with the result of obtaining one absolutely achromatic.

For the first year they kept the secret of this wonderful lens, and then gave it to the world. With such a lens and by staining the objects under the cover-glass, it is easy now to detect the presence and study the ways and habits of these micro-organisms.* It is undoubtedly to Pasteur that the credit is due for the early discoveries in bacteriology. He proved that not only did water teem with life—this had long been known—but that the air around

* *The Realm of the Microbe* (Priestley).

us was also filled with germs of every kind. Without laying claim to being the first discoverer of germs in connection with disease, he was the first to recognise the vast importance of these minute organisms in the economy of nature. It was while working at molecular physics that the germ theory took root in his mind, and caused him to pursue the studies on fermentation, which eventually led to the investigation of "ferment" diseases affecting human beings and cattle. He divided micro-organisms into two great classes: the aërobic and the anaërobic—those which require free oxygen for their existence and those which are killed by it. The former begin their work on the surface of things, their mission being to clear the earth, by a process of slow combustion, of all that is dead; the latter, working simultaneously, spring into activity underneath the surface of putrescide matter and die on exposure to the free oxygen of the air, and are in their turn swept away by those on the surface.* Thus the two great classes of minute living organisms co-operate towards the fulfilment of a common end, the one beginning work which the other takes up and completes. They also attack our plants as parasites, bringing degeneration and death to their hosts, and at times cause the severest diseases in the lower and higher animals, and threaten even man himself with fatal epidemics. On the other hand, but for their united efforts, we should cease to live, for the earth would be littered with fallen *débris* and organic matter of every kind, all of which it is their duty to transmute into the very elements which are essential to life.

Pasteur had discovered that diseases affecting beer or wine were due to microbes which reached the vats through the air, and that no disease appeared if the air were filtered or sterilised. He had established the vitality of all ferments, harmless or noxious, and had proved that the spores were carried to their respective breeding-grounds through the pollen of flowers or dust of the air. This was the first true indication of how disease travelled, and that it was intangible, invisible, and belonged to the lowest forms of vegetable life. He then turned his attention to the disease affecting silkworms, which was then threatening to extinguish a vast industry, and by a series of the most careful experiments

* *Realm of the Microbe* (Priestley).

proved that this disease was of two kinds, both depending upon micro-organisms. In his investigations, he came upon some very interesting points as .to the mode of infection, which threw much light upon all diseases, whether affecting animals or human beings. First, that in the eggs laid by a diseased moth the germs could maintain their vitality when thus enveloped and protected from the outer air, and so could transmit the disease by heredity. Secondly, that climbing over each other, after having crawled over infected leaves, they would inflict occasional pricks with the sharp hooks on their legs, thus causing direct inoculation of the disease; and thirdly, that in a perfectly healthy state the digestive functions of silkworms were so active that the germs of the disease were carried away quickly or destroyed in the same manner as the leaves in the process of digestion : but that if by any cause the digestion of the worms be impaired, the germs were able to multiply rapidly, and the worm was doomed to perish.

By destroying the infected eggs and the worms suffering from hereditary weak digestion, and by improving the hygienic conditions of the environment, he was able completely to stamp out the disease, and to restore the silkworm industry to its former prosperity.

Lemaire, proceeding upon Pasteur's lines, after proving that the presence of carbolic acid was inimical to the life of higher plants and animals, carried his researches further, and found that the lower organisms were similarly affected by the same material; and that the addition of a small quantity of carbolic acid to fluids in which putrefaction and fermentation would ordinarily take place prevented the incidence of these processes ; and reasoning that disease processes, such as pus formation, were the result of fermentations or decompositions brought about by the action of germs, he concluded that they might be prevented by similar treatment, and he actually applied this antiseptic treatment successfully on the wounds of the human subject and of the dog.

Lister independently came to the same conclusions ; but owing to the difficulty of killing the germs after they had once made their way into the tissues, he recognised the absolute necessity of preventing such organisms gaining access to the wounds at all, and founded his well-known antiseptic treatment for the attainment of this end.

Koch, Klein, Klels, Nicolaier, Pasteur, and many other noted men have been at work upon this branch of pathology, and specific microbes have been recognised as the cause of disease in anthrax, relapsing fever, actino-mycosis, thrush, ophthalmia, enteric fever, cholera, diphtheria, gonorrhœa, glanders, tetanus, tuberculosis, influenza, cerebro spinal fever, dysentery, diarrhœa, erysipelas, foot and mouth disease, hospital gangrene, hydrophobia, malarial fever, measles, mumps, scarlet fever, yellow fever, pneumonia, phagedæna, puerperal fever, and pyæmia. In chickenpox, cowpox, smallpox, plague syphilis, typhus, and whooping-cough, no true specific microbe has yet been discovered, but there is no reason to doubt that they exist. Probably, Koch's researches and discovery of the cause of tuberculosis in 1882 and of cholera in 1884 are the most important. In fact, so much importance has been deservedly attached to Koch's work, that when a few years ago he propounded a method of treating tuberculosis by the sub-cutaneous injection of a preparation composed of the toxines produced by the action of his bacillus, his proposition was received with universal enthusiasm. Unfortunately, this treatment has not fulfilled its promise; but the attention thus drawn to tuberculosis has added immensely to our knowledge of the disease.

Owing to the development of bacteriology, surgeons especially have had to extend and enlarge their ideas of infection, and make many alterations in their pathology. Every true inflammation, various forms of necrosis and suppuration, the abscesses, the phlegmonous and purulent inflammations, boils, carbuncles, osteo-myelitis, suppurative arthritis, endo-cranial suppuration, and brain abscess; empyema, suppurative pericarditis and peritonitis, septi-cæmia, pyemia, erysipelas, tetanus, hydrophobia, and the multiple forms of tuberculosis, anthrax, and glanders, are now known to be infection-diseases due to the presence of microbes. These microbes are in the air, the water, and the soil, in our immediate surroundings, in our dwellings, and in our food; present as our constant companions and at times as our dangerous foes.

The entrance of these germs is most easily effected when there is any abrasion, and thus it was in wounds that the septic influence of these microbes was most apparent, and it was to their exclusion from wounds that attention was first turned.

Lister, with a combination of experimental resource, patience, and brilliancy, almost unparalleled in the history of surgical science, step by step, built up a theory and practice of antiseptic surgery—a theory and practice which rapidly revolutionised the treatment of wounds and the routine of ward management. He introduced a system which has affected the practice, not only of those who believe in its accuracy, but of those who cannot accept all the details, yet have nevertheless adopted its principles. He proved conclusively, and his proofs have been confirmed by many others, that ordinary cleanliness in the dressing of wounds, either made or treated by the surgeon, was not sufficient; but that the use of some kind of antiseptic was absolutely necessary to avoid the risk of septic infection. Surgery and its subjects owe to Lister a debt which can never be repaid.

These germs may also find an entrance through the skin and hair follicles, and give rise to localised centres of infection, such as boils, carbuncles, and different forms of cellulites, or they may be swallowed or inhaled, be taken up and circulated by the leucocytes, and deposited at any seat of injury or irritation, and as these germs are constantly present under ordinary conditions of life, their effects would certainly be much more constantly in evidence were there not some natural safeguarding influence. It has long been known that certain cells in the living body are capable of removing absorbable aseptic substances, such as cat-gut ligatures, etc. Metschnikoff introduced the term phagocytosis to designate the process by which the leucocytes and other cells destroy or digest pathogenic micro-organisms, and the cells which perform these functions he calls phagocytes. The leucocytes are the ordinary police intrusted with the duty of taking up unauthorised intruders, but in moving them on sometimes they let them escape. But the other cells—the mucous corpuscles, connective tissue, cells, endothelia of blood-vessels, and lymphatic vessels, alveolar epithelium of the lungs and the cells of the spleen, bones, marrow, and lymphatic glands—may all be enrolled as special constables to repel an invasion in force. When the struggle between a microbe and a phagocyte turns out in favour of the latter, the microbe does not multiply in the protoplasm, or ceases to do so before the protoplasm is destroyed, and as the microbe cannot leave without dissolution of the cell, it remains within its narrow confines, and is destroyed either by some as yet unknown chemical substance or dies from starvation. In either event the vitality of the cell is not impaired, and the microbe disintegrates or disappears. If the conditions for the growth and development of the microbe in the protoplasm of the cell are more favourable, intra-cellular multiplication of the microbe takes place ; the ptomaines which are eliminated produce coagulation-necrosis in the protoplasm ; the cell disintegrates ; and the intercellular culture is liberated in an active condition.

In cases of unsuccessful warfare of the phagocytes against invading micro-organisms, the mechanical obstruction composed of emigration corpuscles and embryonic cells is broken down ; and the rapid increase of micro-organisms at the seat of inflammation gives rise to extensive local and often general infection. From a practical standpoint, it can be said that all therapeutic measures which influence favourably the process of phagocytosis in the broadest meaning of this word are calculated to exert a potent influence in arresting or limiting the infective processes. For example, the application of counter irritants around an inflamed centre would cause an active immigration of leucocytes, and it is probably upon this that their well-known usefulness depends.

In addition to this *vis medicatrix naturæ* just pointed out, we can do a very great deal for our own protection, and the old adage that prevention is better than cure points out to us the lines upon which we should proceed. It is in the field of preventive medicine we have done most and have still much to do. The personal care of the health ; the notification and isolation of disease ; the disinfection of disease areas; the preventive and curative inoculations ; the inspection of our food supplies ; the drainage and ventilation of our soils and houses ; the establishment of laboratories in connection with our Public Health Departments for studying the history of infectious diseases in man and animals ; in fact, the improvement of every condition of life leaves a large field still open for our work.

The author of *Civilisation : Its Cause and Cure*, says :—" If the extent of the national sickness is such that we require 23,000 medical men to attend to us, it must surely be rather serious! And they do not cure us." Well, if all these medical men will believe in the germ theory of infection, that these germs do not exist only in the brains of those who discover them, and if they will thoroughly act up to such a belief, the national condition may be greatly changed for the better—if not by doing away with all the ills that flesh is heir to, at least by considerably diminishing the number of those which are due to infection.

Notes and Reflections, by Sir Henry Holland; *Von Ziemssen's Cyclopædia*, Vol. I.; *Bacteria and their Products* (Woodhead); *The Realm of the Microbe* (Priestley); *Public Health Problems* (Sykes); *Surgical Principles* (Lenn.); *Micro-Organisms and Disease* (Flugge).

Technology of the Diatomaceæ.

BY M. J. TEMPERE.[*]

Chapter II. (continued).

SPECIAL TREATMENTS.—SOUNDINGS.

SOUNDINGS are certainly the materials that offer to Diatomists the greatest interest and at the same time all the chances of disappointment. I say the greatest interest, because we have there to deal with species actually living in our seas and lakes; all those who have attempted the cleaning and examination of Soundings will realise what I mean when I say there is much that tends towards disappointment. I advise the beginner, therefore, to lay in a stock of patience, and not to be discouraged if the result of his attempts should be absolutely nothing, for unless the materials on which he is about to operate have been warranted to contain Diatoms, it will very likely prove that seven or eight times out of ten there is nothing to find, and that his time has been lost.

A rapid, off-hand examination of a sounding, especially if you have only a small quantity to operate on, may very often show no trace of Diatoms, though the material does really contain them,

[*] Translated from *Le Diatomiste*.

for it is only after a series of manipulations that it is possible to ascertain its richness or sterility as regards organisms of this kind. I have many times had at my disposal one or two killogrammes of marine deposit, in which repeated examinations under the microscope showed nothing, but which, after having been subjected to the needful manipulations, have given excellent results, not as regards quantity, but from the specially interesting species that were extracted.

Soundings are of many kinds and are obtained by different methods. Formerly, it was and even sometimes now is, the practice to cover the lower side of a leaden weight (held by a slender cord) with fat or hard soap, to the surface of which a small quantity of the mud which forms the sea-bottom adheres.

Nothing can be more detestable to the diatomist than this kind of material—in the first place, because he has to get rid of the grease or soap employed ; and then because it is only possible by this means to place at his disposal so small a quantity, and that it cannot give an idea of the richness of the bottom whence it comes.

The fatty or resinous matters are removed by washing with ether or benzine, and the soap by hot water. The sediment then having been well dried, is treated with a series of acids and the usual reagents.

In general, it is by means of special sounding-leads or drags, or by means of nets, that this material is obtained in any quantity and appears as argillaceous or sandy mud, more or less calcareous, mixed up with fine *débris* of all sorts belonging to the three kingdoms, brought up by the lead or the drag.

All the differing materials require to be treated according to their nature. I will therefore pass them in review, and indicate the special treatment that each requires.

Treatment of argillaceous or earthy Muds and the extraction of the contained Diatoms by " floatage."—The argillaceous or earthy muds, produced of marine soundings, are rarely very rich in Diatoms ; often, as I have said, they contain none at all. You may assure yourself of their presence and even withdraw the greater part by the method of *floatage.* For this purpose the material must be perfectly dried, either by exposure to sunshine or in a slow oven, taking great care to avoid all contact with any greasy matters.[*]

When perfectly dry, the mud is broken into fragments, if it has agglomerated in drying, and thrown into a vessel sufficiently large and filled two-thirds of water (a tumbler-glass or a celery glass, for instance). The whole is rapidly mixed, and then after one or two minutes of repose carefully remove, by means of a spoon, a kind of scum that will be seen to float on the surface. It is specially on the edges and against the sides of the vessel that the scum is most abundant, and one may use with advantage a large soft brush to remove it, pressing lightly all round the vessel. This operation, if the mass is considerable, may be repeated twice or even three times.

If the mud contains Diatoms, an examination under the microscope will show a great abundance of frustules, filled with air, which has caused them to float on the surface of the water. If nothing is seen after a treatment of this kind, it is a proof that the sounding is sterile, and may then be rejected. (For fossil earths of the same nature the case is different, and the absence of Diatoms in the floatage is not to be taken as a proof of sterility.)

The collected scum, placed in a porcelain capsule with a little water, is brought to the boiling point for a few minutes to drive the air out of the Diatoms, which on cooling will collect at the bottom of the vessel and may then be treated with acids.

Extraction of the Diatoms remaining in the aforesaid floated materials.—Many valves and frustules of Diatoms may be extracted from the mud already floated by the following means —:

1st.—Tamise the whole with ordinary muslin of sufficient solidity, and having meshes of a millimetre, which will allow all the Diatoms to pass through, retaining only the more bulky rubbish—shells, stones, etc.—which may be thrown away.

2nd.—The deposit thus obtained may be again tamised, employing the No. 1 or 2 filter, according to the size of the Diatoms contained. When the water passes through clear, you must treat the part that remains on the filter with a series of acids and re-agents till it is perfectly cleaned.

[*] Mr. W. H. Shruhole has made known this process, which he has employed, for removing the Diatoms contained in the mud of the Thames, in *The Leisure Hour*, 1891, p. 387.

3rd.—The sediment resulting from this second operation should be treated in the same manner, using filter No. 3, which will usually, in the case of marine diatoms, give a final result; that is to say, that all the organisms contained in the soundings will be found in the filter.

The filter No. 4 need only be used if the mud contains very minute species, and sufficiently interesting to call for its use, for the work with this number is very slow and takes up a great deal of time and patience, when a large quantity is to be operated upon.

Extraction of Diatoms contained in the larger debris of various kinds collected by the drags and nets.—The materials, more or less bulky, collected by the drags and nets, ought to be carefully tested; whether fresh or dry matters little. You should put apart the Algæ, the Sponges (these ought to be themselves divided into classes, as I have already shown),* the *débris* of fishes and echinoderms, and, lastly, the stones composed of calcareous concretions, tubes of annelids, fragments of rock, etc.

The preliminary washing of each of these different kinds should be done separately, mixing ultimately those that appear alike.

I have indicated in previous numbers the course to be followed for the cleaning of Diatoms derived from Algæ and from Sponges. The stones, if they are too large, must be broken up and treated as shells, etc.

(To be continued.)

M. Treby's studies of the planet Jupiter for the last thirteen or fourteen years have satisfied him that the conditions existing there are more stable than astronomers have of late years been supposing. Even if the phenomena of the spots and bands are atmospheric, their permanency and regularity point to some fixed cause on the real surface of the planet controlling them. Besides the "red spot," which has now attracted attention for many years, he finds permanent spots, even on the equatorial zone, having a movement of rotation corresponding with that of this object. The supposition may be legitimately drawn from this fact that this period of rotation agrees with that of the planet itself.

* See this Journal, III., Vol. IV. (1894), p. 307.

Predacious & Parasitic Enemies of Aphides
(Including a Study of Hyper=Parasites).

By H. C. A. VINE.   Plates III. & IV.

Part II.—continued.

IN studying the anatomy of the Aphidivorous Syrphidæ, perhaps no point will more strongly fix the attention, even of a very casual observer, than the vast difference in the structure of the alimentary systems of the larva and the imago.

The exclusively animal food of the larva is obtained and assimilated by organs whose special adaptation for the purpose is remarkable, even in a section of the animal kingdom teeming with remarkable and extraordinary structures. In the imago the wholly vegetable diet is taken into the system by the most complex and delicate suctorial organ known, and applied to the maintenance, and more especially to the reproduction, of life by equally elaborate alimentary and excretory organs, which present many points of resemblance to those of graminivorous animals.

I am quite aware that similar changes of food and structure prevail throughout the diptera, and in a less degree among many other insects; but I do not know of any group (unless among the Neuroptera) in which the contrast is so complete, for the larvæ of Aphidivorous Syrphidæ are not only carnivorous, but they hunt and kill the creatures they require for food.

The Mouth Parts of the Larva.

In the Aphidivorous Syrphidæ the mandibles are represented by a pair of strong hooks attached to a chitinous foundation, situated on either side of the ventral folds of the epidermis, behind the opening of the mouth, and immediately behind and beneath the frontal papillæ. When the larva is in a state of repose, they are hidden within the body, and appear when the evagination of the mouth parts has proceeded sufficiently for the papillæ to become visible (Pl. III., Fig. 5, *g, g*).

These hooks (Pl. III., Fig. 7) are broad at base, short, hooked slightly, and sharply pointed. They are unarticulated in any way, and appear to derive their power of movement from the flexibility of the derma, of which they are a process. A muscular band

INTERNATIONAL JOURNAL OF MICROSCOPY AND NATURAL SCIENCE.
THIRD SERIES. VOL. V.

D

may sometimes be observed behind them; but I am inclined to think that it has no special function as to the hooks, but probably assists in the general movements of the skin.

As the larva extends its head by continuing its evagination, folds become visible representing another segment, almost entirely dorsal, which bears on either side a fleshy protuberance carrying two papillæ, which are no doubt organs of special sense. These are shown on Pl. III. at Figs. 5 and 9, and will be further described. Midway between these protuberances an implication of the epidermis, parallel to the length of the larvæ (for at this stage of extension the dorsal surface is considerably in advance of the ventral) forms the upper margin of the mouth, from which the points of the piercing and sucking organ may generally be seen protruding. They are shown at *e* in Fig. 9.

This latter organ has been figured by Reaumur, De Geer, and other writers, but scarcely in sufficient detail for the figures to be useful. It may and probably does represent the first two segments of the head.

In general shape it may be described as resembling the bill of a flat-billed bird, such as a duck. It consists, as to its upper part, of an arched, chitinous framework, with posterior extensions forming two elongated blades or plates (*d, d,* Fig. 2). Forwards, a long spear-headed or lancet-shaped process extends to form the upper half of the "bill," having its extremity slightly rounded and with a central groove, somewhat like a carpenter's gouge. This is opposed on the underside by a plate of somewhat similar shape, but deeply indented on either side, and furnished at its extremity, which is pointed obliquely, with a thin, hard edge of great transparency (Pl. III., Fig. 6). This half of the organ is not articulated in any way, but its base is lost in the mass of muscles (*e, e,* Fig. 4), which, surrounding and inserted into the flat plates before mentioned, give appropriate movement to the two halves of the "bill."

These latter are continuous with, and merge into, the upper and lower surfaces respectively of the pharynx, which, protected by the chitinous plates and strengthened by a series of slight rings of this material (occasionally the rings seem to become plates), passes through the muscle, as is indicated at *f, f,* Fig. 4, and continues as the œsophagus (Fig. 1, *o*, Pl. III.). The transverse muscles shown in Fig. 4, Pl. III., are sheathed by longitudinal bands, as shown at *e*, Fig. 1. Together they control the opposition of the upper and lower blades of the piercing organ, and by causing rhythmic contractions at the base of the pharynx cause it to act as a powerful suctorial organ. On either side of the head are three long muscular bands, arising from the epidermis and inserted into the framework of the piercing organ, as shown at *d, d*, Fig. 1, Pl. III., which control the forward and retractile movements with great completeness.

The Digestive Organs of the Larva.

An outline of the alimentary system has been already given in the last section in explanation of the first figure of Pl. XV. (Vol. 4).

The most notable features are the extensive salivary glands and the great capacity of the stomach. The anterior portions of the former are shown at *n*, Pl. III., Fig. 1. Each consists of a blind tube of delicate, unorganised membrane, lined with large, nucleated, epithelial cells. The tube is greatly enlarged at its remote or blind end, and after doubling upon itself for half the length of the larva is lost at its narrow end among the muscles surrounding the pharynx. The outer surface is smooth, and the nature of the fluid secreted differs from that secreted by similar organs in the imago, inasmuch as it does not coagulate in water. The secretion of the larva is probably non-albuminous, or at most only very slightly so.

The structure of the stomach presents no peculiarities sufficient to occupy our attention here. Under an amplification of 1,000 diameters, it presents no appearance of organisation beyond faint traces of striation here and there, showing the existence of a muscular coat.

One other point may be noted, which may possibly be of value in the future classification of the larvæ—that is, the colour of the hepatic vessels. Thus, in certain larvæ, and especially in the larva of *Syr. luniger*, these vessels are yellow to brown. In another group of larvæ, distinguished by a clear pink stripe down the centre of the back, they are green, while in some very transparent larvæ, which I have not succeeded in rearing, they are of a deep crimson.

The Tracheal System.

The general plan of the apparatus fulfilling the important functions of respiration has been exhibited at Fig. 2 of Pl. XV. (Vol. 4).

Two large and exceedingly delicate tubes (there marked *b, b,* and in Pl. III., Fig. 1, marked *h, g*), containing in their walls the usual spiral elastic thread, traverse the whole length of the larva, one on either side. Towards the extremity of the head, at a point well invaginated when the larva is in a state of rest, these tubes protrude, carrying with them a conical extension of the epidermis, and presenting in some aspects (and especially when the insect is in the act of extending itself) the appearance of a pair of palpi. They terminate each in a narrowed, rounded point, furnished with chitinous slits for the admission of air, the opening or closing of which seems determined by a circular membranous surface situated obliquely at the base of the point.

Towards the posterior extremity the main tracheæ approach one another, and as at the head, pierce the epidermis, but together, and in the central line, frequently forming a double wart-like protuberance. Each tube here terminates with a circular or oblong ring of chitin, within which a drum-like membrane is partially surrounded by three chitinous ridges set in the ring radially. Each of these ridges, being pierced by a slightly serrated slit, forms an efficient valve for the admission or exclusion of air, regulated by the tension of the circular membranes, and giving the larva complete control over its respiration. The volume of air included within the main trachea and its extensive branches is sufficient to enable the functions of life to be sustained with but little inconvenience, even when immersion in water precludes the possibility of obtaining fresh supplies.

If, instead, the larva be placed in a cell only partially filled with water, it will soon raise either the posterior or anterior valves to a level with the surface, and thus resembles, in its method of breathing, the larvæ of the genus *Eristalis.*

A curious adaptation of the breathing process of the larva may be observed if a young and transparent specimen be kept in a shallow vulcanite cell, covered by a loose glass, for several days. It will do perfectly well without food, but should be exposed to the

air by removing the cover each day for a few moments, and dropping upon it a single drop of water from a glass rod. The larva
will then extend itself, and if viewed under a magnifier will be
seen to draw in air *through the piercing organ,* bubble after bubble
passing down the œsophagus and accumulating in the stomach.
In this way only can the pharyngeal and œsophageal passages be
traced in the living larva.

From each of the main tracheal trunks large branches are
thrown off to every segment, and especially to the surface of the
flask-shaped posterior end of the circulatory vessel; to the head,
numerous loops and curves being here interposed, in order to
allow of the invagination of the parts without stopping the flow of
air; and lastly, to the nervous ganglion (i, Fig. 1, Pl. III.), situated
on the ventral side behind the head.

To ensure the continual supply of oxygen to this important
organ, each main branch from the tracheal trunks sends off at a
point not far from its commencement a small tube which passes
direct, generally following the course of a nerve thread, and without
branching, until it reaches the surface of the ganglion, where it
breaks up into numerous fine vessels. These tubes are shown at
k, k, Fig. 1. About the head also a complex and duplicate system
of looped and connected tubes (l, m, Fig. 1) provide against any
check or hindrance to the supply of oxygen from the movements
of the larva.

The Organs of Special Sense.

So far as my observations go, these are to be found in the pair
of double palpi situate above the mouth, and, in a pair of minute
organs, in the edges of the upper part of the mouth.

The palpi have already been mentioned as placed on two protuberances above the upper end of the opening of the mouth.
Two are found on each side, placed obliquely side by side, and
although the structure of the lower part of the organs appears to
be identical, the organisation of the corresponding apices on each
side is so different as to suggest considerable difference of function.

The outer papillæ on either side is rounded at its extremity,
and provided with five minute circular openings, each encircled
by a chitinous ring, and I think provided with a transverse mem-

brane. The effect is that of minute tubercles set irregularly on the papilla.

The inner papilla on each side terminates in a single tubercle, of less diameter than the papilla on which it is set, and in shape like the extremity of a sausage. In some of my specimens these tubercles appear to be set with a few short spines; but in others three or four very minute bladder-like structures are visible, projecting from the tubercle, so close together as to impinge on one another.

It is impossible to predicate, in the slightest degree, as to the precise function of these organs. I have some reason to think that one or other of the pairs is of an olfactory nature, and analogy would lead us to expect that one or both pairs should be endowed with a sense of touch. That they are the seat of some active perception of importance is shown both by their position and by the nerve-fibres, which under favourable conditions may be observed to pass into the base of the papillæ.

Alimentary System of the Imago.

The entire alimentary system, from the proboscis to the rectum, with the various glands forming its immediate appendages, is shown on Pl. IV. at Fig. 1. The drawing from which this plate is taken was made from dissections of *Eristalis tenax*, this genus presenting many advantages for the purpose in comparison with the more delicate *Syrphus luniger* or others of the same genus.

The internal structure of the former insect differs so little from that of the aphidivorous genera, that it may be fairly taken as conveying to the reader all the important characters of the latter, while the larger size and tougher nature of the various organs enables them to be dissected and represented with a completeness almost impossible of attainment if the excessively minute and delicate viscera of *S. luniger* or similar species had been employed.

The principal difference in structure exists in the proboscis, and in order to enable the reader to form a correct idea of that organ as it is found in the genus *Syrphus*, an outline drawing on a small scale is given on the same plate at Fig. 2.

The proboscis of *Eristalis* is longer and of more substantial and horny structure than that of *Syrphus*. As in the *Muscidæ*—the anatomy of which the Syrphidæ nearly approach—the organ consists of three joints, corresponding to, or rather modifications of, the three anterior segments—*a*, *b*, *c*. The foremost, *a*, consists of a pair of membranous lobes of somewhat triangular shape, each of which is divided, on its outer side, about the middle of its length by a deep cleft. Each lobe is provided with a series of channels, known as "pseudo," or "false tracheæ," and consisting of a number of parallel membranous grooves, to which a vast number of incomplete rings give a semblance of tracheæ. These channels arise posteriorly from a transverse channel of similar structure, larger in size, and placed at only a slight obliquity to the central line of the proboscis.

Each lobe is supported by a curved chitinous arm, taking its rise in the mass of muscles at the base of the second or middle joint, and between these supports a pair of oblong chitinous plates attached one to either lobe, from the extremity of the longitudinal hollow on the upper side of the second joint, in which the *ligula* (*d*), *labium* (*e*), and *maxillæ* (*f*, *f*) lie concealed.

The second joint (*b*) consists of two more or less substantial chitinous processes, which support the muscles on the ventral aspect, and between which, on the dorsal surface, is the channel above mentioned for the reception of the tongue and its appendages. These are attached by membranous and tendonous connections at the base of the second segment, the apodèmes of the maxillæ extending to the ventral side of the pharynx in the third or basal joint.

The slender, acute, chitinous tongue, *d*, forms with the labium, *e*, which is furnished at its extremity with several comb-like processes, a hollow tube, which is continuous posteriorly with the pharynx, and which receives and passes on the food obtained by the suctorial lobes. This will be further referred to in the next section, when the food will be considered. Near the extremity of the tongue is the opening of the salivary duct, which, arising from the salivary glands, is continued through the strong, triangular chitinous structure, *n*, of the third joint, which encloses the pharynx. The second joint and the suctorial lobes are amply supplied with the necessary air by extensions of the delicate tracheal sac, composed of a crape-like membrane, and situated on

the ventral side of the basal joint.   Arising from the thickened
folds at the base of the maxillæ, and belonging to the basal rather
than the intermediate joint, are a pair of labial palpi, *g*, presenting
a somewhat lanceolate appearance in *Eristalis*, but in *Syrphus*
more club-shaped, and resembling the same organs in the *Mus-*
cidæ.   They are covered closely with minute bristles and sparsely
with longer hairs, one or two of especial length and size arising
from the apex.

The third or basal joint, *c*, consists of the folds of integument
surrounding the triangular chitinous structure forming the foundation of the whole proboscis.   These folds are continuous with the
integument of the fourth or antennal segment, which is in immediate proximity to the facial opening, in which the proboscis lies
hidden when at rest.   This segment contains the muscular
apparatus by which the chief movements of the proboscis are
accomplished, consisting mainly of five pairs of muscles attached
to the above-mentioned chitinous framework.   It also contains the
tracheal sacs, or air reservoirs supplying the organ, and gives
passage to the prolongations of the cephalic ganglion which inner-
vate the muscular system.   The triangular framework of chitin, *h*,
consists of a pair of thin plates, thickening on their ventral edges
to solid ridges, between which passes the pharyngeal tube.   The
whole structure is anchored by means of two projecting processes,
or *apodèmes*, in oval muscular masses, and between them issues the
continuation of the pharyngeal tube forming the œsophagus *(i, i).*

The œsophagus consists of an extended narrow tube, which,
bending sharply at its commencement, pierces the cephalic
ganglion, and ends in an enlargement just above the proventriculus or gizzard (*m*), where are situated the muscular valves controlling the passages to the sucking stomach or crop (*k*) and to the
digestive stomach *(l)* respectively.

The appearance of the œsophagus, when viewed in dilute
glycerine under a ¼-in. objective, is sufficiently remarkable.   The
internal surface is covered with bristles, somewhat irregularly
placed and generally pointing downwards.   These take their rise
in the third or membranous coat, which is insoluble in potash,
and is continuous with that of the pharynx and the proboscis.
The two outer coats of the œsophagus consist of circular and

longitudinal muscular fibres respectively, giving it exteriorly a rugged aspect, and the membranous coat bears on its interior surface a lining of epithelial cells. The passage towards the crop or sucking stomach is shown at k, and here the circular fibres have become few, while the longitudinal and membranous coats are developed, until the crop itself is reached, when the muscular bands interlace in all directions, especially about the centre. The organ shown in the figure is collapsed or empty, but when occupied with food it presents the appearance of a seamless, distended bag, around the middle of which a broad constricting band divides it into similar hemispherical parts. Its surface when distended is smooth and free from corrugations, but displays the numerous crossing bands of muscular fibre, by means of which the food is regurgitated, to find its way to the proventriculus, m, and the "chyle stomach," or digestive stomach, l. The proventriculus is embraced by the upper portions of a pair of sacculated glands, n, n, which secrete some gastric or other fluid, to assist in the digestive operation.

The digestive stomach, l, which presents no points of special interest, opens by a muscular neck more or less developed into the upper intestine, p, which in the figure is distended by food, and which at its lower end receives the secretions of the four elongated hepatic glands, g., g. These glands are objects of great beauty. The arrangement of the nucleated cells on either side of the tube, which constitutes their main structure, gives them the appearance of a regularly knotted rope, and frequently the whole organs are of a brown or beautiful purple colour.

At the extremity of the intestine a powerful muscular valve controls the passage of food to the lower intestine, r. This intestine is much stouter in its structure, the rugose outer surface presenting at times an almost sacculated appearance, and at its lower end it passes into the rectum by an internal muscular contraction not readily observed, as the external wall is often not contracted. The rectum (v, v) is noticeable for the great development of the renal organs (t, t), which are found, four in number, in an ovoid enlargement at s. These organs are of great interest, and may readily be obtained from a recently killed insect. They will be fully described and illustrated in the next section, together with the organs of generation, and observations upon the food found in the viscera of some species.

EXPLANATION OF PLATES III. & IV.
PLATE III.

Fig. 1.—Details of anterior segments of larva of *S. luniger*. *a,* Piercing organ. *b,* Pharynx. *c,* Palpi. *d, d, d,* Muscles controlling the movements of mouth parts. *e,* Mass of muscular fibres surrounding pharynx. *f,* One of the hooks. *g, g,* The anterior inlets of the main tracheal trunks, *h, h. i,* The nervous centre, consisting of many combined ganglia, and giving off numerous nerve-threads (*k, k*) to the different segments. *l,* Tracheal loops to provide for circulation of air during the invagination of mouth parts. *m,* Cross tracheæ connecting main trunks with tracheæ of nerve centres. These are duplicated. *n,* Salivary glands. *o,* Œsophagus.

,, 2.—Chitinous portions of piercing organ.

,, 3.—Points of same, enlarged.

,, 4.—Muscle surrounding same, the pharynx being shown by dotted lines.

,, 5.—View of sucking organ, with blades opened.

,, 6.—Point of piercing blade, much enlarged.

,, 7 and 8.—Posterior terminations of tracheal trunks, showing structure of chitinous valves.

,, 9.—Under view of front parts of larva partially evaginated. *a, a, b, b,* The frontal palpi with their papillæ, *c, c. d,* The cleft of the mouth. *e,* The extremity of the piercing organ, just visible. *f,* Pair of palpus-like organs on the borders of the mouth. *g,* Border of mouth. *h, h,* Chitinous hooks.

PLATE IV.

Fig. 1.—Alimentary system of *Syrphidæ* as illustrated by *Eristalis tenax.*

a, b, c, The three joints of the proboscis. *d,* Tongue. *e,* Labium. *f, f,* Maxillæ. *g,* Labial palpi. *h,* Pharynx. *i,* Œsophagus. *k,* Passage to crop. *k',* Crop empty of food and collapsed; a few grains of pollen are visible within. *l,* Stomach. *m,* Proventriculus or gizzard. *n, n,* Secretory organs delivering gastric or other digestive fluids to the proventriculus. *o, o,* Salivary glands, which have their ultimate outlet in the tongue. It has been impossible to trace the ducts throughout. *p, p,* Intestines distended with food. *q, q,* Hepatic vessels discharging into lower end of intestines. *r, r,* Lower intestines. *v, v,* Rectum. *s,* Enlargement of rectum containing the renal organs, *t, t. u, u,* Tracheal vessels supplying air to renal organs.

,, 2.—Proboscis of *Syrphus luniger,* with lobes extended and viewed obliquely. The same lettering applies as to Fig. 1.

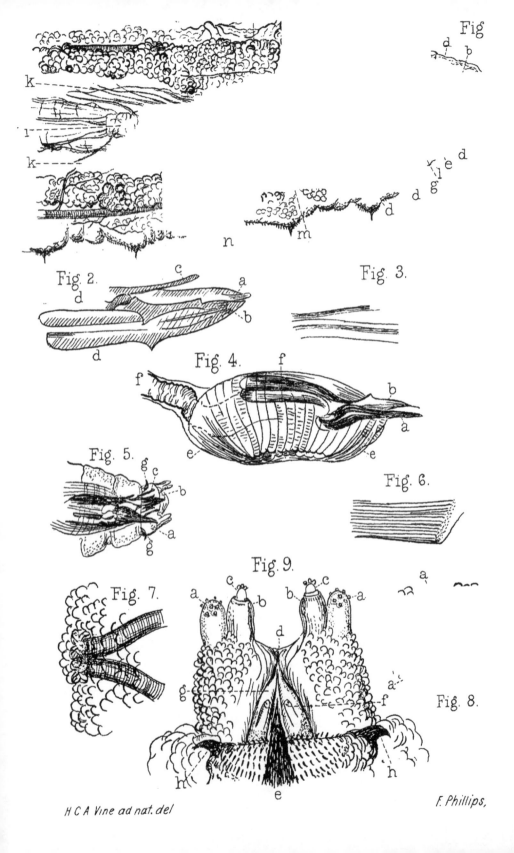

Fig

Fig. 2.

Fig. 3.

Fig. 4.

Fig. 5.

Fig. 6.

Fig. 7.

Fig. 9.

Fig. 8.

H C A Vine ad nat. del

F. Phillips,

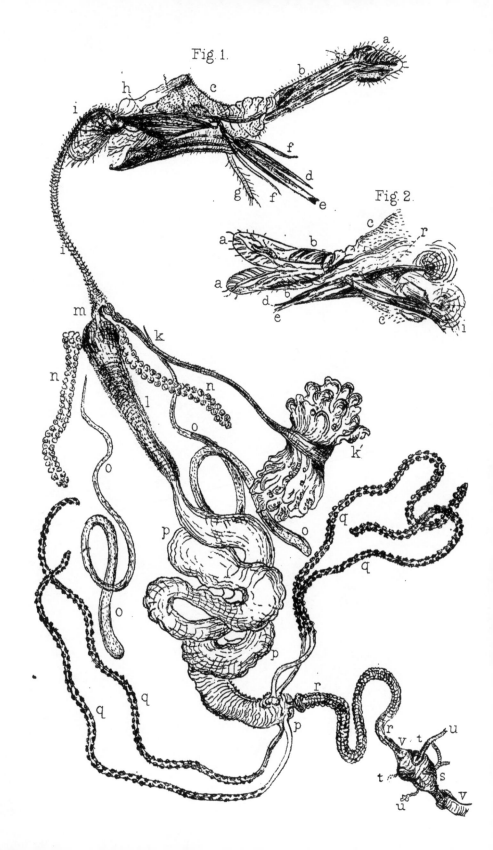

Fig. 1.

Fig. 2.

PLATE V.

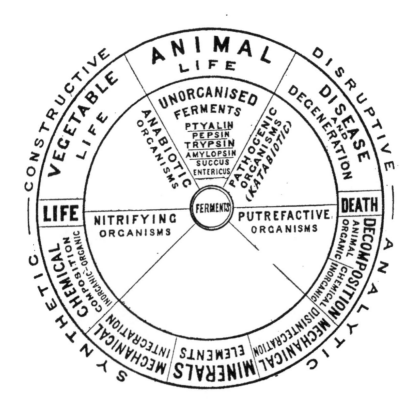

From Dust to Dust:
A Cycle of Life.

BY J. SIDNEY TURNER, M.R.C.S., F.L.S.*

A S medical men, we are ever face to face with Nature's problems, and are constantly being called upon to unravel some of her deepest mysteries. It is our task to repair her working machinery when it becomes rusty and out of gear—to deal with the abnormal rather than the normal. I have therefore chosen a subject which contemplates the natural order of things, and which may, at least, afford us some change of thought. It will necessarily be but a brief sketch of what might well be made to fill a volume.

Man has been tersely defined as a " cause-seeking animal." The Alpha and Omega of life have ever been fascinating problems for speculative thinkers, but the truest philosophy is that which seeks only to discern the knowable.

The Book of Nature, written for all time and all people, is ever open to us, and, although it at first seems to be written in an unknown tongue, like some richly-illuminated old missal, wherein we can only admire the form of the letters and the beauty of the colouring, it requires no translation but that which the earnest student, if guided by love of truth, may accomplish for himself. Thomas à Kempis says, " If thy heart be right, then will every creature be to thee a mirror of life and a book of holy doctrine." No two persons see exactly the same world ; it is greatly in our power to make it a Heaven or a Hell. Happy indeed is he who seeks the pleasures of life—not the fleeting ones that perish, but those intellectual pleasures that have in them germs for future development. Can we do better than inculcate in our youth a love for Nature? Be sure that, where such exists, tender chords of sympathy will be found ever ready to vibrate in harmony with the Eternal Truth.

Who amongst us is not the better for having some hobby? Who may not, in some humble way, add to the growth of that

* Address given at the Fiftieth Annual Meeting of the South Eastern Branch of the British Medical Association, June 13th, 1894.

great tree of knowledge, planted as a seed by primeval man, tended and trained through long ages by the loving hands of those devoted to its culture, and watered, aye, sometimes by the blood of martyrs—for science. has had its martyrs—until it has grown into that sturdy giant which it is now our privilege and duty to tend ! May we guard this sacred inheritance with jealous care, and add new growth to that good old stem which enshrines the labour of many great men, who though long gone from us, are ever present in our life's work !

Aristotle, the father and creator of Scientific Biology, with the clear insight of his philosophic mind, discerned in organic nature an ascending complexity from the vegetable kingdom up to man. Strange indeed that this bud of thought should have remained dormant for about 2000 years, until Lamarck, Goethe, Erasmus Darwin, and, above all, Charles Darwin and Alfred Russell Wallace gave it vigour to unfold and bloom as the fairest flower of modern philosophy. Great and far-reaching as this Theory of Evolution has been, it yet failed to explain many of its own phenomena. It traced, step by step, the gradual development of the higher from the lower organisms—or from ancestors which were common to both—but, it hardly took sufficient account of the work effected and still being done by micro-organisms, without which all life would be impossible, and which have certainly been amongst the principal agents and factors of Evolution.

For a knowledge of these micro-organisms, and of their power for good and evil, this and all future ages will owe a deep debt of gratitude to Davaine, Pasteur, Lister, Koch, Metschnikoff and others, who have explored for us this hitherto unknown world of life ; a world within a world, and, may be, the very centre where life began, and whence it radiates, connecting all life together as in an endless chain. It seems that we have got at least one step further towards the solution of the mystery of Life.

The thoughts of many minds seem for a time to float about, as nebulous theories, in the atmosphere of the scientific world, until some one with master mind condenses and crystallises them into a shape and form, to which the breath of his genius gives energy and life. When any new and great discovery is made, it at first seems to explain so much and to open up so many new vistas of thought, that we are apt to imagine we have found the master-key to unlock all the secret drawers of Nature, but she still keeps many of them closed to us.

We are accustomed to speak of the different sciences, though really but one science exists, which, like a civilised state or a complex organism, requires inter-communication between all its parts to preserve its corporate life. Science, like all other organised complexities, tends more and more to become specialised in its several parts ; so that we get many segregated facts and inductions, until a Darwin or a Lister arises to aggregate, collate, and deduce some general principle from them.

Natural Science seeks to show not only what a thing *is*, but what it *does*, and deals with all the conditions of life, as well as life itself. Geology treats of the formation and inorganic development of the earth, that is to say, of the conditions of life existence. Anatomy, Morphology, Histology, and Embryology are so intimately connected with Physiology, and the latter with Chemistry and Physics, that it is impossible to define their boundary lines. These Sciences, combined with a knowledge of the external requirements necessary to existence, teach us to know the *positive* conditions under which life is carried on ; whilst Pathology tells us the *negative* conditions, which are inimical to healthy life, productive of an alteration of vital properties, and death. Amongst all these Sciences, Chemistry and Physics alone contemplate things in a state of rest, and bring us nearer to the ultimate analysis of matter. The Physicist shows us how gravity, attraction, pressure, and the vibrations of light, heat, and magnetism influence matter ; and the Chemist, by analysing and combining its elements, teaches how inorganic matter can be built up, and brings us to the very threshold of organic life. Well may these sciences be wedded and work together as Chemico-Physics ! Such unions between all the Sciences, with increasing knowledge, will more and more take place, as their action and interaction are better understood. An advance and stepping forward out of line takes place, first in one and then in another science, so that, eventually, each Science is enabled to cast side-lights upon others, and to receive light from them in return.

Bacteriology, the youngest of all the Sciences, has of late years

made such rapid advance, and has shed so much light, that it may
be said to have illumined all other Sciences connected with life
upon our planet; to have elucidated some of the chief phenomena
of Agriculture ; and to have placed many of our manufacturing
industries, such as those of silk, wine, etc., upon a rational footing.
It furnishes us with many friends, as well as many foes, but happily
the former predominate.

Nature seems not only anxious to produce, but to destroy.
There is a Shiva as well as a Vishnu, as the Hindu philosophy
teaches; and it is probable that every animal and vegetable organism has a Shiva near at hand, when the scales, which Brahma
holds, trend their beam towards that side which points to the
setting of life's sun. The maintenance of the equilibrium between
the force which integrates and builds up, and that which disintegrates and pulls down, is the mystery of life ; its loss is the
mystery of death.   Between these two forces lies the path of
progress and the way of endless evolution, the purpose of which
is the inscrutable secret.   Life and death are equally necessary to
the fulfilment of that end for which the former was brought into
existence.   Death is necessary for rejuvenescence of species and
progressive evolution ; according to Professor Weismann it is an
acquired characteristic of the Metazoa ; and the only acquired
character which he concedes to be hereditary.*   Amongst unicellular organisms—Protozoa and Protophyta—there is no natural
death, whilst the conditions favourable to their lives are present ;
they can multiply by fission, so that, as all the body material is
used up in the process, there is nothing left to die ; or, as Weismann puts it, "There is no death because there is no corpse."
These Protozoa and Protophyta (or, as Haeckel calls them, Pro-
tista) produce in the Metazoa both life and death.

.The origin of life in the eternal past, as well as the end of life
in the eternal future, is sealed to us ; it is not, however, beyond
the aim of students of Biology to study the only eternal that our
finite sense is capable of understanding—the Eternal Present—
and to see how the wheel of life revolves.

* In a letter from Professor Weismann, dated August 21, 1894, he says :—
"You seem to believe that I take death for an acquired character, but death
is certainly not a somatogenous character, but a blastogenous one. So there is
no doubt that it must be transmitted," etc.—J. S. T.

How or when life first originated is not for us now to consider; suffice it to say that even Tyndall, who fought such fierce battles with Bastian and Pouchet on the question of Abiogenesis, has told us * "that, as a believer in the nebular hypothesis, he is logically bound to believe that spontaneous generation did come about some time;" and Haeckel† thinks that it may still be taking place in such organisms as *Bathybius Haeckelii.* This moneron, found at a depth of 2,000 to 4,000 fathoms, was discovered and described by Huxley, but of the organisation of this "protoplasmic slime" there is some doubt, though Dr. Emil Bessels,‡ an excellent observer, states that he has studied it in the living condition. Under any circumstance, whether it be organised or only a gelatinous precipitate of sulphate of lime, its spontaneous generation must be an assumption.

The question *where* life first originated is more easily dealt with, and on this point most naturalists are agreed that the ocean was the original home of all life. During a long period of the earth's history there was practically only pelagic life, the conditions of which are so simple that it is easy to believe that the first dawn of life appeared there. Modern science has discovered quite a new fauna and flora in the organisms of the surface, such as *Trichodesmium, Pyrocistis, Protococcus,* the *Coccospheres, Rhabdospheres,* the free-swimming algæ and bacteria. Whether or no the bacteria in sea-water perform the function of fixing the nitrogen of the air, so that it can be utilised by pelagic algæ, in the same way as nitrifying organisms are known to do on land, has yet to be discovered; but it is clear that the teeming myriads of marine animals, the most conspicuous of which are all carnivorous, must get their nitrogen from somewhere. The small amount of *débris* coming from the land, and the very limited supply of vegetable food which the littoral shore can afford, could never support so vast a fauna, far larger indeed than that of the land.§ On the earth's surface we

* *Natural Science,* Jan., 1894. *Tyndall,* by Prof. Huxley.

† E. Haeckel, *The History of Creation,* v. I., p. 344.

‡ *Jena Zeitschrift,* v. IX., p. 277.

§ *Natural Science,* October, 1894, p. 242. Review of Prof. W. K. Brook's paper in *Journal of Geology,* August, 1894 :—

"In a few vivid pages we are shown how the ocean is almost destitute of

have a vast vegetable kingdom, whose grand and mysterious function it is to turn inorganic matter into organic life, by the action
of rootlets and spongioles upon nitrogen, salts, and minerals ; and
by that of the chlorophyll of the cells upon carbon, under the
influence of the sun's light.　　What a mystical moment it must be
when a particle of inorganic matter suddenly, in an instant,
becomes part of an organic existence !　　Could we but follow
any single particle on its journey through life and back again to
its former state as an element, we should know not only its own
history, but that of the greater part of the universe.

Herbert Spencer* in " Principles of Biology " says—" As in
all cases we may consider the external phenomena as simply in
relation, and the internal phenomena also as simply in relation,
the broadest and most complete definition of Life will be :—*The
continuous adjustment of internal relations to external relations.*　It
will be best, however, commonly to employ its more concrete
equivalent—to consider the internal relations as 'definite combinations of simultaneous and successive changes ;' the external
relations as ' co-existences and sequences ;' and the connection
between them as a ' correspondence.' "

† " Each advance to a higher form of life consists in a better
preservation of this primary correspondence by the establishment
of other correspondences." . . . " Perfect correspondence
would be perfect life."

For life to have existed at all, the *conditions* of life-existence
must have been first present.　　We are accustomed to use the

plant-life, how its flower gardens are only stocked with animals, and how its
very herbs and lichens are corallines and sponges ; we learn how the vast
animal armies of the sea attack and devour other animal armies, but have never
a plant for their forage ; and thus by descending steps we are brought to the
conclusion that ' the basis of all the life in the modern ocean is found in the
micro-organisms of the surface,' lowly animals and plants so abundant and
prolific that they meet all demands.　These minute pelagic creatures must
have been the first to exist, and where they first appeared, there they have
since remained, undisturbed by varying environment or the stress of competition, and therefore retaining their primæval simplicity."

* Herbert Spencer, *Principles of Biology*, § 30.

† Herbert Spencer, *First Principles*, § 25.

terms "adaptation" and "the survival of the fittest." We probably all know what is implied by these terms, but it will be as well to bear in mind that organisms do not *adapt* nor *fit* themselves to their environment, but they are compelled by the conditions of their environment to harmonise with it, or to die : in short, they are *conditioned by* their enviroment.   A better term than "survival of the fittest" has been suggested in "the survival of the least unfit."   It is not always, therefore, the *fittest* organisms which survive, but those which are *best adapted* to their external relations by the conditions which environ them.

When we speak of Life we must not forget that, besides the life of the individual, there is a *continuous* life from individual to individual, stretching back to its most remote first appearance upon our planet.   The Evolution of Life, in its broadest sense, is in a chain, the links of which are composed of cyclical existences. This chain may be compared to the rolling wave, made by the impetus of separate particles of water, each of which moves in a cycle and returns to its former position of rest.

* "An entire history of anything must include its appearance out of the imperceptible, and its disappearance into the imperceptible."   In order to have a beginning we must therefore assume that the Earth was composed of the primary rocks.   Upon these solid rocks the action of heat, cold, rain, and frost has been and is continually causing degradation, erosion, grinding, denudation, and alluviation.   To their weathered products we apply the name of "Soil."   The various component parts of the earth-surface are thus being continually mixed up, so that we have a *Mechanical Integration*, which very quickly passes into a *Chemical Composition*.

Recent Science has shown that this "soil" is incapable of sustaining vegetable life, even in its lowest forms, until it has undergone some change by exposure to the air, or, rather, to what the air contains.   If we take soil and sterilise it—a very difficult thing to effect completely, except by burning it—nothing, with the exception of certain micro organisms or nitro-monads, will grow in it.   It is true that certain spores and seeds will germinate, but only thrive for so long as they can live upon the nourishment

* Herbert Spencer, *First Principles*, § 93.

which has been stored up, within the seed itself, by the mother plant. These nitrifying bacteria, on the contrary, seem to thrive *best* in a medium devoid of all organic matter ; and this is of the greatest importance in the economy of Nature. Prof. Frankland[*] has kept them alive for four years in a purely mineral solution ; and Winogradsky proved that they not only flourish, but multiply and build up living protoplasm, in a solution from which organic matter has been rigorously excluded. Other bacteria transform iron and iron-peroxide into oxide of iron ; whilst others oxidise sulphuretted-hydrogen, and transform it into sulphur and sulphuric acid, but as these are inimical to vegetable and animal life we need not speak further of them.

Prof. Stoney asks :—[†]" Whence comes the vital energy of the innumerable bacilli which are excluded from the direct influence of sunlight, and why are these organisms all extremely minute ? " There are some bacilli (*e.g.*, the nitrifying bacilli of the soil) which seem to thrive entirely upon mineral food, and not only so, but perform their functions while completely removed from the sun's rays. The manufacture of protoplasm and other complex compounds from inorganic matter involves a considerable amount of energy, and the bacilli must somehow obtain this from the surrounding gases and liquids; Prof. Stoney regards it as conceivable that the energy may be imparted to the organisms directly by the impact of the more swiftly moving molecules of these gases and liquids ; and if that be the case, then the necessity for the excessive minuteness of the bacilli—scarcely more than molecules— is explained.

Nitrogen is quite as necessary for plant-life as for that of the animal. It exists in the soil, even when it is not manured, and largely, as we know, in the atmosphere. It is only recently, however, that we have any certain knowledge of the means whereby the free nitrogen of inorganic nature becomes part of a living body. That plants do not obtain their nitrogen from the air by the action of their leaves has been conclusively proved, and it has now been clearly established that those which are not insectivorous

[*] Prof. Percy Frankland, *Our Secret Friends and Foes*, p. 87.

[†] Sci. Proc. Roy. Dublin Soc., n.s., v. VIII., pp. 154, 156, quoted in *Natural Science*, v. III., p. 252.

obtain it by means of nitrifying bacteria. Darwin has shown that *Drosera, Dionæa muscipula, Utricularia*, and other insectivorous plants obtain their nitrogen from insects, larvæ, and other small organisms, which are caught in their traps.

Schloesing and Müntz, in 1877, demonstrated that a *living* ferment was necessary for the production of nitric acid and nitrates in the soil; but it was due to the researches of Prof. Percy Frankland and Mrs. Frankland, Warington, and Winogradsky, that the complete knowledge of the process of nitrification was obtained, and that *two* different bacteria were necessary to accomplish it, one of which converts ammonia into nitrous acid, when the other takes up the work and produces nitric acid in a form which can be assimilated by plants. These have now been completely isolated, and it is remarkable that the nitric-acid-forming bacterium was first found in a sample of soil taken from a region near the great saltpetre layers of Chili. European soils contain much smaller quantities of nitrates than those of South America or Africa, but this has been accounted for by the much greater rainfall in Europe, which removes the soluble salts from the surface of the soil, whilst the dryness of the tropical climates of Chili and Peru allows them to accumulate. It is quite conceivable that the rapid growth of vegetation which occurs in an Arctic or Alpine summer, may be due to the accumulation of nitrates in the soil, which has been kept dry and comparatively warm by the snow for so many months.

We all know what an immense benefit the importation of nitrates has been to agriculture, and also that manufactured nitrates have been of some use. It remained, however, for Science to show that, besides importing the nitrates, we also imported the nitrifying organisms, which *continue to fertilise* our lands (even after the nitrates themselves have been used up) by acting upon the ammonia in the manures, which we afterwards supply to them. Hitherto our methods of cultivation have been empirical, and one of the great maxims has been "*to expose the soil to the air.*" Empiricism, the result of well-observed experience, has nearly always proved right knowledge. Science explains the knowledge obtained by empiricism, develops it, and by deducing general laws and principles gives it wider scope of action.

That certain plants, such as peas, beans, vetches, sanfoin, and clover, instead of impoverishing the soil, possessed the power of enriching and rendering it more productive for other crops, has been a well-recognised fact for many centuries. In the earlier Rothamstead experiments, previous to 1861, it had been ascertained that while the higher plants, as a rule, absorb no nitrogen from the air, except the small amount to be obtained from nitric acid and ammonia after thunderstorms—the leguminous plants were able in some way to obtain it. Dr. Wilfarth and Professor Hellriegel discovered that their roots, grown in fertile soil, were covered with nodules, and that these nodules were caused by the presence of certain bacteria (*B. radicicola*), which, however, were not parasites, but lived with their plant-host in true symbiotic relationship. They borrow from the plant the necessary hydrocarbons and supply it with nitrogen, which they assimilate from the air circulating in the soil. It has been proved, moreover, that these bacteria are the *cause* of the nodules appearing on the roots of the plants; for, when the plants are grown without the bacteria, they cease to produce these swellings on their roots, which are so remarkably rich in nitrogen and swarming with bacteria.

Prof. Nobbe confirmed these results, and found that a similar symbiosis exists between microbes and *Robinia*, and he further discovered that each plant has its own particular microbe, and did not thrive so luxuriantly when the microbe of another plant was substituted for its own. Thus, besides a *general* nitrification of the soil constantly occurring for the maintenance of all plant life, there appears to exist a *special* arrangement between certain plants and certain bacteria, for their mutual benefit. Whether or no these microbes can be made to adapt themselves to other hosts, in course of time, is being made a matter of experiment by Dr. Schneider, of Illinois.

Numerous instances are mentioned by Sir John Lubbock, Frank, and others, of symbiosis or commensalism of certain fungi and some of our forest trees, such as the Beech and Pine, which thrive only when their roots are in contact with *Mycorhiza* and other fungi; and, although it was formerly supposed that the fungi were attacking the roots, it is now considered that the tree and the fungus mutually benefit each other.

In the well-known instance of lichens, the algal cells are nourished by the hyphal tissue of their fungal overgrowths, which weave themselves round and enclose them, and supply nitrogen and inorganic materials; whilst the algal cells, in turn, convert the carbonic acid of the air into organic compounds, such as starch, which the fungus makes use of. The fructification in lichens is produced by these fungal overgrowths; and it is interesting to observe that in the proximity of large towns they do not attain to this stage, because the smoky air is inimical to the growth of the fungi. No doubt many of us have observed the (so-called) "fairy rings" on our hills and downs; these are produced by the better and more vigorous growth of the grass, owing to the excess of nitrogen afforded by the fungi, which composed the ring of the previous year.

· According to the recent experiments of Schloesing and Laurent,* various mosses and minute algæ (Confervæ, Oscillatoriæ, etc.), which usually develop on the surface of the soil, also absorb nitrogen from the air, which, after having been assimilated to the soil, goes to feed the higher plants. Whether or no these lower plant organisms take the nitrogen direct from the air, or are indebted to bacteria for their supply,† we do not at present know, but that the higher plants are dependent upon bacteria and the lower fungi for their nitrogen is clearly established. There is every reason to believe that nitrifying organisms supply the lower fungi, and that these, in turn, associate with algæ to form lichens, and afford nitrogen to them.‡ Thus we ascend by the gradual increase

---

* *Comptes Rendus*, 1891, t. 113, p. 777.

† See *Nature*, July 19, 1894, under Notes commencing "Frank and others."

‡ *Nature*, July 19, 1894, p. 276.—"Frank, and afterwards Schloesing and Laurent, showed that soil containing bacteria and algæ can fix free nitrogen in large quantities; their experiments, however, did not decide whether algæ alone are capable of doing this. In order to answer this question Kossowitsch has estimated the amount of nitrogen present in a nutritive soil before and after the growth of pure cultures of two kinds of algæ—Cystococcus and Stichococcus. In neither case was any sensible increase of nitrogen detected; so that it appears that neither of these algæ alone have the power of fixing free nitrogen. Cystococcus, even when mixed with pure cultures of the bacteria which enable the Leguminosæ to assimilate free nitrogen, was found powerless in this direction; whereas a mixture of soil-bacteria and Cystococcus, which

of available plant food to the higher forms of vegetable life, from which the *Animal World* derives its sustenance.

There is, however, really no boundary line between the vegetable and animal kingdoms; the distinction which Cuvier and others made between them all break down under examination. Some organisms, such as Mycetozoa, appear to belong, at different stages of their existence, to both kingdoms. Broadly speaking, however, plant life is nourished and sustained from without; whilst animal life is supported by digestion of food from within. In other words, plants have their roots outside, and animals have theirs inside themselves; except those lower organisms which obtain their food from the fluid media in which they are bathed, and others which lead an entozoic and parasitic existence. It became necessary for those animals which were attaining a higher complexity of organisation, and had to seek food by moving about from place to place, to take their roots with them. This commenced by what is known as "gastrulation," passing through every stage, from mere invagination of the ectoderm to the gastro-intestinal system of the higher mammals. It also became necessary that the higher animals should be, to a great extent, independent of the immediate assistance of external ferments; consequently they have acquired the power of digesting and assimilating the organic proteids, carbo-hydrates, and fats, by means of their own ferments, such as Ptyalin, Pepsin, Trypsin, Amylopsin, and the Succus Entericus. These are unorganised ferments, which were formerly supposed to exert merely a chemical action. They, however, possess the character of living things, except that they cannot be identified as living organisms.

also contained a small amount of other algæ, had the power of fixing free nitrogen to a large extent. The same author also describes a number of experiments with heterogeneous mixtures of algæ and bacteria, and shows how in each case the capability of fixing free nitrogen is greatly increased by the addition of dextrose to the nutritive substratum. From this, and also from the fact that such mixtures of algæ and bacteria, which are capable of fixing free nitrogen when exposed to light, cannot be shown to assimilate it in the dark, he concludes that although in no case has it been proved that algæ by themselves possess the power of fixing free nitrogen, yet they are in a symbiotic relationship with the nitrogen-fixing bacteria, and he regards it as probable that these latter draw on the assimilation products of the algæ to supply the carbon they require in growth."

Prof. Halliburton, in his recent work, *The Essentials of Chemical Physiology*, says—" The gastric juice is antiseptic ; the pancreatic juice is not. .. . . It is often difficult to say where pancreatic action ends and bacterial action begins, as many of the bacteria that grow in the intestinal contents, having reached that situation in spite of the gastric juice, act in the same way as the juice itself. Some form sugar from starch, others peptone, leucine, and tyrosine, from other protéids ; while others, again, break up fats. There are, however, certain actions that are entirely due to these putrefactive organisms." It is a pity that the word " putrefactive" is employed here, as it implies more than is really meant. I need not enter into the details of the action of these intestinal Anabiotic bacteria upon carbo-hydrates, proteids, and fats ; they are carefully described in his book. Dr. Halliburton also tells us how we are protected from an alkaloidal poison, called Choline, by the bacteria which break it up into carbonic acid, methane, and ammonia.

Dr. Alexander Houston,[*] in his report upon the Scott-Moncreiff sewage system, states "that bacteria exist in sewage, which are capable of peptonising solid organic matter, or, in other words, of preparing it by a process comparable to that of digestion for its final destination.

Thus far we have dealt with bacteria, which, as factors of change, are friends rather than foes, and we have now attained to the summit of life's wheel—where "perfect correspondence" takes place by the " continual adjustment of internal relations to external relations "—perfect metabolism—perfect life.

The processes of evolution have been synthetic, constructive, ever and ever becoming more complex as we ascend. The links which form the chain of life are made up of one or more species, whose component members are continually filling the gaps left by their predecessors fallen by the way. What appear to be the weak links are made up of the greatest ; while those which are stronger are made up of the smallest numbers. No single link must long remain too weak or too strong, or the perfect work of the whole chain will be impaired. If a link should become too strong in

* *Nature*, Report on Scott-Moncreiff Sewage System, by Dr. A. Houston, 1894.

numbers, its average is adjusted by diminished supply of food on the one side, or by increase of its enemies on the other. There is thus a constantly shifting equilibrium, to which we are accustomed to apply the term—the balance of Nature; and, by this, every species is maintained at a working point. I must not, however, pursue this interesting problem any further, but proceed to the "facilis descensus," which we find all too easy.

We have now to consider micro-organisms in regard to their nature as foes to health, as *Pathogenic* organisms producing *Degeneration* and *Death*. Leaving accidents out of consideration—and these may be often the cause of the entry of pathogenic germs—we may roughly define *disease* as a disturbance of vital phenomena, using the word "vital" as symbolic of perfect life. It may be some disarrangement of the alimentary or mechanical systems, which, in many instances, will be found to depend upon some remote bacterial action—such as, diarrhœa produced by souring of milk, or ptomaine poisoning by preserved meat or vegetables, or tubercular inflammation of a joint. We have also many diseases caused by parasites of all sorts and sizes, from *Taenia* to *Trichina spiralis* and *Filaria*. However, a great and ever-increasing number of diseases are now perfectly well ascertained to be the direct result of pathogenic organisms. There is no need for me to mention their names, as you are perfectly well acquainted with them. I would like, however, to emphasise this fact—that disease is not an "affliction," but, in most cases, the result of an ignorance of, or a contempt for, those methods, which, for want of a better name, we call "the laws of Nature." In other words, disease of zymotic origin is "preventable;" and it is a blot upon our so-called advanced civilisation that it is not more prevented. Dirt has been well described as "matter in the wrong place." Zymotic disease is *life in the wrong place*.

After the stage of disruption which we call *Death* has been reached, *Putrefactive* organisms are the factors of change, causing organic to be broken up into inorganic matter, in the process called *Animal decomposition*. We must nöt forget that the world has many scavengers by sea and land, such as sharks, vultures, hyænas, and many others; but these serve but temporarily to arrest organic matter in its downward course, and leave the work but very incompletely effected ; and our old friend *Bacterium termo*, or *Bacillus proteus*, as he is now called, ultimately finishes the business. Were it not for this, certain substances would assume a state of rest, and be incapable of being again used ; whilst the earth's surface would be littered and clogged with dead and useless organic matter, and all life, as we now know it, would cease to exist. Animal decomposition comes to an end only with the change of the last particle of organic matter into the inorganic, when chemical decomposition and mechanical disintegration restore the mineral and gaseous elements to the condition of being capable to live again. It is quite possible that in some instances a "short circuit" is made, but our present teaching tends more and more to show that all life—as first life must have been—is built up from elementary particles, such as nitrogen, carbon, silicon, water, nutrient salts or mineral food stuffs, such as the soluble sulphates, phosphates, nitrates, and chlorides of calcium, magnesium, sodium, potassium, and iron ; whilst for plants living in the sea, iodine and bromine are of especial importance. Fluorine, manganese, lithium, and various other metals, such as alumina, found in the ash of lycopodium, also enter upon their cycle through life by means of plants.

Cyclical change thus appears to be the key-note which dominates all the harmonies of Nature.

SUMMARY.

I have made the attempt to show how the study of Nature—having especial regard to the phenomena of life—to which we give the comprehensive title " Biology," embraces and binds all other Sciences together.

How the thoughts of many minds have been crystallised into that priceless pearl of knowledge, "The Theory of Evolution."

How the discovery, in our own times, of the hithero unknown world of micro-organisms had thrown its light on this, and many of life's deepest mysteries.

How the conditions of life existence—the formation of soil—preceded all evolution of life.

How organisms turned inorganic material into protoplasmic life, but

That we were quite unable to conceive the first beginning of life—how or when it originated—but that it was probable

That the ocean was the first home of life.

That the earliest differentiated forms of life appear as a rod and a simple cell.

That the rod-form, or bacterium, was capable of thriving in inorganic matter, and building up protoplasm for higher forms of life to exist.

That—leaving the small amount of nitric acid and ammonia, which some plants may, at intervals, obtain from the air after lightning, out of consideration—these nitrifying bacteria are necessary to produce a supply of nitrogen (which is a necessary constituent of all protoplasm), for the growth and development of plant life.

That inorganic matter by the action of the rod and cell forms of life was started on its cycle through the vegetable kingdom.

That vegetable life supports animal life by means of unorganised yet vital ferments, provided by the animal economy; but that the conversion of carbo-hydrates, proteids, and fats, is assisted by many bacteria, to which I have given the generic name of Anabiotic, or builders up of life in contradistinction to the Katabiotic bacteria, or destroyers of life, which we are accustomed to call Pathogenic bacteria.

That bacteria were friends and foes; and that many diseases were produced by life in the wrong place.

That in such disease a conflict takes place between the rod and the cell forms of life, in which, happily for us, the cell is often victorious, though sometimes the reverse occurs, causing degeneration and death.

That putrefactive organisms are beneficial to the world of life, by restoring the mineral and gaseous elements to a form which can be again utilised to create another world of life.

And, finally, that these bacteria are the factors of change both in the evolution and dissolution of all organic matter.

Such is a brief outline only of this most interesting subject, and no one more than myself realises how inadequate it is to explain what we desire to know. I have striven to state as facts only those that have been well proven, and to base deductions upon them.

Aristotle said, "We must not accept a general principle from logic only, but must prove its application to each fact, for it is in facts we must seek general principles, and these must always accord with facts." "Prove all things ; hold fast that which is good."

It seems to me, however, that micro-organisms are the great *factors of change* in the economy of Nature ; commencing their work in the inorganic world, carrying it through the organic world, and back again to the inorganic, where their task is recommenced, and cycle rolls on cycle "down the ringing grooves of change."

These wonder-working micro-organisms, which render such subtle service in the unseen world around us, would be well compared to the Erdgeist, or Earth-spirit, in "Faust," who called Nature "the living, visible garment of God."*

> "In Being's floods, in Action's storm,
> I walk and work, above, beneath,
> Work and weave in endless motion !
> Birth and Death,
> An infinite ocean ;
> A seizing and giving
> The fire of Living.
> 'Tis thus at the roaring Loom of Time I ply,
> And weave for God the garment thou seest Him by."

Thus may we all work and weave
"Ad majorem gloriam Dei."

In a collection made by Captain W. G. Thorold in Thibet of plants growing at elevations between 15,000 and 19,000 feet, fifty-seven, or one-half, were found between 17,000 and 18,000 feet, five between 18,000 and 19,000 feet, and one, *Sansurea tridactyla*, at 19,000 feet. A large majority of the plants hardly lift themselves above the surface, the characteristic type being a rosette of small leaves closely appressed to the ground with a central sessile inflorescence. Judging from the fact that many of the species are found in the most widely separated parts of the country, there must be very few local species ; and the circumstances indicate that the distribution marks the remains of a probably much richer flora.

* Carlyle, *Sartor Resartus.*

Address to the Members of the Bath Microscopical Society.

By the Rev. E. T. Stubbs, President.

AFTER a few preliminary remarks, Mr. Stubbs said :—
"The first object that we have in belonging to a Microscopical Society is *work—actual work*. It is not to bring people together for the sake or showing them marvellous sights, and things they have never seen or dreamed of before ; it is not that they should open their mouths in stupid amazement, or that we should fill them wfth ignorant wonder, and thus gather to ourselves a few moments of empty, unreasoning applause. But it is for the sake of work ; for the sake of seeing more ourselves ; and therefore of learning how little we really know as we advance in the study of those works of nature in the animal, vegetable, or mineral world, of which so small a part is really known.

There is an enormous untrodden field lying open for fresh explorers and for new discoveries, and we may render our humble aid in the work.

Just consider for how brief a period the microscope has for any purpose of real and accurate research been in the hands of man.

It is, I believe, to Roger Bacon, who flourished in the 13th century, that mankind is indebted to the first conception of the microscope. Roger Bacon was one of the most eminent men which England has ever produced ; he was a native of Somerset, and was born at Ilchester in 1214. This is a quotation from one of his works :—' We can give such figures to transparent bodies, and disperse them in such order respecting the eye and the objects, that the rays shall be refracted and bent towards any place we please. And thus from an incredible distance we may read the smallest letter, and may mark the smallest particles of dust and sand, by reason of the greatness of the angle under which we may see them.' Roger Bacon was accounted a necromancer because when at Oxford he showed people things that were out of sight ; and we are told that during his life he spent a considerable sum in experiments amounting to £2,000 ; *i.e.,* £18,733 present value.

But there is no record of any definite step having been taken in the direction of making microscopes until the year 1618, when Cornelius Drebbel, of Alkmaar, in Holland, is said to have invented the microscope.*

But the instrument as described in the works of that and succeeding times was cumbrous, ill provided with illumination, difficult of manipulation, and with mechanism so coarse and so imperfect, and the methods of examination so few and ill-arranged, that very little progress could be made until the beginning of this century. However, within the last fifty years, the mechanism of the microscope has been gradually improved in every direction: in the penetrating power of the lenses; in the employment of special glass, with high refracting index; in the consequent increase of the visual angle; in the mechanism of the tube; in sub-stage illumination; in stage motion; and in many other particulars. The instrument in our hands to-day is very different from what it was fifty years ago, and if our ancestors in microscopical investigation did good work then, how much more easy must it not be for us to work now, who have the way made so smooth and easy before us!

And although I am well aware that progress in microscopical knowledge depends much more upon the operator than upon his instrument, still we must recognise the fact that the present improvements in our microscopes are a very great help to the faithful and zealous worker.

But, no matter of what sort or excellence may be the instrument a man has, it will be of no avail—nothing will avail—if he has not had *practice* and *facility in manipulation*. Whatever may be the instrument a man has, he ought to *know it thoroughly* and to be able to use its powers to the utmost.

A friend, an old General Officer, himself an ardent microscopist, told me the following story:—A lady he knew had attained to great eminence in Natural History and especially in microscopic research. Her work was particularly in Marine Natural History. She had permission to put several letters after her name, as Fellow, or Associate, or Corresponding Member, or Honorary

* Is not Mr. Stubbs in error here? We believe that Drebbel only copied and sold Janssen's instruments. See Brewster, Quekett, etc.—[*Ed.*]

Member of several learned societies. She was the wife of a captain in the merchant service, and thus had special opportunities for prosecuting her studies and furthering her research ; and as she had no family, she was in the habit of accompanying her husband on his various voyages. My friend asked to see the microscope by means of which she had made so many discoveries, investigated so many secrets of nature, and gathered to herself so many honours ; and he told me he would not give five shillings for the entire instrument ! Some of the parts were held together by thread, india-rubber bands, twisted wire, and so forth ! Yet with such a poor microscope she had been able to observe and to add to the stores of knowledge in such a way as to deserve all those honours I have mentioned. This shows how necessary is constant practice, perfect familiarity with your instrument, steady perseverance, as well as single mindedness in your work.

In all investigations it is of great importance to ascertain, in some degree at least, what is the order of nature. For then we may know to a certain extent where impossibilities lie ; where the way of further advance is really barred against research ; so that no time may be wasted in that direction. Just as the searcher after coal will not waste his time in boring through those rocks which experience or a knowledge of geology tells him never overlie the carboniferous strata, so a knowledge of the order of nature will tell you in what localities the gnat may be found ; and in what waters *Plumatella*, *Melicerta*, *Stephanoceros*, *Lophopus*, or *Floscularia* may abound. And a knowledge of the order of nature will enable us to see that the means we employ are adequate ; or at least not opposed to attain the end in view ; and also to show the easiest, simplest, shortest, most efficient method to reach that end.

But there is a vast field of investigation almost untrodden lying open before us. And it seems to me that there are several directions in which investigations may be made with advantage, and new facts brought to light. Now I am speaking only of what is easy and within the reach of any earnest worker.

For instance, the whole question of polarised light is very imperfectly known, and no very real profit has as yet been derived from its study. Is it not the case that tourmaline, and, I believe, all polarising substances, are more or less electric ? Now, if it is also true, as we have lately learned, that electricity and ether are only terms for the same thing, may not a deeper study of polarising substances account for this strange property? I saw a photograph of a flash of lightning, ribbon shaped; was this because the light was made to vibrate in one plane? If so, that light was polarised!

How little is known about the structure and uses of the antennæ! Kirby and Spence, which we have in our library, will give us valuable information. Sir John Lubbock, in No. lxv., International Scientific Series, and Linnean Transactions, 1862, Vol. xxii., p. 283, treats on the subject, and a very interesting paper was given to this Society some years ago on this matter, but after all much more remains to be known. Just think of the variety in shape of antennæ: the leafy-branched antennæ of *Melolontha vulgaris*, the common cockchafer, seven-branched in the male, six in the female; all the family of the Lamellicornes: the oak egger moth, *Lasciocampus quercus*, with its branched antennæ; the plumose antennæ of *Liparis dispar;* or of the emperor moth, *Saturnia carpini;* the incrassated antennæ of *Sphinx ligustri*; and the immensely long antennæ of *Adila frischella*, eight times longer than the body, with 197 joints, each joint equal to the other, and having two hooks curved and placed on the eleventh and twelfth joints. Of course, we are all familiar with the swelled joint on one of the antennæ in the *Cyclops quadricornis;* the hook in the antenna of the male of *Lipeurus baculus*, the pigeon louse, is also noticeable.

Again, in the microscopic world of Rotifera there is much that remains to be done. That splendid work on *Rotifera*, by Hudson and Gosse, the last and best on the subject, will give great help. Prichard on *Infusoria* we ought not of course to neglect; it is in our library. Only let us bear in mind that he wrote when it was thought that all Rotifera (or as then called Rotatoria) were produced by infusions, which is very far from being the case.

And I cannot too strongly recommend those whose holiday may lead them to the seashore, to study on the spot the beauties there spread out before them. Strange it is how very delicate are those microscopic forms which are found in abundance on the

INTERNATIONAL JOURNAL OF MICROSCOPY AND NATURAL SCIENCE.
THIRD SERIES. VOL. V.

F

most exposed coasts, lashed by the fiercest waves in the rocky pools left by the sea water. To know anything of their life and beauty one must study them on the spot, if you wish to see them alive, remembering that of all fluids sea water corrupts the soonest and becomes unfit to sustain marine life. But every stream and every pond will yield materials for abundance of study to him who keeps his eyes open to these things.

Here we have the Kennet and Avon Canal, where most of the species of Rotifera and Entomostraca may be found in abundance, as well as various kinds of water plants, Potomagetons, etc. It needs some experience in the search, and some knowledge of the localities, to discover these treasures, but it is well worth the trouble. With such advantages before them the members of the Bath Microscopical Society may well be expected to do something in the way of investigation.

ABSORPTION OF ORGANIC MATTER BY PLANTS.—In a communication from Prof. Calderon, of the Institute of Las Palmas, Canary Islands, he contests the ordinary view that the nitrogen of the tissues of plants is derived entirely from the nitrates and ammoniacal salts absorbed through the roots. He does not, however, adopt the old theory that the source is the free nitrogen of the atmosphere, but rather the nitrogenous organic matter which is always floating in the air. The nitrogen of plants he divides into three classes :—*Necrophagus*. the absorption of dead organic matter in various stages of decomposition ; *plasmophagus*, the assimilation of living organic matter without elimination or distinction of any kind, between useful and useless substances, such as the nutrition of parasites ; and *biophagus*, the absorption of living organisms, such as that known in the case of insectivorous plants. A further illustration of the latter kind of nutrition is, according to Prof. Calderon, furnished by all plants provided with viscid hairs or a glutinous excretion, the object of which is the detention and destruction of small insects. To prove the importance of the nitrogenous substances floating in the air to the life of plants, he deprived air of all organic matter in the mode described by Prof. Tyndall, and subjected lichens to the access only of this filtered air and distilled water, when he found all these physiological functions to be suddenly suspended—*Nature*.

Bacteria of the Sputa and Cryptogamic Flora of the Mouth.

By Filandro Vicentini, M.D., Chieti, Italy.

SECOND MEMOIR.

Translated by Professor E. Saieghi.

Recent Bacteriological Researches on the Sputa; the Morphology and Biology of the Microbes of the Mouth.

Further Remarks on the Bacteria and Bacilli found in the Sputa,

and their relation to the " *Leptothrix buccalis* ;" the result of original observations on the morphology and biology of this parasite, principally on its higher phases, now for the first time examined and described.

(Continuation of the Memoir on the Sputa of Whooping-Cough, Vol. XLIII., of the " Atti Accademicia," Medico-Chirurgica di Napoli.)

" The manner of reproduction and the reproductive bodies of the *Leptothrix* are not known ; but, perhaps, more accurate research will discover them : this hypothesis being based on the fact that. in the interior of filaments magnified to 800 diameters, small, round bodies are seen which might be spores."—Robin.

SUMMARY :

§ 1.—Further remarks on bacteria and bacilli found in the sputa and contents of the mouth—A *résumé* of the preceding work—Some forms of bacteria not therein described—Other forms of bacilli, *ut supra.*

§ 2.—Summary of the present investigations and methods of technique—Collecting and preparing the sputa and the buccal contents.

§ 3.—Previous opinions on the micro-organisms of the buccal cavity, according to Miller—History and general facts—*Leptothrix buccalis (Leptothrix innominata,* M.)—Other buccal micro-organisms—Buccal pathogenic fungi.

§ 4.—Observations on the morphology and biology of *Leptothrix buccalis*—The four phases of *Leptothrix*, particularly the fructification by spores—Production by *points* (male organs, ?) —Fructification of spores reproduced in the sputa—Dissemination of microbes in the posterior organs.

Recapitulation.—Bibliography.

<hr>

§ I.

FURTHER REMARKS ON THE BACTERIA AND BACILLI FOUND IN THE SPUTA AND IN THE CONTENTS OF THE MOUTH.

AT the meeting of 25th November, 1888, I had the honour of presenting to this illustrious Academy a memoir on the sputa of Pertussis, in which I incidentally touched upon sputa of another character.

It is not my intention to re-open all the questions before treated, but simply to refer to the bacteriological aspects. The recent observations refer to all bacteria and bacilli found in normal sputa as well as in those of various morbid conditions, the so-called tubercular bacilli not excluded, it being understood that the many special questions belonging to the latter bacilli will be fully considered later on.

Confining my remarks to the bacteriological researches in sputa, I will briefly repeat the conclusions I came to in the preceding memoir :—" The bacteria generally found in the sputa, and particularly in those of Pertussis, or Whooping-Cough, are derived, in all probability, from germs originating in the mouth and nose, so that (and until irrefutable proof to the contrary) their presence is to be taken as a natural fact, which has nothing to do with the cause of the disease. Such bacteria, although of very varied forms, can be reduced to two sole types, representing as many phases or forms of transition. One group may be identified with *Bacterium termo* and the other with *Leptothrix;* and probably even these two groups belong to a single species."

And elsewhere :—" In summing up the feature of the different types, we may conclude that the many forms of Bacteria delineated in our plate, together with the others simply pointed out in the text, can be reduced to two species, viz. : *Bacterium termo* and *Leptothrix*.

The forms *a, b c*, above (Fig. 2, Pl. VII.*), *d, e, f, f', g, i, k, l, l', p,* and *p'*, apply to *Bacterium termo*; the forms *b', c* (below), *m, n, n', o* (left), *q, r, s, u, v, v', x, x', y,* and *z,* to *Leptothrix*. And it is not improbable that, in their turn, the two species, according to Cohn, Naegeli, De Bary, and others, may form only one. These authors classified the round *Bacterium (monas crepusculum), Bacterium termo* and *Leptothrix*, as three forms or developments of one and the same species.

I am rather inclined to believe that even the forms known under the names of *Vibrio rugula, clostrium,* and *rhabdomonas* may be attributed to the same species, as three simple varieties of its bacillary forms." I base these conclusions upon indirect signs, and especially upon certain morphological characters, common or transitional, on the usual habitat, etc.

Now, having followed up this line of research for the last two years, I think I may be able to give a direct demonstration of my work, by indicating, so to speak, the explanation of the polymorphism of the various microbes lodged in the sputa.

Moreover, in the preceding paper, several types of Bacteria and Bacilli, for brevity's sake, were overlooked, as well as other specimens drawn from further observations. †

* This plate appeared in the April (1894) part of Journal, and Figs. 4 to 16 will be found on Plate VI. of the present part.

† Whilst again touching on Whooping-cough, and in continuation of our bibliographical notices, we may mention a late article from Mircoli on " Renal Alterations in Pertussis " (*Arch. p. le Sc. Mediche,* 1890, p. 63, etc.) Mircoli inoculated, ineffectually, the cultures of bacteria, taken from the larynx of children who had died from Whooping-cough, into the larynx or under the skin of rabbits. He mentions articles from Sseurtschenho and Wendt, confirming the observations of Afanasieff. In the *Journal of Microscopy and Natural Science,* edited by Alfred Allen (January, 1890), there is a Lecture given by Dr. Shingleton Smith before the Microscopical Society of Bristol ("'On Some Recent Developments of the doctrine of a Contagium Vivum "), where (at p. 32) it is said that the efficacy of special micro-organisms in most of the contagious diseases, and particularly in those most common to children, as *Measles, Whooping-cough,* and *Scarlet fever,* is not proved.

Some forms of Bacteria not described in the preceding Memoir (Fig. 4).

All forms of Bacteria and Bacilli herein described (Figs. 4 and 5) are coloured with gentian violet, excluding the filament or serpentine bacillus, *h'* (Fig. 5), which has been stained with iodine solution.

(*a*) In the preceding Memoir I maintained that the typical curved Diplococci—*g*, *l*, *p* (Fig. 2)—which are morphologically identical with the Gonococci of Neisser, never present a capsular appearance, and this proves to be so in most cases. But from further observations I have found exceptions to this rule, as is shown in Fig. 4*a* (magnified to 690 diameters). Here, in fact, the curved Diplococci are intermixed with straight Bacteria, on parallel cross lines, and are enclosed with the same envelope, which contains the whole. I found on December, 1890, similar specimens for the first time in the thick, opaque sputa of a child of seventeen months affected with Whooping-cough, and on the seventh day after the expectoration. The sputa contained, in the midst of common bacteria and bacilli, a fair number of red blood-corpuscles with a few ellipsoidal epithelia and granules of myelin. But, later on, I accidentally met with similar specimens in urine. This is no cause for surprise, as I believe that the impregnation of microbes in the urinary and spermatic passages to be the same as that of microbes vegetating in sputa, as they proceed from the same parasite, *Leptothrix*, which vegetates both in the mouth and in the external genito-urinary passages, as I shall prove later on.

In the third section I hope to show likewise the analogy between this type of Bacterium and the *Jodococcus vaginatus* of Miller.

But, besides these curved diplococci, which are intermixed with straight bacteria in a common envelope, others are to be found coupled only and enclosed in one sheath by pairs (seldom) in the *patina dentaria* (white deposit upon the teeth) around the rich colonies of Diplococci. Their capsular appearance is made striking, owing to the staining with picric acid and successive saturation in glycerine.

b, b'.—This specimen (magnified to 1750 diameters) is taken from the *patina dentaria*, mixed with saliva ; groups of diplococci are seen in it, both large and of medium size, interwoven with strings of very minute cocci, morphologically identical with those of type s (Fig. 2).

Such specimens, as well as diplococci with the capsular appearance of type l' (same figure), reproduced in b' (Fig. 4), are numberless. These diplococci are of two sizes, large or medium of type l' (Fig. 2) and others, surrounded likewise by a clear sheath, but small, as in b' (Fig. 4, right). The small cocci enclosed in b' do not differ materially from those of the beaded forms (with the exception of being surrounded by a sheath). I have observed that these minute diplococci are the heavier, because they are found mostly on the slide, whilst the large diplococci adhere to the cover-glass.

Now, the specimen drawn in b shows that the beaded forms, morphologically identical with the bacilli of Koch, are found in the *patina dentaria*, as well as in the solid and stagnant substrata, viz., within the buccal epithelia or in the morbid secretions, either in the crypts of the larynx or the tubercular products, in so far as the inactivity of the part favours the multiplication of the lineal series of cocci, which first form strings of beads and then small rods or filaments, as I have already demonstrated.

In the second place, from this specimen, I believe there exists a close affinity between those beads and diplococci, either of the buccal cavity or of some other habitat; otherwise I cannot understand why they should be so close together as to resemble fruits fallen from the same tree.

Another example of the tendency those minute cocci have to adhere to the diplococci is seen in i.

(*c*) The couples of dumb-bell bacteria herein shown (magnified to 2,500 diameters) are taken from the nasal mucus in its normal condition. I contend that they do not differ at all from the ordinary dumb-bell and chain-like bacteria found in the sputa upon the types c, d, e (Fig. 2), which remind us of the form and size of *Bacterium termo*.

On examining thick nasal mucus, I met with myriads of such bacteria. In this case the material to be examined must be taken from the dry cavities of the nose by a glass rod.

We shall see that those dumb-bell bacteria are more apt to form chains; in fact, they constitute the remarkable groups or bundles of chain-like bacteria found on the surface of the tongue.

According to my researches, each dumb-bell is the result of two cocci linked together by their heads. For this reason, it is often found, enclosed by a sheath, either inside or outside of the epithelia of the mouth, nose, or urethra. It is this adhesive tendency which forms them into chains or groups, from which in less moist parts will germinate the growth of *Leptothrix*, as we shall see later on.

I must refer to the preceding Memoir for the other forms of bacilli and bacteria, all related to *Leptothrix*. With regard to the nasal mucus, its pavement epithelia are smaller than the same epithelia of the mouth and less softened, perhaps from not being under the dissolving action of the saliva.

The bacteria which invade the pavement epithelia of the urethral mucus, taken directly from the meatus, often resemble dumb-bell forms. However, in the blenorrhœgic flow, are often found rather small diplococci, some in the pavement epithelia and others within pus corpuscles ; but curved diplococci or gonococci of Neisser are rarely found in it. Beside these common forms, I also detected very minute cocci, like those rosaries of type *s* (Fig. 2) or *b* (Fig. 4), and filaments or short articulations of *Leptothrix*.

d, e.—I have said before that the features of pneumococci of Friedländer, or Fränkel, etc., were probably derived from a degree of colouration of some points of their single articulations or from modifications of forms, as types *d* and *e* will show.

In *d* (Fig. 4, magnified to 2,500 diameters) are seen three sheathed diplococci, with sheath partially coloured.

In certain cases of pneumonia, we found some specimens of these, intermixed in the same group or colony.

In *e* the diplococci have no sheath and show points of varied partial colouring. Such specimens were obtained from a pulmonary sputa, in which, amongst other forms of bacilli and bacteria, were pneumococci of types *f, g, h, i, i'*.

f, f', g.—These are specimens of the second kind, taken from the same sputum in a case of croupal pneumonia. In *f* (magnified to 1,750 diameters) the capsular appearances are so well surrounded as to resemble a capsular membrane. The single pneumococci are sometimes in groups of eight, ten, and fourteen. The internal articulations are prolonged and end in points.

In f' (magnified to the same degree) the articulations are also acuminated, but have no halo, and at first are very pale. In successive days they become partially coloured in the preparation if kept artificially moist. These pale *Diplococci*, without sheath, appeared to be in a state of active germination, as we detected therein the more minute and proportionally paler forms, some being composed of unequal cocci, and several were moving about in the medium. The forms surrounded by sheath are firmer or less active, as if the capsule indicated the quiescent state of the microbe. In fact, if it moves, it is by a motion of simple translation, without vibrating, as is the case with bacteria in active germination.*

It appears that the capsule, owing to its great refraction, and also to the difficulty the near small bodies find in crossing it, resembles a fatty substance surrounding the microbe or proceeding from it. This is only an hypothesis.

In g (magnified to 2,500 diameters) is shown a distinct diplococcus, but with the internal articulations nearly cylindrical, with a large and somewhat irregular capsule, resembling the mono-articulated bacillus of type q (Fig. 2). The sheath indicates, even here, a quiescent state of the microbe and the successive secretion of pale substance. Some analogy may be found in it to the *Bacillus crassus sputigenus* of Kreibhom. (See section 3.)

h.—Intermixed with the diplococci e and f' we find, at times, in the pulmonitic sputum, these forms without a sheath (magnified to 2,500 diameters). In one of these forms we infer the process of increase by seeing the pointed articulation at the right part as pale as in f'. The lengthened part of the mono-articulated bacillus is strongly coloured, and, in a good light, exhibits internal punctuations of deeper colour, especially near the edges : a remarkable circumstance that makes them resemble the *Leptothrix*, in which are also these punctuations or future gemmules.

The above forms are met with in the second period of pulmonary affections after the fever has subsided, but may also be ·

---

* In a recent work, Billet states he has met in other species with this form of incapsulated diplococci, and considers it a phase of life common to various species of bacteria. He calls it the zoogloeic state. (This will be again referred to in a note at the end of Section 3.)

found in successive periods. These pneumococci retain tena-
ciously the gentian violet, as I stated in the first memoir.

Several of the preceding forms of pneumococci, particularly
those of types *d, e, f, f'*, and *h*, are analagous with those drawn by
Cornil and Babes in *Les Bactéries et leur rôle*, Pl. XV., Figs. 1, 4,
6, and 13.*

i, i.—The large round diplococcus, *i* (magnified to 2,500
diameters), was also found in the second period of a pulmonary
affection. Its articulations resemble two detached hemispheres,
and round the envelope adhere very minute cocci, not one of them
being able to penetrate the clear zone. This fact clearly indicâtes
the solidity of the sheath.

In *i'* there is a small *chain-like* bacterium, the articulations of
which exhibit a series of similar diplococci, so connected that the
hemisphere of the one adheres to that of the other, leaving a
clear space between the hemispheres of each diplococcus. Such
chain-like bacteria are sometimes found grouped together, although
in small number, in pneumonic sputa.

Other Forms of Bacilli, *ut supra* (Fig. 5).

Some forms of bacilli were omitted in the other memoir for
want of space ; others were collected later on, and these I will
now briefly describe. The new forms added to the first will better
show the identity of the bacilli of the sputa with the articulations
or fragments of the filaments of *Leptothrix* vegetating in the mouth.

In the preceding Memoir the bacillus *n'* (Fig. 2) represented a
form of transition from the young bacillary forms, *n, n, n*, to the
mature forms, *m, m*. I then pointed out other forms of transition.

a, b, c.—In *a* (Fig. 5, magnified to 1,250 diameters) is seen a
bacillus that partakes of three distinct types, divided into three
segments : one internally occupied by four ellipsoidal bacteria, the
clear segment in the middle, slightly coloured, and the other
deeply coloured and opaque. It shows that the clear bacilli
inside—those containing bacteria and those entirely opaque—
belong to the same species. This bacillus was taken from influenza sputum, but similar forms were found in other sputa and in
the saliva.

* Work quoted in the Bibliographical Appendix.

Another analogical proof we have in *b*, where a bacillus, opaque for five-sixths of its length, exhibits the other one-sixth clear and slightly coloured. This specimen was taken from the saliva.

In *c* we find four types—viz. : a short, clear segment above, and on the top of it a bacterium of elliptical form ; in the middle, a long, opaque segment ; and in the lower part a pale, granular segment like the middle feature of type *n'* (Fig. 2) and types *e* and *e'* (Fig. 5). This specimen was also taken from the saliva.

d.—This short but large bacillus, taken from the *patina dentaria* (magnified to 2,500 diameters), exhibits the upper part opaque and deeply coloured ; the other part is divided into two segments. The middle segment is very short, clear, and contains a dumb-bell bacterium placed across ; the lower one is opaque, but not entirely, as there is on the right a clear space between the opaque part and the outline, and presents a kind of indentation. Other bacilli have the same features towards the edges, or near the junctions of their segments.

As regards the enclosed bacterium, there is no doubt about its position in the envelope of the bacillus, from the look of the specimen as well as from the oscillation of the preparation, in which the bacterium was seen to be always inside. In the preceding specimens we have also abundant proof of this fact.

Now in the first place, it shows that not only elliptical, very minute, and linear bacteria, but also *dumb-bell* bacteria, may be found in the envelope of the bacilli ; from which we conclude there is no difference between the *dumb-bell* bacteria and the bacilli. And since the small chains, and the bundles especially, adhering to the tongue, are mostly formed of dumb-bell bacteria, we must infer the analogy of those forms with the bacilli, and consequently of both to the *Leptothrix.*

In the second place, comparing the various specimens of bacilli (as *a* and *d*), containing bacteria with the fertile filaments of *Leptothrix*, we find in the latter a variable disposition of internal granules, or buds, adhesive to the envelope. Some are alternate ; others on the same level at both sides. If we imagine that in those filaments the buds grow up towards the middle line of the stem, it will result that, when the left bud is placed higher than that on the right, the bacteria resulting from their increase will range themselves in a longitudinal line, taking the elliptical form, as in bacillus *a*.　On the contrary, if the two buds are on the same level, we shall have the case of bacillus *d*—viz., the two cocci, resulting from the increase of the buds, will meet in the centre of the stem, and, there joining together, will produce, in a cross line, the *dumb-bell* bacterium.

Therefore, wherever we turn our eyes, the notion of the poly-morphism of the microbes, and their probable derivation from a single species, is gaining firmer ground.

e, e'.—The bacillus drawn in *e* (Fig. 5, magnified to 1,750 diam.) was found in the sputa of a young woman in labour, who was affected with bronchitis.　It was mixed with other similar granular bacilli, clear and of the same length, but more slender, as we see in *f*.　These bacilli, however, are common also in other sputa and in the contents of the mouth; they present a pale, granulated mass, similar to that of bacilli, *n'* (Fig. 2) and *c* (Fig. 5).　They also show, in various places, buds or internal granules, smaller and paler than those found in the fertile filaments of *Leptothrix*.

In *e* the ends are pointed, but in *e'* they are round.　The larger of the two is faintly coloured, but exhibits near one of the heads a clear interstice, as is seen in *a* and *d*.　The other specimen, smaller and paler, exhibits a younger form of the same type.　These two last bacilli were found in pulmonitic sputa, and were also found in the contents of the mouth.

In the uncoloured preparations of the *patina dentaria* there are *nail-shaped* bacilli, like the type *e*, soft, flexible, slightly veined, and having quick motion (Fig. 14, *d*).　These bacilli are originally lodged on some pointed productions, as was observed by me for the first time; their slender points form the bacilli *f*, *h*, and *h'*.

f, g, h, h'.—The very slender bacilli, *f* and *g* (magnified to 2,500 diameters) are also taken from a case of pneumonia.　In *f* we see a mono-articulated and strongly coloured bacillus, as well as another pale and granulated.　In *g*, a very slender one, the ends of which appear to have been recently detached from a longer filament.

By comparing these specimens with the *Leptothrix*, no import-ant difference is noticed.

In *h* another specimen of curved bacillus, resembling a point of interrogation, taken from pulmonitic sputum. These bacilli are not uncommon in the *patina dentaria.* They have quick vital movements. They resemble the *Spirillum sputigenum* of Miller, or the *Comma bacillus* of the saliva described by Lewis, as identical with the cholera bacillus of Koch, *Bacillus virgola.**

I have, however, found that, between these bacilli and the lesser points, which are abundant on certain filaments of *Leptothrix*, there exists a great similarity. These *points* vary in size, but all possess quick movements (Fig. 14). We shall consider later on if these bacilli act like the reproductory filaments (spermatia or antherozoids) of many cryptogams, although in *Leptothrix* they are not originally contained in special receptacles (spermogones or antherids).

In a more advanced hypothesis, we shall have also to consider if these bacilli represent the spirilli, or if the spirilli proceed from such filaments, endowed with the same motile power. It appears more probable that the spirilli may generally be derived from the fragments of certain curved filaments, fertile or not, of *Leptothrix*. Sometimes the spirillum is thicker than the filaments of *Leptothrix*, as in Fig. 1 of Cornil and Babes.†

In *h'* is seen a larger filament (stained with iodine), found in the *patina dentaria*, which is somewhat similar to spermatozoa.

In the uncoloured preparations I have seen the very rapid movements of all these bacilli. In conclusion, all these forms of bacilli are common in the expectorations and contents of the mouth.

---

§ II.
SUMMARY OF THE PRESENT INVESTIGATIONS AND METHODS OF WORKING.

ARGUMENT.

In the fourth section of my Memoir on Whooping-cough, I stated that bacteriological researches on the sputa would be fruit-

---

* Lewis, *A Memorandum on the Comma-shaped Bacillus, etc.*

† Cornil and Babes, *op. cit.*—Vicentini, "Sopra un caso di febbre ricorrente ecc. e sul riscontro degli *Spirilli* nel Sangue," *Atti del. R. Accad. Med. Chir. di Napoli*, 1883, t. XXXVI.

less until the natural history of the pathogenic or non-pathogenic microbes that lodge there is thoroughly known.

In most cases a complete organism grows in length, breadth, and thickness. But the chain-like bacteria and the filaments only grow in one direction (in length) simply by a repetition of particles in a lineal series. We must, therefore, suppose that neither these original particles, nor the *chain-like* bacteria and filaments resulting from them, are complete organisms, but that they are rather rudiments of more perfect ones.

By examining all the different forms of bacteria and bacilli, we were led to believe in a single type, to which all other forms were related. Now, we will show that the existence of this vegetation is to be found in the *patina dentaria*, in certain sputa, and in the mucus of the urethra, all the forms hitherto described being only fragments of it.

If somebody, for instance, not knowing all the phanerogams, should enter a wood-house, looking at the wood, heaps of branches, leaves, etc., he might at first suppose that they were different things instead of parts of a whole. Likewise, if he should see a threshing-floor upon which is spread and beaten Indian corn, he might take the stems, roots, leaves, tufts, and grains for as many different objects. But it will be sufficient to take him out into the field, to convince him of his mistake. Looking through the lines of the corn, he would see that most of the plants (the female ones) are provided with ears, and he would see also others higher than the rest (the male ones), having instead of ears only a tuft.

Then he might suppose that the plants of this second type were, perhaps, of different kind from the others, not thinking that Indian corn is a dioecious plant, with distinct sexes on different individuals.

This is precisely the case with the microbes of sputa. We were led to the study of the morphology and biology of *Leptothrix*, by considering that nothing is useless in nature ; those parts which now appear superfluous to us in living organisms, (as the nipples in man) are rare exceptions.

The illustrations of *Leptothrix* contradict this law. Look, for instance, in Bizzozero, the figure of *Leptothrix buccalis*, reproduced in our Fig. 6, and the query will be asked :—Of what use are those upright, barren shoots resembling grass, which have been eaten close by cattle ?   This want of fruits cannot be accounted for by mechanical injuries.   The figures of Robin, Frey, Miller, and others do not suggest a different conclusion.*

After these reflections, we were induced to investigate this parasite.   Our researches will be particularly demonstrated in Section 4, only a synthesis of them being given here.

From our observations *Leptothrix buccalis* would, therefore, live under four forms, or have four different stages of development, according to the conditions of the alimentary substratum.

In the liquid secretions, as saliva and mucus, its spores – bacteria and cocci—and its gemmiferous sprouts—real bacilli (excluding the Spirilla, the *nail-like*, the *snake-like*, and the *comma*)— increase and multiply like moving spores and germs of fungi. This would be *the immersed vegetation of Leptothrix.*

The bacteria, the yeasts, and the spores have generally the power of reproducing the full organism like the seeds of the phanerogams ; but, between the spores of the bacteria on one side, and the seeds of the phanerogams on the other, there is this difference : that those seeds cannot multiply by division, producing fresh seeds ; whilst moving spores, yeasts, and bacteria multiply abundantly, by fission, even by themselves.

If, however, the liquid is kept quiescent in the crypts or cavities, or if the part is, as on the surface of the tongue, very firm, the second period or low vegetation begins, viz., *beaded, chain-like,* and *bundle-like* bacteria are formed, so common on the patina of the tongue.   But vegetation stops where (as in the mucous parts) the continual friction impedes a further formation.

In the secretions and mucus there is a relative inactivity ; but if it becomes effective, a reproductive growth will be formed, as shown in Fig. 16, taken from pulmonitic sputum, which for several days had adhered in the interior of a closed tube.

The third period, or growth, is when it lodges on a solid part,

* Bizzozero, *Manuale cit. nella Bibliografia*, Pl. II., Fig. 27 ; Robin, *Histoire naturelle des végétaux*, etc., 1853 (see Bibliography), Pl. I., Figs. 1 and 2. Frey, figure reproduced by Perroncito, quoted in Bibl., Fig. 11. Miller, figures quoted later on.

on the tongue or teeth.   But, owing to the continual friction on the tongue and teeth, the growth cannot reach the fourth period, or that of fructification.   Then we have bare growths apparently sterile, as represented above.   The same happens with the tufts of *Leptothrix*, often found on the epithelium scales, on the residue of undigested food, or other corpuscles in urine, sputa, etc., forms which may be called " aërial," although incomplete, " vegetation."

But in many parts of the dental surface, where there is less friction, as between tooth and tooth and near the gums, the fourth phase or fructification takes place.   This fact has not been observed, and is not even suspected, as we gather from the most recent work of Miller.*

I call it aërial vegetation, because the stems of the microphite have a tendency to rise in the air, although immersed like the algæ.   In our case the surrounding liquid is the saliva.   This fructification by spores, first noticed and investigated by me, consists of comparatively long ears (Figs. 10, 11, 12, 13, 16), formed by the exudation of a viscous substance, faintly coloured, round the ends of fertile filaments.   In this substance are entangled small round spores, often brightly coloured, placed in six longitudinal lines, as we shall see later.   In the strongest specimens (Fig. 16, *f*, *g*) as many as 720 spores can be counted in an ear.   All the cocci and minute bacteria are probably disseminations from these spores.

There is another form of special production, or pseudo-inflorescence, which is drawn in Fig. 14, *a*, *b*, *c*, resembling *points*, round a filament or group of fertile filaments.   These *points* having fallen off form the *nail-like*, *snake-like*, or *comma* bacilli. If we admit the analogy of these bacilli with the reproductive filaments of other cryptogams, *Leptothrix* may be compared to a fungus, or *diœcious alga*, with two sexes upon different filaments ; and such pointed productions would exhibit the male organs, the elements of which, spread about the surrounding medium, would, with their great mobility, act as antherozoids in a manner not yet known to us.   In such an hypothesis, the scanty number of such inflorescences, with regard to fructification or female organs, would be fully explained.   Therefore,

---

* W. D. Miller, *Die Mikro-Organismen der Mundhöhle*, etc., 1889.

in summing up the morphological and biological series of *Leptothrix*, beginning with the forms of Cocci or Bacteria, we pass through the chains, the growths, and end in the fructification by spores, in order to begin again another series *usque ad infinitum*. But, beside this cycle of reproduction, we must notice another—of buds and internal impregnation (that already foreseen by Robin), which is seen in the interior of the envelope of old filaments, articulations, and bacilli, and through which, from the same bacilli, are generated other bacteria, and this owing to the reserve buds described above. From those bacteria are, in their turn, reproduced the chains, the tendrils, and their articulations or bacilli, as we have seen in the former Memoir on the subject of bacilli *b'* and *n* (below) and in *u, u* (Fig. 2).

Let us consider what a great cause ·of dissemination this is, placed at the entrance of the digestive and respiratory ·organs. According to our calculations, there would always be found present in the nose and mouth, taken together, from two hundred to three hundred trillions of bacteria, or other elements of the microphite, ready to enter the stomach at the time of deglutition, and to fall on to the respiratory passages at every breath. In comparison with that mass, what are the few germs we inhale from the atmosphere?

But if, at every breath we inhale, a propelling motion sends whole swarms of elements or germs of the microphite into the air-passages, another slower but continuous movement pushes germs or analogous elements on into the genito-urinary passages through contiguity, from one epithelium to another, and onwards along the mucous patina, passing through the urethral orifice, which Pasteur has compared, in this case, to the Thames Tunnel.*

We will briefly treat these questions in Section 4, and demonstrate the identity of *Leptothrix preputialis* with *Leptothrix buccalis*. Now, to enable others who may be willing to repeat and to verify our observations, we will proceed to describe the *modus operandi*.

---

* Schutzenberger Le fermentazioni, ecc. trad., Milano, 1876, p. 204 ; V. also Vicentini Caso di vegetazione di funghi microscopici nell' Uretra ecc, Morgagni, April, 1880.

Collecting and the Preparation of the Sputa.

In the preceding work we dealt with the collecting and the treatment of sputa, putting off to another occasion the diagnosis of derivation of the various materials expectorated, in order that the real sputa might be distinguished from the particles of food ; the sputa from the saliva or throat, and the mucus flowing from the nose, being also differentiated. Now is the moment to touch on this point again, in order to describe improved methods of preparation and colouring.

The materials to be submitted to microscopic observation really do not always proceed from the air-passages. In the case of a boy, 7 years old, affected with Whooping-cough, we remember having received a portion of food ejected from the stomach, and thought to be a sputum. This substance was gelatinous in appearance, the colour of tobacco, and contained a large amount of alimentary residue ; but what surprised us was a very dense mycelium of conspicuous sprouts after the type drawn in former plate,* Fig. 3, *d*, interwoven with other slender ones. The fructification of these mycelia produced forms identical with small heads (capitula) or sporangia, delineated in *k* and *k'*, very vigorous, upon as many as five fertile branches, intermixed with similar vigorous forms of *Aspergillus glaucus*. This would lead us to suppose that the deglutition of the relative germs with the sputa, and their secondary development, is even more easy and vigorous in the stomach.

In some cases, especially with children, the proper expectoration is lost or swallowed ; and when the patient is asked to spit out again, he emits only a little saliva, mixed with small flakes of mucus, from the throat and tonsils. These salivary sputa, however, may be recognised generally on the following day through the clarification, after remaining undisturbed, with an upper liquid stratum (more considerable than in crude sputa, or those mixed with abundant saliva), and a scanty deposit, very rich in buccal epithelia, and, eventually, alimentary residue. Besides, they are recognised from being wanting in ellipsoidal epithelia and granules of myelin.

* See April, 1894.

Another source of error, against which it is difficult to guard, is the reflux of nasal mucus, through the posterior nares, in the back of the mouth, and its successive expectoration mixed with saliva. Naturally, ellipsoidal epithelia and myelin, even in such sputa, are wanting.

In any case the presence of the myelin and the ellipsoidal epithelia induces us to regard the sputum as proceeding from the air-passages. These are two important points when, for instance, a doubt arises whether a sputum with a rusty appearance proceeds from the nose or the chest; as, in our daily researches on sputa, we never detected ellipsoidal epithelia or myelin in spurious sputa; whilst in the genuine expectorations they are very seldom absent, and when not found in the first, they appear in the second preparations.

We come now to the manner of collecting the sputa.

The act of spitting is accomplished in two different efforts : by the first, the genuine sputum, nearly entirely free from saliva, and that which is required is coughed up ; but soon after by a second effort, saliva is emitted with the residue of the sputum left behind. This second sputum (being nearly all saliva) is eliminated. In order to avoid mixing the sputum with saliva, the patient should not keep it in the mouth too long.

Instead of a spittoon, it is better to use a colourless plate or saucer, perfectly clean, where the sputa can be kept separate, and then selected for examination. The sputum should be taken at the most viscid point with a bent needle or forceps, and raised so that the saliva and the more fluid part of the mucus with which it is mixed, may trickle down on the plate. What remains on the instrument is quite sufficient for successive investigations, as it will necessarily contain very little saliva. To preserve it, it should at once be placed in a glass tube, which has previously been cleaned with sulphuric acid, and washed out with alcohol.

For the coloured preparations we have to modify the directions given in the preceding work. It is better to keep the solution of gentian violet separated from the aniline water, so as to prevent decomposition ; but we must avoid putting the glass rod, still wet with aniline water, into the colouring solution, and *vice versa.* Two different rods must be used. With a small rod put first on the slide a drop of aniline water, taken from the relative bottle, and soon after add a colouring solution. For, by placing the solution first and the aniline water afterwards, we obtain a greater precipitation of coloured granules in the preparation.

It is necessary that the particle of sputum should be well immersed in the colouring mixture, so as to have it well coloured on its under as well as its upper surface; because, if the particle is placed on the slide before the solution, the part adhering to the glass cannot be coloured. The particle of the sputum should not be larger than 1/4th to 1/3rd of a grain of millet; otherwise, when pressed between the two slips, it will press out from under the cover-glass, and consequently the best specimens, which are on the edges, will be lost, or the preparation will prove too thick for investigating the most delicate parts.

The particle of sputum, taken with a bent needle, should be immediately put into the colouring mixture, pressed down for a short time with the needle, then left for two hours under a watch-glass; a drop of distilled water may be added to it, if a weaker colouring be desired; or, in hot weather, to prevent hardening or evaporation.

After the required time has passed, take the sputum with a perfectly clean needle, put it in a watch-glass containing distilled water, and, by gentle agitation, it will be freed from the granular deposits. Meanwhile the slide must be well cleansed. Then put upon it a drop of distilled water, in which the particle is to be again immersed. By using glycerine we cannot obtain the proper thinness of the preparation, and many features would appear to be altered and colourless. Glycerine can ultimately be substituted by capillarity to preserve the preparation, the upper surface of the cover-glass, made greasy by the immersion, being washed first with benzoin and then with water.

In the preceding Memoir, we recommended the spreading and thinning of the particle of the sputum with needles; but for our present investigation it would not be advisable to do so. It was, perhaps, owing to this that we then lost some details, especially with regard to the fructifications of *Leptothrix*. In fact, spreading with needles must destroy the relations of continuity and contiguity necessary to preserve intact the said fructifications.

To thin the preparation as much as possible, we must place upon it the cover-glass, well centred, and let it act by its own pressure. In this way the mucus becomes thinned, and spreads slowly without altering materially the elements involved in it. When the preparation has become partially thinned, press down lightly and gently with the rod and leave it undisturbed, and so on until the very slender film of the coloured matter has in every direction neared the edges of the cover-glass; care being taken in these handlings to strain off the redundant liquid without wetting the upper surface of the cover-glass.

The preparations so mounted, in water, are in all parts very transparent, clear, and bright; they do not exhibit superposition of elements or corpuscles, and even in hot weather, keep well under the microscope for several days, if care be taken to wet now and then one of the edges, as we suggested in the first Memoir. Even the bacteria in the centre of the particle of sputum which were not at first reached by the colouring medium become in time gradually coloured, and these weak colourings are often very useful for examining minute details.

In order to obtain fructifications of *Leptothrix* from sputa, it is necessary to take the particle of sputum, not from the bottom, but from that side of the tube where, perhaps accidentally, a vertical streak of the mucus, slender and adhesive, has been left, as we shall see later.

For researches on fructifications, powerful immersion objectives are required, so as to be able to detect all details. We have used a No. 11 objective (1/18th homogeneous immersion) of Hartnack.

Collection and Preparation of the Contents of the Mouth.

We shall especially deal with the *patina dentaria* (surface of the enamel, tartar), etc., saliva and patina of the tongue.

Patina Dentaria.—The deep layers of this patina contain almost exclusively bud-growths and knots of large filaments which take a brilliant colour with the solution of iodine. The fructifications, on the contrary, are simply found on the superficial layers, so that for their investigation we must possibly remove the upper surface of the patina. By scraping the tooth with a small instrument or the nail, too much tartar is removed, which must be disintegrated twice over; first to select a particle, and then to thin and spread it on the slide I have best succeeded with a bent needle, first scaling lightly the labial surface of the tooth, and carrying the very minute particle of the patina on to the slide (on which was placed a drop of aniline water), and then scraping likewise the interstice between the teeth, to place on the slide another similar particle. In order not to spoil the fructifications, the scraping should be made neatly from top to bottom.   This manipulation should be done in the morning, fasting, and cannot give good results in those subjects who habitually clean their teeth with tooth-brush and powders.

Thus we get within the drop of aniline water two tiny islands of *Leptothrix*, but they are yet too thick to be reduced to the thinness required.   We shall break them with the needles, so as to carry on to the slide only a kind of whitish dust, which will occupy a third of the area to be covered.   Having done this, we shall add with the rod another drop of aniline water, and soon after, with a different rod, a drop of gentian violet solution.   We shall speak of the other colourings in Section 4, the treatment being nearly the same.   The preparation is kept for a little while under a glass, and when it appears to the naked eye to be sufficiently coloured, it is mounted without washing, which, in this case, would take away all the fragments, the fructifications will be set free, and the bacilli float about each tiny island.

When we wish to examine together the saliva and the *patina dentaria*, instead of placing a drop of aniline water previously on the slide, we will put on it a drop of saliva in the manner we shall indicate by-and-by ; the rest of the treatment is the same.   However, the preparations without the saliva turn out clearer and thinner ; whilst those mixed with it cannot become thin enough, owing to the resistance presented by cumuli of large buccal epithelia, which place themselves like supports between the slides ; still worse when one of those cumuli happens to be on a bed of *Leptothrix.*   Naturally, in the preparations mixed with saliva, the rich cumuli of *Diplococci* are more abundant, whilst they are seldom found on those of the patina only.

Lastly, the cover-glass is put on and gently pressed down with the rod, until the layer of the *Leptothrix*, spreading in every direction, reaches the edges of the cover, and the redundant liquid trickles down by its side. The preparations, stained with aniline colours, become a little hard (perhaps owing to the alcohol in it); but yet they get thinned enough. Those treated with the iodine, and especially those treated with picric acid, or the colourless ones, being more flexible, become more easily thinned.

In order to see the fructifications in a fair quantity and distinctly, it is necessary to wait some time. Naturally, those on the edges of each tiny island are seen more distinctly, but all do not show them, because many of the islands contain only filaments or bud-growths. Some are wasted, and others overlie one another, mixing their relative edges together. Now, in those edges with exuberant fructifications, these at first are close together; but the preparation in time partially dries up, and on pouring on it a fresh drop of distilled water the cover-glass slightly rises, and then the tufts of ears begin to open; the ears separate one from another, and (in preparations with picric acid) several may be counted in the visual field. Such observations can be continued for several days by adding to the preparation from time to time a drop of distilled water.

To demonstrate the productions by points, the method of working must be modified, as we shall see in Section 4.

Saliva.—To get the saliva as far as possible free from froth and air-bubbles, the following plan should be adopted :—Let the patient before tasting food spit on a well cleaned glass rod, held horizontally; then incline it so that the saliva shall run down towards the hand, parting, in its course, from the small air-bubbles, which, being lighter, adhere to the rod and stop the froth. When a drop of saliva has gathered on the handle, it may be taken off by the point of a second rod and carried directly on to the slide (if we wish to examine it with the *patina dentaria* and then mix them together); or, if we wish to examine them separately, it should be immersed in a drop of colouring mixture, already placed on the slide. As soon as the preparation is sufficiently coloured, it should be covered without washing. If the colouring is very weak, with slight vestiges of colour, the vibrating motion in the salivary corpuscles will be better detected.

The Cuticle of the Tongue.—On an émpty stomach, the surface of the tongue is scraped with a spatula or tongue-scraper. We take a small particle of the product and put it in the aniline water previously deposited on the slide. Afterwards it is carefully divided and coloured like the *patina dentaria.* However, owing to the superposed epithelia, these preparations can never be properly thinned, and therefore are only in a few points rendered clear enough for investigation under high powers.

Snow Crystals.

By J. H.

IT is almost impossible to imagine the extreme beauty and variety presented by snow crystals. During a sudden fall of temperature lasting only a few days in the last winter, very many intensely beautiful forms of crystals were preserved by the aid of photomicrography. Very many were made up of prisms of six facets. Many presented the appearance of being double; that is, two crystals alike in form were united to an axis at right angles to the plane of each. These were generally fine examples of the less minute crystalline bodies; the rays proceeding from the under crystals in most cases filling the intermediate space between those of the upper, and as crystals of a more complex order, and in most cases exhibiting a richness of effect hardly to be exaggerated. It was to me an extremely agreeable study of the various complicated forms presented to the eye, whether observed by a low power, or seen by the aid of the microscope, in all their beautiful details. A country friend engaged in a similar pursuit made a series of sketches, drawn about their natural size, with a crowquill; while another presented me with sketches made about the same time as my own, some of which presented wonderful details, surpassing any of those I had observed.

A collection of snow crystals, accurately recorded, might, it appears to me, be made to present an interesting feature in meteorologcal investigation. At the same time, it is highly probable that the conditions of their formation are more complex than might be imagined, familiar as we are with the conditions relating to the crystallisation of water on the earth's surface.

PLATE VII.

SNOW CRYSTALS.

PLATE III.

Apochromatic Objectives for the Microscope.

M R. J. G. Wright publishes the following paper in the *Journal of the Birmingham Natural History and Philosophical Society* (Vol. I., 1894, pp. 57—61) :—

"A great advance has been made during late years in optical science as applied to microscopical work, through the laborious and exhaustive investigations of the subject by Prof. Abbè and Dr. Schott of Jena. Those gentlemen, after years of patient labour and study, have discovered new vitreous compounds, by the use of which, and aided by the celebrated optical workshops of Herr Carl Zeiss, they have constructed lenses capable of forming images much more perfect than any hitherto seen, and indeed have lifted the art of the construction of microscope 'objectives' into an almost ideal position.

The nature and advantage of these improvements will be understood from the following considerations and explanations. The ideal desideratum in a lens or a combination of lenses is to produce an image of perfect definition, free from all colour inter-ferences. There are, however, some serious difficulties in the way of realising it. When rays of light pass through a lens, those passing through the axial zone are focussed at a point further from the lens than those passing through the peripheral zone, while those passing intermediate zones focus themselves at points intermediate between those two. Hence the image formed by the lens is imperfect in any one of these focal points. This difficulty is known as 'spherical aberration.'

When a beam of light passes through a lens, the various coloured rays are refracted at different angles. Thus, the violet rays are focussed at a point nearer to the lens than the red rays, and the intermediate colours at varying points between those two, so that the image viewed at any one of these foci is more or less imperfect and blurred. This difficulty is known as 'chromatic aberration.' Both these difficulties are partially corrected by a so-called 'achromatic' arrangement of lenses, in which a bi-convex lens of crown glass is closely attached to a plano-concave lens of flint glass, the latter possessing a lower refractive index and a less

dispersive power than the former.   By this means the spherical
aberration is corrected for *one* colour in the brightest part of the
spectrum, and the chromatic aberration is reduced by uniting *two*
of the colours of the spectrum in one point of the optic axis.
The remaining colours deviate and form what is known as the
'secondary spectrum.'   The above-named corrections represent
the highest conditions of achromatism previously attained even in
the very best lenses.

The new German crown and flint glasses, possessing different
relations of dispersing and refracting power, have, in combination
with a lens of fluor spar, made it practicable to destroy the secondary spectrum, or very nearly so, by uniting *three* of the colours in
one focus, and at the same time to correct the spherical aberration
in *two* colours.

This higher range of achromatism Prof. Abbè calls '*apochro-
matism.*'   It was a great advance in microscopic optics, and seems
to leave little room for further progress in the present state of
optical science.

In wide-angled objectives, where, by reason of its shape, the
front lens cannot be made achromatic, there remains still a trifling
colour deviation which interferes with the perfection of the image.
This Prof. Abbè corrects by the construction of what he calls
'compensating' eyepieces or 'oculars,' in which the under-correc-
tion of the objective is balanced by the over-correction of the eye-
piece, and so a perfect image is obtained.   And in order that these
eye-pieces may be used with all his objectives, Zeiss makes them
all under-corrected, so as to be balanced by the compensating
eye-pieces.

Another advantage accrues to these eye-pieces by the new
method of their denomination.   The old and almost meaningless
denomination by letters—(A, B, C, D, etc.)—is discontinued, and,
instead, they are designated by numbers—(4, 6, 8, 12, 18, and 27)—
each indicating the initial magnification of the ocular.   By this
means the total magnification of the object viewed is most readily
ascertained.   It is only necessary to multiply the initial magnification of the objective by that of the eye-piece (as ascertained by its
number) to give the combined amplification.   The power of the
objective is readily found by dividing the length of the microscope

tube (10 in. in the standard English instruments) by the nominal length of focus of the objective. Thus : a 1-inch objective magnifies 10 diameters, and if used with a 12 eye-piece the amplification is $10 \times 12 = 120$ diameters. Similarly, a ¼-inch objective magnifies 40 diameters, and if used with an 18 eye-piece the total amplification is $40 \times 18 = 720$ diameters.

Another very important advantage in the new system of construction arises from the fact that, whereas in the old achromatic arrangement the image formed by the lenses was so imperfect that deep-eye piecing could not be resorted to without almost certainly breaking down the image; in the apochromatic combinations amplification may be accomplished by the use of the deepest oculars without at all impairing the definition. This enables the worker to take advantage of the longer working distance, which the focal length of the lower power objective gives him, and also secures the greater amplification with less costly apparatus.

The advantages of the apochromatic system may be conveniently summarised as follows :—

In the Achromatic System.

The sharpness of the image is limited to *one* colour of the light rays—the green-yellow in ordinary lenses ; the blue-violet in lenses for photographic purposes ; the other rays giving confused images.

Complete colour correction is effected in one zone only of the objectives, the peripheral and central zones showing a marked imperfection. And only *two* colours are united in one focus in that zone.

In the Apochromatic System.

The images are very nearly equally sharp for all colours of the ⋆spectrum, and therefore scarcely affected by the colour of the illuminating light.

Colour correction is uniform in all the zones, while *three* colours are united in one focus. By this means the amount of focal difference for the various colours of the spectrum is reduced to less than 1/8th of its original amount, which is a practical elimination of the differences.

A considerably increased concentration of light, and the reproduction of the natural colors of objects, even of delicate tints, as well as the ability to use very high power *eye-pieces*, without breaking down the image, are advantages which will be keenly appreciated by the scientific worker.

All Zeiss's apochromatic objectives are made with relatively wide angles of aperture. A few years ago the contentions among microscopists as to the relative merits of high- and low- angled objectives were very strong, and even fierce ; but later experience from the extended use of the former has much modified the views of those whose contentions were in favour of the latter. The present opinion of the best experts seems to be that for critical *ocular* work the wide-angled glasses are greatly superior to the low, while for photographic and some other purposes the low-angled glasses with their longer focus have an advantage.

Prof. Abbè has made a very necessary and useful reform in the designation of the aperture of lenses. The aperture is the measure of the defining power of an objective. The wider the angle the greater the number of image-forming rays passing into the objective ; and conversely, the smaller the angle the fewer the rays. The old term, ' angular aperture '—which means the angle contained between two lines drawn from the focal point to the opposite sides of the pencil of light emerging from the back lens of any objective was sufficient to indicate the relative defining power of objectives, so long as *air* was the medium through which the object was viewed. But when ' immersion ' lenses were introduced, where water or oil became the media in place of air, the differing refractive powers of these fluids rendered the term inadequate. Obviously, the refractive index of the medium intervening between the cover-glass of the object and the front lens of the objective, must be a factor in the determination of aperture.

These considerations led Prof. Abbè to adopt the term ' Numerical Aperture ' (N.A.), and to establish for its determination the formula, $n \sin. u = \text{N.A.}$, where n represents the refractive index of the medium between lens and cover-glass, and sin. u

represents the natural sine of half the angle of aperture of the objective in air. The refractive index of air being taken as 1, water is 1·33, and the kind of oil usually employed is 1·52. On this basis the greatest possible theoretical angle in air—180°—is equalled by 97° in water and 82° in oil; that is to say, as many rays pass into the objective through oil with an aperture of 82°, and through water with 97°, as can be received through air in a " dry " objective with an angle of 180°.

The advantage, therefore, of the immersion system is obviously very great. The highest results which the present condition of optical science has placed within the reach of the microscopist are obtained, so far as the present writer is aware, by the use of the apochromatic system of homogeneous immersion lenses construct- ed by Carl Zeiss, of Jena, from formulæ of Prof. Abbè, and specially by an objective with an aperture of 1·60 N.A, with which an immersion fluid of higher refractive index than oil is used. There does not appear to be any scientific bar to the con- struction of objectives of a yet increased aperture. The difficulty consists in the discovery of a *suitable* immersion liquid, which should have an index of 1·8 or 1·9 to make any real advance on present attainments. Glass for the front lens and cover-glasses of the corresponding index could be made without difficulty so as to preserve the homogeneity.

Regarding the relative merits of the *best* German and English made objectives, the present writer is not aware of any superiority in any way in the English make, nor of any inferiority. The German glasses are first-rate in performance and in workmanship throughout ; whilst in the case of some of the lower powers (half- inch to quarter-inch) the cost is from £2 to £3 each less than their English rivals. The cost in all cases increases rapidly as the numerical aperture approaches the limits. A one-inch Zeiss' apochromatic objective with o·30 N.A. costs £6 ; a half-inch of o·65 N.A., £7 ; and the special 1/10 lens, above referred to, of 1·60 N.A., costs £40 : while the best English lenses cost £10 for a half-inch of o·64 N.A., and £50 for a 1/10th with 1·50 N.A. These high prices are a secondary consideration where scientific investigations demand their use ; but they are, without doubt, a bar to the extensive use of the new lenses among ordinary workers.

The Aquarium.

WE have often heard microscopists express a wish to start an aquarium ; but they always seem to be deterred from doing so under the erroneous impression that it will entail too great a tax on their spare time. We have assured them that this is not the case ; but our remarks have been received with a *Thomas*-like belief. We are, therefore, very glad to have received the permission of the Editor of the *British and Colonial Druggist* to reproduce the following article, which has just appeared in the *British and Colonial Druggists' Diary*, an extremely elegant and useful work :—

"There must be many . . . . who have often desired to follow some inexpensive 'hobby' with which to beguile the casual leisure they may have at their disposal. To those . . . . we can recommend the starting of an aquarium as an unfailing source of amusement and recreation. Those situated in large towns, even in the heart of the metropolis, need not be debarred from this pleasure, for all that is needed is one or two visits to a well-stocked pond to procure an ample supply of material, both animal and vegetable. For Londoners we know of no happier hunting-ground than that known as the 'Leg of Mutton Pond' at the bottom of the West Heath at Hampstead. From it we have obtained almost all the animal life mentioned in this paper, including many rare species ; most of the plants are also to be procured in it, or in the ponds in the near neighbourhood.

First, as regards the position of the aquarium. This should be placed in a good daylight, diffused, but not too strong ; otherwise, confervæ will grow too rapidly. . . .

In arranging our stock, we have always preferred the natural objects that are to be obtained from any pond or stream to the conventional plants and animals . . . which are, to our mind, uninteresting and commonplace. One friend [a pharmacist] for whom we installed an aquarium containing, *inter alia*, a score or two of lively tadpoles, had the pleasure of hearing one of his juvenile customers remark that 'Mr. X had a lot of his pills swimming about in water with tails to 'em.' . . . . The first point to be attacked is the bottom soil, in which to grow the plants ; i possible, this should consist of a pound or two of mud or clay from the bottom of a pond. This will probably contain many ova or larvæ, which will develop in due course into interesting objects. If this be not procurable, a good layer of silver sand will suffice. This should be covered to the extent of about an inch with red gravel, which in its turn should bear a stratum of about half-an-inch of larger stones and a few empty clean shells. A little tap-water should then be syphoned on, and the aquarium is ready for its plants.

Of all plants, none, in our experience, excels the common American pond-weed, *Elodea canadensis* (formerly called *Anacharis*). This plant will grow anywhere with the greatest profusion; it need not be rooted; a few sprigs thrown into the water will speedily send down roots of their own accord. So luxuriantly does it flourish, that it will generally require a little judicious pruning to keep it within bounds. Another excellent plant is *Ceratophyllum submersum*, the hornwort. This should he taken with the roots and planted; it will then live for years. It frequently has adhering to the leaves the interesting rotifer, *Melicerta ringens*, and also sometimes the lovely *Stephanoceros eichornii*, which, to the owner of a microscope, will be an unfailing source of pleasure. The common pond-weed, *Potomageton heterophyllus*, the perfoliate pond-weed, *P. perfoliatus*, also *P. pectinatus*, are useful and attractive plants. All may be found in the canals round London or in the country. A root or two of the water crawfoot, *R. aquatilis*, will also grow well, and give a beautiful effect with its finely divided, fennel-like, submersed leaves. If taken when in flower in the early summer, it will continue to bloom for weeks, the elegant white flowers being as attractive in appearance as those of a rare exotic. A few plants of one of the duckweeds may be usefully included; but too many should not be introduced, as when their location suits them they increase by budding very fast. The great duckweed, *Lemna polyrhiza*, and the ivy-leaved duckweed, *Lemna trisulca*, are preferable to the other species, being neater and not so prone to decay and leave unsightly withered fronds.

Having introduced the plants, rooting such as have roots, and placing a few small stones over the soil to prevent them from dragging their anchors, the aquarium should be placed where it is

intended to remain, and the rest of the water should be syphoned on. This is best effected by means of a piece of feeding-bottle tubing, a water-can full of tap water being placed at a higher level than the aquarium, such as on the top of a pair of steps, the shorter end of the tube being in the can, the longer in the aquarium. By this means the minimum disturbance of the contents is effected, and the water more readily settles down bright.

This done, the aquarium is ready for its animal life ; and in the selection of the various members of the happy family no small degree of care is requisite, otherwise the many are apt to suffer for the benefit of the few, for numbers of aquatic insects are most voracious and carnivorous, even cannibalistic in their habits. The first selection should be directed to molluscs. A few water-snails are indispensable ; not only do they keep the plants within bounds, and by their excreta form a suitable soil for continued healthy growth, but by browsing on the glass sides clean away the confervoid growths, which would otherwise speedily obscure the view. The best for the purpose we find to be *Limnea auricularia*, which may be found crawling on the bottom of most muddy ditches. It has a very wide-mouthed shell, and is often prettily blotched and marked ; its food consists almost entirely of the minute confervæ, which are apt to grow too luxuriantly if the aquarium is exposed to strong daylight. It is preferable to the larger *Limnea stagnalis*, which is generally recommended, for the latter will attack the growing plants. *L. stagnalis* is known at once by its larger size and longer corkscrew-pointed shell. Another genus of useful snails is *Planorbis*, known by its wheel- or "ammonite"- like shell. Two species—*P. corneus* and *P. marginatus*—are most frequently met with, and both are excellent for our purpose. Another snail is interesting, but less often come across—*Paludina vivipara*—known by its resemblance to the familiar periwinkle, and, like it, possessing a horny scale or shield, called the operculum, with which, when the animal retires within, the mouth of the shell is closed. It also has a peculiar elephant-like proboscis, at the extremity of which the mouth is placed. This snail is generally found upon wooden stakes or posts which stand in deepish water, and may be captured by running the hand along the surface, when they may be felt and detached where they cannot be seen.

Among the bivalves, a specimen or two of the cockle-like *Sphærium corneum* and half a-dozen of the pea-shells (*Pisidium*) may be included. At the outside not more than a couple of dozen molluscs altogether should be introduced at once, since, if their surroundings are suitable, they increase very rapidly. The tendency with beginners is to overstock with animal life. To prevent disappointment, it is better to err on the other side, for excess of vegetable life is less likely to give bad results than too many animals. When the crop of confervæ grows so fast that the snails cannot keep it within bounds, we have found an old tooth-brush fastened to the end of a glass rod a very useful weapon in cleaning the glass. After snails, a few beetles should be taken, but only the smaller kinds, since several of the larger ones are very voracious. This is particularly the case with *Dytiscus marginalis*, the common large pond beetle, which both in its larval and perfect state of development, should be avoided, for if introduced, it would speedily devour every other living animal. Another voracious beetle, smaller than *D. marginalis*, is *D. (Ilibias) ater*, a shining black species, which should also be given a wide berth. Exception in the bannings of the large beetles must be made in favour of the largest, the great water beetle, *Hydrophilus piceus*, which is entirely harmless. It is not very common, but occurs in the neighbourhood of London. It may be known by its appearance and size, resembling a female stag-beetle, and by the beautiful play of colours shown on the elytra when swimming in the water. Another harmless and very active beetle, constantly on the move, somewhat resembles a flattened acorn in size and shape, is *Agabus agilis*. A large flattened beetle, *Acilius canaliculatus*, is also to be recommended, for although carnivorous, it will not generally attack living prey if it can find dead animal matter. It therefore becomes a very useful scavenger. Most of the smaller beetles may be included; among others, *Agabus maculatus*, which has blackish elytra with white spots, and also *Laccophilus minutus*. The small oval *Hyphidrus ovatus*, always on the move, is also an attractive insect in captivity; with it may be classed *Cnemidotus cæsus*, which may be known by its grooved and dotted wing-cases of yellow colour. A few whirligig beetles—the little black *Gyrinus natator*—should be captured; they are quite inoffensive. It should be borne in mind that all these beetles are not entirely aquatic; consequently, if kept in an uncovered aquarium, are apt in summer to take flight during the night. They should be kept in a vessel provided with a cover.

A very common but curious insect is the water scorpion, *Nepacinereus*, a member of the *Hemiptera*. Although far from being a vegetarian, it may be safely kept in an aquarium with other things if occasionally fed with flies. These it catches with its first pair of legs, and thrusts it upon its sharp-pointed snout, through which it extracts their blood. Its appearance resembles a small flattened leaf, grey above, and often reddish beneath, with a long pin-like tail. It scrambles about in the water upon the weeds or rests on the surface with its fore-legs extended on the watch for flies. There is another water-scorpion, *Nepa lineatus*, but much rarer than this, which is a very close approach to the exotic stick insect in appearance; in fact, when drawn from the water and feigning death, it is difficult to recognise a living insect. This is similar to the common water-scorpion in its habits, and in attitude recalls the well-known 'praying mantis' while laying in wait for flies. Both these insects are very subject to the attacks of parasitic acari, which will seem adhering to them like minia-ture sheep-ticks.

No aquarium should be without a couple of water-boatmen, *Notonecta glauca*. They may be known at once by their curious action when swimming on the back, recalling an out-rigged sculling-boat. These, too, are not above staying their stomachs with their fellow captives, but will not do so if fed on flies, which they will even seize from the fingers, clasping their prey as though nursing it, and keeping it beneath the surface until it is drowned. There are three common species—one large, and with hard, coloured wing cases; the others flatter, and with elytra reticulated, *Conixa striata* and *affinis*. Unfortunately, the first preys greedily upon the latter two, so that they cannot be kept together for long.

Among larvæ we have a wide selection of curious forms, which may be successfully reared, but in collecting we should be cautious not to include any with large sickle-shaped or knife-like jaws; such are certain to be carnivorous and predatory. A little

practice will soon enable the collector to distinguish between the
murderous crew and the innocent ; but not always, for some of
the former positively wear a moveable mask, like the highwaymen,
behind which they hide their weapons of offence. An eye should
be kept on the aquarium, and any aggressive member of the stock
removed by means of a very small muslin net, or a wide glass
tube used as a pipette ; an inverted thistle-funnel is useful for this
purpose. A few caddis worms (*Phryganeidæ*) should be taken for
the quaint appearance of their cases, different species constructing
their curious abodes with various material ; only one or two
specimens should be included however, otherwise the stock of
plants will suffer. Another animal, the water-spider, *Argynotea
aquatica*, deserves a place. It may be found in any pond, running
on the surface, and then disappearing with the metallic sheen of a
globule of mercury. When captured, it should be carried home
in a box with moist weed, and not in water, for if shaken about in
water, it speedily drowns.

Among the more highly-developed creatures, a few tadpoles
deserve a place. In early spring the spawn may readily be found,
that of frogs in masses, toads in strings, and newts wrapped in
leaves. A few eggs should be placed in the aquarium ; from this
the development of the young batrachians is most interesting to
watch.

If it be desired to include fishes, a single specimen, or a pair,
male and female, of sticklebacks may be added. If, however,
the aquarium be intended as a nursery for microscopic pond
life . . this addition is hardly to be recommended.

For collecting the minuter forms, which in the limit of this
note cannot be noticed *seriatim*, a brother pharmacist adopts the
following plan, which we have followed with great success. A
small filter bag is made of fine cambric, and fastened to a ring of
copper wire, which is afterwards twisted out on each side of the
ring, and attached to two long meat-skewers. Arrived at the
pond, the skewers are stuck in the bank, supporting the filter between them, and the latter is kept charged with pond water, ladling
with a wide-mouthed bottle attached to a stick. By this means,
myriads of minute forms of life, such as Hydræ, Desmids, Cyclops,
Daphnia, etc., are captured in a few moments ; the ' concentrated

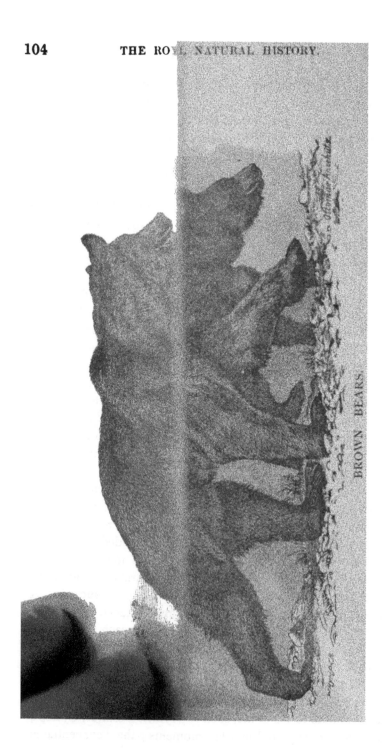

BROWN BEARS.

Microscopical Technique.

COMPILED BY W. H. B.

Hodgson's Compressor.—This simple form of parallel compressor has been devised by Mr. J. V. Hodgson, of Birmingham, to meet the requirements of those who wish for an effective instrument at a moderate cost, and at the same time permitting the use of dark ground and oblique illumination of wide angle. It consists of a plate of brass, 3 in. by 1⅜ in. with a central hole ¼ in.

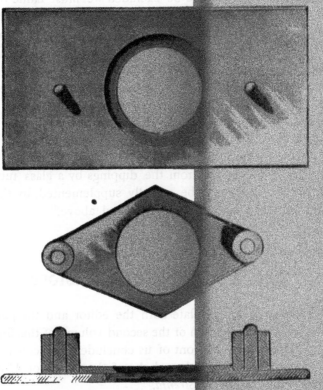

in diameter, over which is cemented a cover-glass. On either side of the hole is an upright peg. The upper plate is diamond shaped, and the central hole corresponds to that in the lower plate. The two ends of the diamond carry stout tubes fitting closely to the upright pegs of the lower plate, and enable it to be adjusted to a nicety. The upper plate cannot be raised by one hand alone. Mr. Hodgson also makes a form in which the cover-glass on the lower plate is sunk below the level, while the upper plate is infusion' is then transferred to a larger bottle, which is emptied into the aquarium with as little delay as possible. By this means food is supplied to the inmates, and numerous curious additions obtained to the permanent stock. The Hydras are very attractive creatures, of which two kinds may be expected in most localities, *H. fusca* of a dull greyish or pinkish tinge, and *H. viridis*, a bright green ; the latter is interesting, being an animal which owes its colours to chlorophyll.

Among the creatures to be excluded we must name the common water wood-louse, *Asellus aquaticus*, and the fresh water shrimp, *Gammarus pulex*, which, although said to be carnivorous in their habits, we find to be very destructive to plants, biting through the growing stems and thus cutting them adrift.

With regard to apparatus for collecting, the only indispensable article is a good pair of eyes. A vasculum holding a few quinine vials, and a small muslin net, which can be tied to the end of a walking-stick, is all that is needed ; the net may be supplemented with a bottle similarly attached, for the smaller forms of life, which may be picked out from the dippings by a glass tube used as a pipette. These may be usefully supplemented by the filtering arrangement which we have described above."

"The Royal Natural History." *

W E have to congratulate both the editor and the publishers on the completion of the second volume of this fine work.

The volume in front of us concludes the account of the Carnivora, and includes the whole of the order of Ungulates, finishing with the order *Sirenia*.

Where all is so good, it is difficult to quote any especial part without citing so much that the publishers would be down upon · us for pirating. But we cannot resist giving the following anecdote relating to " the lasting and pernicious effects of even a drop of skunk secretion." The story is told by Mr. W. H. Hudson about

* Edited by Richard Lydekker, B.A., F.R.S. (London : F. Warne and Co. Published monthly. Price 1/-) Vol. II. (Parts 7—12), 1894.

a settler who started one evening to ride to a dance at a neighbour's house :—" It is a dark, windy evening, but there is a convenient bridle-path through the dense thicket of giant thistles, and striking it he puts his horse into a swinging gallop. Unhappily, the path is already occupied by a skunk, invisible in the darkness, that, in obedience to the promptings of its insane instinct, refuses to get out of it, until the flying hoofs hit it and send it like a well-kicked football into the thistles. But the forefeet of the horse, up as high as his knees perhaps, have been sprinkled, and the rider, after coming out into the open, dismounts and walks away twenty yards from his animal, and literally smells himself all over, and with a feeling of profound relief pronounces himself clean. Not the minutest drop of the diabolical spray has touched his dancing-shoes. Springing into the saddle, he proceeds to his journey's end and is warmly welcomed by his host. In a little while people begin exchanging whispers and significant glances. . . . Ladies cough and put their handkerchiefs to their noses, and presently begin to feel faint and retire from the room. Our hero begins to notice that there is something wrong, and presently discovers its cause. He, unhappily, has been the last person to remark that familiar but most abominable odour, rising like a deadly exhalation from the floor, conquering all other odours, and every moment becoming more powerful. A drop has touched his shoes after all."

Our space is now exhausted, but we must call attention to the descriptions of the general characteristics of the several orders, which Mr. Lydekker has made a special feature.

The work is profusely illustrated with most excellent woodcuts. We give on the opposite page a figure representing Brown Bears on the march, in which the awkward and ludicrous movements of the animals are admirably depicted. The plates are also extremely good, with the exception, in our opinion, of the plate of Fallow Deer, which does not come up quite to the standard of the others, and we cannot say that we quite admire the plate of the Indian Elephant, whose forequarters appear to be a little too thin for his height. But these, perhaps, may be mere trivialities. The highest praise we can give to Mr. Lydekker is that we consider the work should be on every naturalist's table.

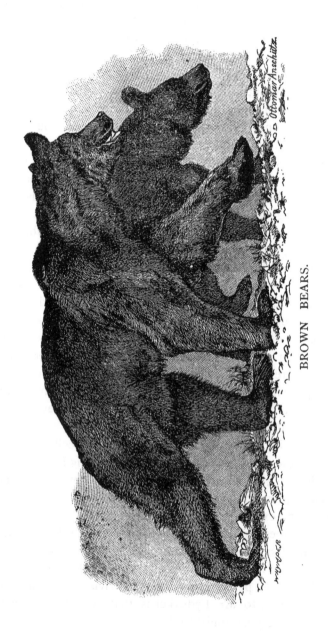

BROWN BEARS.

Microscopical Technique.

COMPILED BY W.H.B.

Hodgson's Compressor.—This simple form of parallel compressor has been devised by Mr. J. V. Hodgson, of Birmingham, to meet the requirements of those who wish for an effective instrument at a moderate cost, and at the same time permitting the use of dark ground and oblique illumination of wide angle. It consists of a plate of brass, 3 in. by 1⅜ in., with a central hole ¾ in.

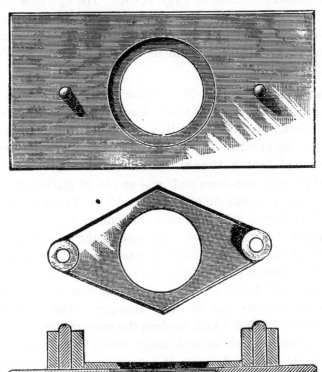

in diameter, over which is cemented a cover-glass. On either side of the hole is an upright peg. The upper plate is diamond shaped, and the central hole corresponds to that in the lower plate. The two ends of the diamond carry stout tubes fitting closely to the upright pegs of the lower plate, and enable it to be adjusted to a nicety. The upper plate cannot be raised by one hand alone. Mr. Hodgson also makes a form in which the cover-glass on the lower plate is sunk below the level, while the upper plate is bevelled to admit of high powers being used close to the edge, as in illustration. The compressors can also be obtained thickly nickelled, thus preventing tarnishing, which to some persons is so objectionable.

Filtration of Agar-Agar.*—Dr. W. S. C. Symmers finds that the ordinary methods of filtering agar-agar takes too much time. He employs the method used at the Pasteur Institute. The important requisite in this method is the filter-paper known as " papier chardin," made by Cogit et Cie. The agar-agar is heated in an autoclave to 120° C. and poured at once on to the filter-paper in a cold funnel. The agar-agar filters as rapidly as nutrient gelatine does in the ordinary method, and a litre may be obtained in half-an-hour.

Embedding Medium.—M. Camille Brunotte describes in the *Journal de Botanique* a new mode of utilising gelatine as an embedding medium, by which the action of heat upon tissues containing water can be avoided. Thin sheet gelatine, 20 gm., is dissolved by the aid of heat'in 100 gm. of distilled water. The solution is filtered through fine linen and 30—40 cm. of glacial acetic acid, and a gram of bichloride of mercury added. This addition keeps the gelatine liquid at the ordinary temperature (15° C.) and of the consistence of thick syrup. To embed the material, a small quantity of the liquid gelatine is poured into a little mould made of thick absorbent paper (papier buvard), and the object placed in it. The liquid being transparent, the position of the object by immersion in water can be easily arranged. The whole is then immersed in alcohol, which hardens the mass of gelatine.

In cases in which alcohol might render cloudy the cell contents, picric acid, bichromate of potassium, or chrome alum may be used for hardening the gelatine, but these re-agents require a much longer time to act. The sections, as soon as cut, can be mounted in gelatine or glycerine, or can, if required, be quickly freed from the thin coating of gelatine that surrounds them.

Fixing Celloidin Sections.—Prof. H. E. Summers, of Cornell University (U.S.A.), recommends the following method of fixing celloidin sections :—

" The sections are first placed in 95 per cent. alcohol for a

* *British Medical Journal*, No. 1765, 1894, p. 951.

minute or two ; they are then arranged on a slide, and the superfluous alcohol drained off by tipping the slide, after which sulphuric ether-vapour is poured over them from a bottle partly filled
with liquid ether.   Under this treatment the celloidin softens and
becomes transparent.   The slide may then be immersed in 80 per
cent. alcohol, and the section afterwards stained, anhydrated, and
mounted in balsam by the usual process, as the ether treatment
serves to fasten the sections firmly to the slide.—*Journal Recon-*
structives.

Styrax for Mounting.—Styrax is being recommended in England and America as a mounting medium in certain cases.   As
the balsam of the shops is always full of dirt and impurities, it
requires preparation before it can be used in microscopy.   The
best plan is to to dissolve and filter it.   To do this, dissolve a
portion of the styrax in sufficient benzol to make a liquid of a thin
syrupy consistence, and filter through two thicknesses of Swedish
filter-paper.   Use a large filtering-funnel, fitted with a cap to
prevent evaporation, and let the filtering-paper reach not much
more than midway up the sides.   Should the medium become too
thick to pass through, add more benzol.   The product will be too
thin to use at once, and may be set aside, properly protected from
dust, until sufficient benzol evaporates.

Wheat and Rye Starch.*—E. Guenez points out that these
starches possess very similar characters, and it is difficult at times
to say decidedly that a given sample consists of one rather than
the other.   To distinguish the two kinds, he recommends that a
little of the material be mounted in water for examination with
the microscope.   The wheat-starch will then be seen to contain
comparatively few split grains, which possess an isolated fracture
situated near the edge or proceeding from the centre to the circumference.   In the case of rye starch the split grains are more numerous, and possess a star-shaped fracture with three or four branches,
apparently originating in the centre of the grain and rarely reaching the edge.   Some grains may also be found which have only a

*Bull. de Pharm. de Bordeaux, XXXIV., 289, in Pharm. Journ., Nov. 3, 1894,
p. 356.

linear crack ; but this will be larger in the centre of the grain than towards the edges, just the reverse of what occurs in wheat.

Iron-Hæmatoxylin and Centrosomes.*—Iron-hæmatoxylin has been used by Heidenhain in the study of the centrosomes and astrospheres. The original process, which is also repeated in the new modification, was the following :—

Fine sections of preparations in sublimate are fixed on the slide by means of distilled water, dehydrated with alcohol containing iodine, and exposed to a $1\frac{1}{2}$ per cent. solution of ammonio-ferric alum. The crystals of this salt should be clear violet in colour; if they are yellowish and opaque, they have suffered from exposure to air and are no longer fit for use. The solution must be made cold, as the salt is decomposed by heat. The slide is next washed with distilled water and then placed in a $1\frac{1}{2}$ per cent. solution of *Hæmatoxylinum purissimum* (Grübler). The over-stained sections are then again treated with the iron-alum solution used before, in order to remove the superfluous colour. The process of extraction must be followed under the microscope and continued until the cell protoplasm is completely decolourised, and the chromatin network of the nucleus becomes clear. One may interrupt the differentiating process any moment by washing with fresh water, and then continue it. When the extraction of the stain has been carried far enough, the slide should be washed fifteen minutes in fresh water and mounted in the usual way in balsam.

Heidenhain noticed that when the differentiation was effected quickly, the centrosomes were stained in greater number than when the process occupied a long time. It seemed, therefore, that the defects of the method might be corrected if a way could be found by which the decolouring process could be hastened. How could the cytoplasm be freed from the stain in the shortest time? Assuming that a stain acts by chemical combination, it seemed probable that the process of extraction might be hastened, if the receptivity of the cytoplasm could be at least partially saturated before the application of the hæmatoxylin. Accordingly, Heidenhain selected as *preliminary* stains (" Vorfarben ") such as affect the cytoplasm and

* *American Naturalist*, XXVIII. (1894), pp. 976—977.

the nucleus, and leave the centrosomes unstained.   Thus, the chemical affinities of the centrosomes for the hæmatoxylin would remain at full strength; while those of the cytoplasm and nucleus would be more or less saturated, and to the same extent weakened for the hæmatoxylin.   In this way the process of extraction was brought under some control, and the method greatly improved.

Stains reached in this way are called " subtractive."   Bordeaux R., Anilin blue and Methyl-eosin were employed as preliminary stains.   Bordeaux R. proved to be the best.   In preparations that have been successfully differentiated as to the centrosomes, the nucleus and its chromatin are almost colourless, so that the centrosome may be easily studied, even when it lies behind the nucleus.   The *nucleoli* remain strongly stained.

The Chromatin.—Heidenhain shows that there are two kinds of chromatin to be distinguished—namely, an *oxychromatin* brought out by *acid* anilin stains (*e.g.*, Rubin S.), and a *basichromatin*, which is brought out by *basic* anilin stains (*e.g.*, Methyl green).   The "basichromatin " is the chromatin of Flemming and authors in general.

The differentiation of the two chromatins can only be accomplished when the nucleus is exposed at the same time to both *acid* and *basic* anilin colours, as is the case when Biondi's solution and Ehrlich's triacid are used.

If one mixes ammonium vanadate with hæmatoxylinum pur (Grübler), a blue stain is obtained which stains *cytoplasm* and *oxychromatin* strongly, while the *basichromatin* is often left nearly colourless.   The two chromatins probably differ only in the amount of phosphorus present, basichromatin containing more, oxychromatin less.

———————————

ALUMINIUM has the property, when used as a pencil, of leaving an indelible mark on glass or any other substance having a siliceous base.   A deposition of the metal takes place, and, while this may be removed by a suitable acid wash, the mark itself cannot be removed by rubbing or washing.   Magnesium, zinc, and cadmium have a similar property, but the mark of magnesium is easily removed, the application of zinc requires a wheel, and zinc and cadmium tarnish ; while aluminium is permanent and remains bright.

Notes.

A new material for pathological casts is advocated by Dr. C. W. Cathcart in the October number of the *Journal of Pathology and Bacteriology*. It is obvious that if it is of service for pathological reproductions, its usefulness can be extended to many other subjects. In searching for some material for casts, Dr. Cathcart turned his attention to the papier-mache methods. His principal difficulty was to get surface detail, and after trying many papers he found none so satisfactory as the " Robosal" blotting-paper, which is made of wood pulp, and while very strong and tough while dry, is exceedingly absorbent, and when wet becomes very soft and pliable. It is made of various thicknesses and almost any shade of-colour, although white is preferable, because it may be coloured afterwards according to the specimen.

The surface detail on casts made of " Robosal" is quite fine enough for ordinary purposes. Thus, the skin markings on the back of the hand can be brought out distinctly. It is much cheaper than plaster of Paris, and being only a hollow shell of paper weighs in ounces what the other weighs in pounds. It is tough and can be knocked about with impunity. The interior of the *moulds*, which of course are of plaster of Paris, must be well smeared with vaseline, tallow, or other greasy substance to which paste will not stick.

The method is described as follows :—"The casts are made of ' Robosal' blotting-paper (Messrs. Robinson, Liverpool), about the thickness of ordinary writing-paper, and ordinary flour-paste. After the mould has been prepared as above, a piece of ' Robosal' is laid on a flat plate and the paste is rubbed into it on both sides till the paper is limp and soft. While the softened paper is still lying flat on the plate, scraps are torn from it so as to leave their edges frayed and thin. If necessary, the paper can be torn into thin films, where finer markings are to be cast. Beginning at the margin of the mould on one side, the operator lays on a piece of the torn softened paper, and presses it with his finger or a stiff brush firmly into the irregularities of the mould. Overlapping this he puts on another piece, and another and so on, until the whole of the mould is covered, being careful to avoid wrinkling the soft paper. Over the first layer a second is laid, and sometimes a third. For small casts this will suffice, but for larger casts an additional layer of ' rope,' brown paper, or of muslin, soaked in paste as before, is to be applied in the same way. It is advisable to put on all the desired layers at once, so that in drying they may shrink uniformly. It is also important to let the paper stick to the margin of the mould by leaving them ungreased. This insures a good shape. When all the required layers have been put on, the cast is dried. Slow drying in free air and warmth is preferable, and takes from twenty-four to forty-eight hours ; but quicker drying in an oven may be used if need be. When the cast is dry, the inside is painted over with a coat of spirit varnish. The paper is then cut from the mould at the margins, and the cast will be found to have shrunk slightly from the surface, and thus be loose. With care, it can then be worked out from the crevices and corners. The rough edges are next trimmed with scissors and bound with blue paper. A thin layer of size is then painted over the cast. As soon as this is dry, the necessary painting can be carried out. Oil colours should be used, mixed with turpentine if a dry surface is wanted, or mixed with plenty of megilp, and afterwards varnished, if a moist surface is to be imitated. The time required for making each cast is about the same with papier-mâché as with other materials."

How are Young Spiders Fed ?—In my rambles for botanical specimens in the last three years, many new and curious things have been thrust upon my attention in the insect world, and these I have recorded for future use. One fact in particular struck my attention, and I herewith submit it to the readers of *Science*, partly to record the fact, and partly to ask if any other readers have ever observed a similar fact. We have been taught by the best works on spiders, that the young of spiders derive their food mostly from the atmosphere. The " Encyclopædia Britannica " confirms this view.

On the 19th June, 1891, I discovered in a ploughed field an enormous spider, of the Lycosidæ species, which was 1½ inches long. She presented a very curious appearance, being covered with scores of tiny spiders from one end of her body to the other. When I touched her with a weed stem, the young spiders scampered off at a lively rate, only to return when left to themselves. The spinneretts and abdomen of the mother spider were greatly distended. Suddenly there was a copious flow of white liquid, which the young greedily devoured. Examining the fluid under my microscope, I was fully convinced that this was veritable milk, and that this spider at least nursed her young, instead of bringing them up on atmospheric moisture. I shall be glad to know if any reader of *Science* has ever observed a similar occurrence.

Naples, N.Y. John W. Sanborn.

Purification of Water by Green Plants*.—The prevalent idea that green plants (flowering plants and algæ) growing in

Pharm. Journ., Nov. 5, 1894, p. 356.

running water add to its impurity and unfitness for drinking purposes, has long been shown to be erroneous. As long as they are in a growing condition, they can only tend to purify the water by giving out oxygen into it. In a paper in the *Archives für Hygien* (1894, No. 2), Dr. T. Bokorny maintains that aquatic bacteria also have a share in the purification of water, so long as it contains a considerable quantity of organic matter. A series of experiments carried on by the author showed the algæ are capable of decomposing fatty acids, such as butyric and valerianic, as also glucose, leucin, and tyrosin.

CANADA BALSAM FIR.—The following note from the *Drug Reporter* will be interesting to microscopists :—

"While there has been little or no increase in the consumption of Canada balsam within the past five years, the supply has been gradually diminishing, partly as a result of natural conditions, and partly because the work of gathering has been, to an extent, neglected. The collection of balsam fir is not a regular industry, but has been prosecuted by lumbermen and labourers in other fields, who devoted their leisure to it. So long as the balsam could be obtained near by the markets to which the gatherers bring it, the supply was ample and regular; but with the cutting-down of the Canadian forests for timber the source of supply has been further and further removed from commercial centres, and the collection of the balsam has not proved profitable enough, for a number of years past, to encourage those who heretofore engaged in it to continue to bring it to market.

In some years the supply has been larger than in others, owing to the scarcity of other employment, and in view of the widespread distress among the labouring classes, including the Canadian lumberman, during the past eighteen months or more, it would be natural to expect that these people would have turned their attention to the gathering of balsam as a means of livelihood. Such, however, does not appear to have been the case, for according to reliable reports the quantity collected this year was very small, not exceeding fifty barrels. This small yield was partly due to wet weather toward the end of the gathering season—late August and early September—but the chief reason for it is that the gatherers found it unprofitable to go so far into the interior for the balsam as they are now compelled to go.

The Grand Falls of Labrador.*

A N account of his visit to the Grand Falls of Labrador has been given by Henry G. Bryant, of Philadelphia, in the *Century Magazine* and in a *Bulletin* of the Geographical Club of Philadelphia. They are situated on the Grand or Hamilton River, which rises in the lakes of the upland region of the peninsula and flows in a general south-easterly direction into Hamilton Inlet—the great arm of the sea which, under various names, penetrates into the interior a distance of one hundred and fifty miles. No scientific explorer has advanced far into the country, and all that is known of it is derived from vague information furnished by Indians, a few missionaries, and the Hudson Bay Company's men. The first white man to visit and describe the falls was John McLean, of the Hudson Bay Company, in 1839. They were visited twenty years afterward by Joseph McPherson. These are the only white men who are known to have seen the Grand Falls till the summer of 1891, when Mr. Bryant and an expedition from Bowdoin College reached them independently of one another. Mr. Bryant, accompanied by Prof. C. A. Kenaston, of Washington, and a Scotch and an Indian assistant, left North-west River Post, at the head of Hamilton Bay, on August 3rd, to proceed up the stream by canoe. On the 27th they reached the point where the further navigation of the stream is obstructed by rapids, whence they proceeded overland and reached the Falls Sept. 2nd. "A single glance showed that we had before us one of the greatest waterfalls in the world. . . . . A mile above the main leap the river is a noble stream four hundred yards wide, already flowing at an accelerated speed. Four rapids, marking successive depressions in the river bed, intervene between this point and the falls. . . . An immense volume of water precipitates itself over the rocky ledge, and under favourable conditions the roar of the cataract can be heard for twenty miles. Below the falls, the river, turning to the south-east, pursues its maddened career for twenty-five miles shut in by vertical cliffs of gneissic rock, which rises in places to a height of four hundred feet. The rocky banks above and below the falls are thickly wooded with furs and spruces, among which the graceful form of the white birch appears in places." The height of the falls was found, by as accurate a measurement as could be made with cord, to be three hundred and sixteen feet.

* *Popular Science Monthly.*

INTERNATIONAL JOURNAL OF MICROSCOPY AND NATURAL SCIENCE.
THIRD SERIES. VOL. V.

Reviews.

A NEW ENGLISH DICTIONARY, on Historical Principles, founded mainly on the materials collected by the Philological Society. Edited by James A. H. Murray. D to DECEIT, commencing Vol. 3 ; F to FANG, commencing Vol. 4, by Henry Bradley, Hon. M.A. Oxon. (Oxford : The Clarendon Press. 1894.) Price 3/6 and 2/6.

These two parts are compiled with the same care as has marked all the previous issued parts. The past and present method of spelling, with illustration of its use, are given of every word from the 10th century. The letters A, B, C and E have already been published, and we are pleased to notice that one section, at least of each of the letters D and F, will be published quarterly during 1895, at 2/6 each part. It is a grand work.

McDOUGALL'S HIGHER GRADE ARITHMETIC in Theory and Practice. Cr. 8vo, pp. vi.—199. (London : McDougall's Educational Co., 4 Ludgate Square.)

In this book, the author has reasoned out and demonstrated to pupil teachers and others so much of the science of arithmetic as is needed for examinations in the Education Department. Commercial Arithmetic has received special attention, and a chapter is devoted to the Metric System.

A NEW SCHOOL METHOD. Part 3. How to teach the class subjects—Geography, Grammar, History, and Elementary Science. By Joseph H. Cowham. Cr. 8vo, pp. 376. (London : Westminster School Book Depot and Simpkin, Marshall and Co. 1894.) Price 1/6.

On a former occasion, we noticed the first two volumes of this series. The object of the author is to show how the teacher may combine the highest training, with the best instruction. We are struck with the simplicity of the method employed, and with the possible satisfactory results. There are a number of helpful illustrations.

A MONOGRAPH of the Land and Fresh-Water Mollusca of the British Isles. By John W. Taylor, F.S.S., with the assistance of W. D. Roebuck, the late Chas. Ashford, and others. Part I. October, 1894. (Leeds : Taylor Bros.) Price 6/-, or by subscription 5/- (5/3 post free.)

The object in issuing the present work is to bring together, as far as practicable, all reliable information bearing on the study, and thus form a standard work of reference as well as a reliable text book upon British Land and Fresh water shells. The first volume will be devoted to the general treatment of the subject, the different forms and characters of the shells, the morphology of the animal and description of the structure and functions of the various organs of its body.

The work is profusely illustrated, there being 138 wood cuts and one coloured plate in the first part. We are desired to state that the author will be pleased to receive assistance from Microscopists and Micro-photographers on the subjects of dentition, microscopic structure, and other peculiarities. The work promises to be a very important one, and we hope to refer to it again in our next.

EUROPEAN BUTTERFLIES AND MOTHS. Cr. 4to. (London Cassell and Co.)

Parts 1 to 7 of this very important work have reached us. Each number contains a superbly coloured plate showing the butterflies in all their stages, together with the food plants of the larvæ. No. 1 contains also a plain plate describing very thoroughly the anatomy of the Lipidoptera. The parts are published monthly at 6d. each. ———

THE BIRDS ABOUT US. By Charles Conrad Abbott, M.D. 8vo, pp. 288. (Philadelphia, U.S.A. : J. B. Lippincott Co. 1895.)

A thoroughly interesting book is now before us. In the introduction, comprising about a dozen pages, the author gives a pleasing history of bird-life generally, and hopes that the book may add something towards the growing disposition to cultivate rather than persecute our feathered friends. There are 25 chapters, in which the different families of birds are very interestingly described. We find, also, there are 24 plates, principally showing nests and homes of birds, and 50 illustrations in the text.

———

A DICTIONARY OF BIRDS. By Alfred Newton, assisted by Hans Gadow and others. Part 3. 8vo, pp. 577—832. (London: Adam and Charles Black. 1894.) Price 7/6.

This part reached us just as we were going to press. It covers from MOA to SHEATHBILL, and gives besides descriptions of those birds whose names fall between these letters, very good articles on the Muscular and Nervous Systems, Nidification, etc. With the publishers' permission we shall hope to refer to this interesting work again. ———

THE EARTH: An Introduction to the Study of Inorganic Nature. By Evan W. Small. Cr. 8vo, pp. viii.—220. (London: Methuen and Co. 1894.) Price 2/6.

One of the University Extension series contains the substance of a course of lectures delivered in various centres during the winters of 1884—90. They are well adapted for the general reader who wishes to gain a knowledge of the more striking phenomena of the earth. The subjects treated of are, The Earth as a Planet; The Materials of the Earth; Work and Energy; How the Materials of the Earth's Crust were formed; and the Evolution of the Earth. There are a number of illustrations. ———

PONDS AND ROCK POOLS, with hints for collecting and the management of the Micro-Aquarium. By Henry Scherren. Cr. 8vo, pp. 208. (London: Religious Tract Society. 1894.) Price 2/6.

The chapters in this book have been considerably enlarged from the pages of *The Leisure Hour*, in which they originally appeared. They treat of Pond and Rock-Pool Hunting; The Beginning of Life; Sponges and Stinging Animals; Worms; Star-Fishes, Arthropods, and Molluscs; and The Micro-Aquarium. There are 66 illustrations.

———

A POCKET FLORA OF EDINBURGH and the surrounding District. By C. O. Sonntag. Fscap. 8vo, pp. xii.—246. (London: Williams and Norgate. 1894.)

We have here a collection and full description of the Phanerogamic and the principal Cryptogamic plants classified after the Natural System, with an Artificial Key and a Glossary of Botanical terms, besides a folding map of the County of Edinburgh and surrounding district.

This little book has been compiled with much care, a great number of localities favoured by the individual plants are given, the author only specially indicating his own in the case of rare plants. Botanists, especially those living in the locality, will find this an invaluable pocket companion.

Die Naturlichen Pflanzenfamilien.   By A. Engler and K. Prantl.   Parts 106, 7, 8.   (London : Williams and Norgate.   Leipzig : Wilhelm Engelmann.   1894.)

We have here the completion of the third and parts of the fourth volume. The families treated of are Cactaceæ by K. Schumann (concluded); Geissolo-maceæ, Penacaceæ, Oliniaceæ, Thymelacaceæ, and Elæagnaceæ by E. Gilg ; and Borraginaceæ by M. Gurke ; Gesneriaceæ (concluded); and Columelli-aceæ by Karl Fritsch ; and Bignoniaceæ by K. Schumann.   There are 49 fine illustrations, composed of 369 figures.

Half-Hours with the Microscope.   By Edwin Lankaster, M.D.   Fšcap. 8vo, pp. 130.   (London : W. H. Allen.   1894.)

The nineteenth edition of this useful little book is now before us. Con-sidering the present advanced state of Microscopy we should have liked to have seen an entirely new written and up-to-date book, but the volume before us has the appearance of having been printed from the stereo-plates of a former edition. There are one coloured and eight plain plates, besides other illustrations.

Directions for Laboratory Work in Bacteriology.   By Frederick G. Novy, Sc.D., M.D.   8vo, pp. 209.   (Ann Arbor, Mich., U.S.A.: George Wahr.   1894.)

The subject matter of this book has been arranged with reference to pro-gressive work in the laboratory, and is intended to cover a course of daily afternoon work for twelve weeks.   No illustrations are given, but blank pages are inserted that the student may add them.   Very full and valuable formulæ and instructions are given.   We consider the book a most useful one.

Practical Lessons in Elementary Biology for Junior Students.   By Peyton T. B. Beale, F.R.C.S., etc.   P. 8vo, pp. vii.—136. (London : J. and A. Churchill.   1894.)

In the useful lessons which have been used with much success in the biolog-ical section of King's College, the author treats in a practical manner those points in the structure and life history of some organisms which are of great importance, and in which the student frequently experiences a difficulty.

In each lesson the author indicates what may be seen and how important parts may be practically demonstrated.

Outlines of Biology.   By P. Chalmers Mitchell, M.A., F.G.S., etc.   Post 8vo, pp. xv.—297.   (London: Methuen and Co.   1894.) 6/-

This book is arranged in accordance with the syllabus of the Royal College of Physicians and Surgeons for the guidance of candidates preparing for exam-ination in Elementary Biology.   It it divided into 18 chapters, commencing with Protoplasm and the Building up of Protoplasm to The Dog Fish and the Frog ; the last chapter being on Embryology.   There are 74 illustrations in the text.

Practical Physiology of Plants.   By Francis Darwin, M.A., F.R.S., and E. Hamilton Acton, M.A.   Cr. 8vo, pp xxviii.—321. (Cambridge and London : The University Press.   1894.) Price 6/-.

Part I. of the book before us deals with general physiology, and Part II. treats in a special manner of certain departments of physiology, viz.—Che-mistry and Metabolism, and presupposes a greater amount of knowledge on the part of the student.

SCIENCE PROGRESS. No. 10. (London: The Scientific Press.) Price 2/6.

The December Number of this valuable Journal has reached us. It contains the following articles :—On the Artificial Hatching of Marine Food-Fishes; The Molecular Weight of Liquids ; The Origin of the Vascular Plants ; Recent Researches in Thermal Metamorphism (Part 2) ; Continuous Current Dynamos (Part 2) ; On the Morphological Value of the Attraction-Sphere (Part 2) ; Kew Thermometers. There is also an Appendix of Chemical Literature for October.

COMMON THINGS and Useful Information. Cr. 8vo, pp. 256. (London : Thomas Nelson and Sons. 1895.) Price 1/-.

One of the series of Royal Handbooks of General Knowledge ; gives in Dictionary Form the more important points in Natural History, Elementary Science, Physiology, and Common Things. These are all treated in full and readable manner ; the book is nicely illustrated.

THEOPHRASTUS ON WINDS AND WEATHER SIGNS. Translated by Jas. G. Wood, M.A., LL.B., F.G.S., and edited by G. J. Symons, F.R.S. 8vo, pp. 97. (London : Edward Stanford. 1894.)

This is a translation of the work of Theophrastus of Erebus, the favourite pupil of Aristotle, on Winds and on Weather Signs, to which is added an Appendix on the Direction, Number, and Nomenclature of the Winds in classical and later times. There are 5 plates showing Maps of Greece and the Country round Athens ; Horologium of Andronikos ; Aristotle's Diagrams ; and Table of the Winds in the Museo Pio Clementino.

THE DIVINING ROD: Its History, with full instructions for finding subterranean springs. By J. F. Young and R. Robertson. Cr. 8vo, pp. viii.—138. (Clifton: J. Baker and Son. London: 25 Paternoster Sq. 1894.) Price 1/6 net.

The authors explain here various methods of using the rod, and show how it can be verified. The book also contains an essay, entitled "Are the claims and pretensions of the Divining Rod valid and true?"

LECTURES ON THE DARWINIAN THEORY. By the late Arthur Milnes Marshall, M.A., M.D., B.Sc., F.R.S., etc. Edited by C. F. Marshall, M.D., B.Sc., F.R.C.S., etc. 8vo, pp. xx.--236. (London: D. Nutt. 1894.) 6/-

This volume consists of a series of lectures delivered in connection with the Extension Lectures of the Victoria Univ., 1893, and are on the following subjects:—History of the Theory of Evolution ; Artificial and Natural Selection ; The Argument from Palæontology ; from Embryology ; The Colours of Animals and Plants ; Objections to the Darwinian Theory ; The Origin of the Vertebrated Animals; and the Life and Work of Darwin. There are 36 illustrations and a photo-engraved Frontispiece of the Archæopterix.

WITH THE WOODLANDERS & BY THE TIDE. Cr. 8vo. pp. 305.

ANNALS OF A FISHING VILLAGE. Cr. 8vo, pp. viii.—261.

FROM SPRING TO FALL, or When Life Stirs. Cr. 8vo, pp. 244. By "A Son of the Marshes." Edited by J. A. Owen. (London : William Blackwood and Son. 1894.) Price 3/6 each.

It is with no small degree of pleasure that we read these exceeding interesting little volumes from the pen of "A Son of the Marshes." There is a peculiar fascination in the author's style, which makes it impossible to find a dry page in any number of his books, turn where you will.

WOODSIDE, BURNSIDE, HILLSIDE, AND MARSH. By J. W. Tutt, F.E.S. Cr. 8vo, pp. v.—241. (London: Swan Sonnenschein and Co. 1894.)

Mr. Tutt, the author of *Random Recollections of Woodside, Fen, and Hill*, which we were pleased to notice a short time ago, records in the volume before us, in very readable language, a few more of the interesting phenomena which are to be observed everywhere around us by those who take the trouble to look for them, and to give such explanation of their causes as may easily be understood even by those whose scientific knowledge is small. The book is nicely illustrated and thoroughly interesting.

THE PARISH COUNCILS' ACT, 1894. A Popular Handbook and complete guide to Elections of Parish Councillors, Rural and Urban District Councillors and Guardians. By J. Wallis Davies. Cr. 8vo, pp. xv.—236. (Bristol: J. Wright and Co. London: Simpkin, Marshall and Co. 1894.) 2/6.

The author, who is Solicitor and Secretary to the Parish Councils' Association, gives his information in the form of questions and answers, his object being to explain and elucidate the somewhat complicated act in an interesting and pleasing manner. We believe the information is very concisely given.

ALBUM OF 24 ENGLISH AND WELSH CATHEDRALS. 4to. (London: *Church Bells* Office.) Price 1/- net.

The 24 views of the various Cathedrals are nicely executed, and opposite each is a short descriptive history. In the Preface is a short history of Cathedrals generally.

MONISM, as connecting Religion and Science : The Confession of Faith of a Man of Science. By Ernst Haeckel. Translated from the German by J. Gilchrist, M.A., B.Sc., Ph.D. Crown 8vo, pp. viii.—117. (London: Adam and Charles Black. 1894.) Price 1/6 net.

The author declares his purpose to be twofold. First, to give expression to that rational view of the world which is being forced upon us with such logical vigour by the modern advancement of our knowledge of nature as a unity ; and second, To establish a bond between religion and science.

HEREAFTER AND JUDGMENT : The Satan of the Old Testament ; The Satan of the New Testament. By the Rev. W. H. Tucker. Cr. pp. 237. (London : Elliot Stock. 1894.)

We have read this book very carefully, but cannot say that we are pleased with it. The author makes assertions which, to our lay mind, he does not prove.

AN OUTLINE of the Principles of Modern Theosophy. By Claude Falls Wright. Cr. 8vo, pp. 192. (Boston [U.S.A.]: New England Theosophical Corporation. 1894.)

Theosophy is to us an incomprehensible subject. The author here professes to lay open the system which furnishes the key to every religion wherein is buried the truth about our nature and destiny.

CARBON PRINTING. By E. J. Wall. Cr. 8vo, pp. 69. (London: Hazell, Watson, and Viney. 1894.) Price 1/-.

This is Vol. 8 of the Amateur Photographers' Library, its object being to furnish historical as well as practical notes on the subject of this most permanent of all printing processes. The instructions given are concise and to the purpose.

PHOTO-MICROGRAPHY. By Dr. Henri Van Heurck. Re-edited and Augmented by the Author from the fourth French edition, and translated by Wynne E. Baxter, F.R.M.S., F.G.S. Cr. 4to, pp. 41. (London: Crosby, Lockwood and Son. 1894.)

This work is extracted from Van Heurck's larger work on the Microscope. There are 18 illustrations and a frontispiece showing *Amphipleura pellucida*, × 2,000 diameters.

THE BRITISH JOURNAL PHOTOGRAPHIC ALMANACK, 1895. Edited by J. Traill Taylor. (London: H. Greenwood.) Price 1/- and 1/6.

A bulky volume of 1343 pages, containing a good deal of useful information to the photographer, whether professional or amateur, and a large number of advertisements, to which he will doubtless very frequently refer.

THE ANNUAL OF THE UNIVERSAL MEDICAL SCIENCES : A yearly report of the progress of the general sanitary sciences throughout the world, edited by Charles E. Sajous, M.D., and seventy associate editors, assisted by over 200 corresponding editors, collaborators, and correspondents. 5 Vols. (London : F. Rebman. Philadelphia, New York, and Chicago : The F. A. Davis Co. 1894.)

We have before us the seventh series of this extraordinary annual. It is in our estimation the most important work of the kind published, as every article of interest published throughout the world is epitomised here, the exact reference being given so that, if desired, the entire article can be studied by those desirous of doing so. Each volume has its own index, and in Vol. 5 is a general index to the entire work, together with a list of all volumes and periodicals consulted. It is illustrated with plain and coloured plates, woodcuts, etc. No physician's or surgeon's library is complete without it.

THE FLOWER OF THE OCEAN : The Island of Madeira ; A resort for the Invalid ; a field for the Naturalist. By Surgeon-General C. A. Gordon, M.D., C.E. 8vo, pp. x.—110. (London : Bailliere, Tindall, and Cox. 1894.)

An interesting account of Madeira is given here. We find Historical Notices, Meteorology, certain points relating to Hygiene and Medicine, the Fauna of the Island, Agriculture and Viticulture, and some Medicinal Plants.

LANDMARKS IN GYNÆCOLOGY. 2 Vols. By Byron Robinson, B.S., M.D. Fscap. 4to, pp. 220. (Detroit, Mich. : George S. Davis.)

Two volumes of The Physician's Leisure Library, containing abstracts of the author's lectures at the Chicago Post Graduate Medical School during the past three years.

THE VACCINATION QUESTION. By Arthur W. Hutton, M.A. Cr. 8vo, pp. 128. (London : Methuen and Co. 1894.) Price 1/6.

The volume before us is a Letter addressed by permission to the Right Hon. H. H. Asquith, Q.C., M.P., H.M. Principal Secretary of State for the Home Department ; to which is added by way of Postscript some remarks on Dr. Klein's Alleged Discovery.

THE CHEMIST'S AND DRUGGIST'S DIARY, 1895. Crown 4to. (London : 42 Cannon Street.)

Besides the Diary, of which a page is given for each week, it contains a number of articles of special interest to the Chemist and Druggist ; 42 pages being devoted to Perfumes and how to make them. The Diary is interleaved with blotting-paper. The volume is nicely bound.

THE BRITISH AND COLONIAL DRUGGIST'S DIARY, 1895. 4to.
(London : 42 Bishopsgate Without, E.C.)
The fourth annual issue of this useful work contains a great deal of information specially collected for, and particulary useful to all engaged in pharmarcy. The diary is interleaved with blotting-paper and the whole is strongly bound.

———

MEMORIES OF GOSPEL TRIUMPHS among the Jews during the Victorian Era. By Rev. John Dunlop. Royal 8vo, pp. 510. (London ; S. W. Partridge and Co. and The British Society's Office, 96 Great Russell St. 1894.) Price 5/-.
This very handsome book is the Jubilee Volume, prepared by the accomplished and genial Secretary of the British Society and Editor of the *Jewish Herald*; it is a wonderful record of arguments, facts, and illustrations in favour of missions to the Jews.
It contains biographical sketches of the founders of the society and early missionaries, together with an account of the splendid work accomplished by the society during the last 50 years. There are 250 portraits and illustrations, and is splendidly bound. We are informed that the Queen and Prince of Wales have each accepted a copy. ———

THE CHILD'S PICTORIAL. Crown 4to, pp. 192. (London : Society for Promoting Christian Knowledge. 1894.) Price 2/-
Just the book for a present. It is full of plain and coloured pictures and good tales, besides a series of illustrated articles on the Zoo by the Rev. Theodore Wood, F.Z.S. ———

STIRRING TALES OF COLONIAL ADVENTURE. A Book for Boys. By Skipp Borlase. Cr. 8vo, pp. 376. (London : Frederick Warne and Co. 1894.)
A series of eight stirring and instructive tales, such as must please every boy who reads them. There are also some good illustrations.

———

A SALT-WATER HERO. By Rev. E. A. Rand. Cr. 8vo, pp. 330. (New York : Thom. Whittaker.) Price $1·25 (5/-.)
This is an exceedingly interesting and nicely illustrated book for boys. The author not only writes a thoroughly readable tale, but he manages at the same time to instil some very manly thoughts into the mind of the reader.

———

THE LONE INN : A Mystery. By Fergus Hume. Cr. 8vo, pp. 265. (London : Jarrold and Son. 1894.) Price 3/6.
A good tale, in which the mystery is well sustained. A shorter tale at end of the volume is entitled "Professor Brankel's Secret," which we must leave the reader to find out. ———

GAMES FOR WINTER EVENINGS.—Messrs. A. N. Myers and Co., 15 Berners St., Oxford St., W., have sent us a parcel of amusing games, of which we enumerate the following :—The Manchester Ship Canal ; Cat and Mice ; Ten Little Niggers running from Slavery to Freedom ; In Egypt with the Pyramid ; The Owl Party ; and The Tower Bridge. This last is an interesting model made in perforated cardboard, in many pieces, to be built and fixed together, and forms, when properly arranged, a pretty model of this celebrated structure.

Predacious and Parasitic Enemies of the Aphides
(including a Study of Hyper=Parasites).
By H. C. A. Vine.   Plates VIII. and IX.

AS it is only among the larval forms of the Syrphidæ that the aphis-eaters are to be found, it is of great interest to identify if possible the respective larvæ, in order to ascertain in what species the aphidivorous habit prevails.   The principal difficulty encountered in this study is the fact that fully 95 per cent. of the imagines, reared at great expenditure of time and trouble, prove to be the common *Syrphus luniger*, or some nearly allied variation.   Another difficulty is caused by the irregular periods of abstinence which more or less affect all these larvæ, and during which they not infrequently die from the attack of some species of fungus.   I have, however, succeeded in placing beyond doubt the position of three aphidivorous larvæ, and these, with the addition of one species which I have as yet failed to rear to maturity, comprise all which I know to be aphis eaters.

The eggs or young grubs have been obtained in different localities around Bath and near Radstock, Somerset, and by the kindness of Mr. C. J. Watkins, numerous specimens have been sent me from Painswick, Gloucestershire.   Hence the specimens observed may be taken to be fairly representative of English species.

A vast proportion of the larvæ, wherever collected, are more or less rough in skin, brown, or very dark grey, finely speckled with black, sometimes beset with small spines, and having a more or less black, or occasionally coloured, lines, visible through the transparent outer skin, down the centre of the dorsum.   These larvæ form pupa cases, of a dark grey brown, coffee brown, or brown black, which retain the finely sprinkled black specks of the larvæ, and frequently exhibit slight oblique marks of lighter or darker colour, together with five dark lateral spots, and in some instances a slightly marked black line along the dorsum. The fly which emerges is invariably *Syrphus luniger*, and is figured on Pl. XVI. of Vol. IV., together with the larva and pupa.

A larva occasionally met with, and more especially in fruit and kitchen gardens, is, I think, smooth, save for the creases of the skin, plump, and of the colour of parchment. It is frequently marked with blotches, which are generally situate on either side of the medial line of the dorsum, and which vary in colour from dull yellow to dark brown and black. These patches are due to masses of pigment underneath the skin, through which they are visible. The pupa-case is similarly parchment-like in appearance, free from speckling, and sometimes slightly ribbed, especially after the escape of the fly, which belongs to the same genus as the last described, though to a different species. It is *Syrphus topiarius*.

The third variety which I have succeeded in identifying is a smooth skinned grub, far from common, of a pale green colour, and having a more or less marked stripe of pink or carmine down the centre of the back. This larva eventually produces the fly, *Catabomba pyrastri*.

Besides these, I have made repeated attempts to rear an aphidivorous larva, small, greenish, or grey in colour, nearly or quite smooth, with very delicate markings, and generally having fine lines of light carmine on the dorsum and around the lateral edges. These larvæ habitually undergo long fasts and periods of inactivity, and the specimens which I have succeeded in keeping alive for several months have in each instance succumbed to the attacks of a fungoid growth, which seems to affect the mouth parts. But I have also reason to think that the larva of *Platychirus albimanus* is aphidivorous in its habit, and if so, it would seem to me to be exceedingly probable that the larva last described belongs to that species.

It is impossible to say, from only a few seasons' experience of the counties of Somerset and Gloucester, whether all the aphidivorous varieties have been observed. One, *S. balteatus*, is known to be an aphis eater, and I have on one or two occasions reared it from larvæ closely resembling those of *S. luniger*, but have failed to identify the grub from the others. The larva of *S. ribesii* is also no doubt aphidivorous.

It may be said, then, with some certainty, that the following species prey upon Aphides :—*Syrphus luniger, S. topiarius, Catatomba pyrastri, S. balteatus, S. ribesii*, and in all probability *Platychirus albimanus.* Mr. G. B. Buckton, in his *British Aphides*, mentions the genera, *Scæva* and *Cheilosia*, as containing species which are aphis eaters; but, although I have frequently *taken* flies of this genera, I have never found any of my aphis-eating larvæ *develop* them. This may be accounted for by the fact that many species of insects are very local. It should perhaps be added, by way of warning, that notwithstanding the particulars which I have been able to give, the recognition of these larvæ requires considerable experience, and cannot be considered as certain until the appearance of the pupa-case fully confirms the diagnosis of the larva.

I now come to a very interesting point in the economy of the Syrphidæ. In the course of numerous dissections of the flies which I have made during the past two years, my attention has been frequently arrested by the pollen-grains which are invariably present in all parts of the alimentary system, from the œsophagus to the rectum, and even in the excreta.

The idea that the mouth-organs of the Diptera are only adapted to the imbibition of fluids or juices must be considerably modified, if it be agreed that pollen forms a considerable, if not the greater, proportion of the food of the Syrphidæ in a state of nature. And that such is the case I am quite satisfied. My observations of the stomachs of several species of *Syrphus, Eristalis*, and *Rhingia* leave no doubt that in all these, pollen constitutes the bulk of the food ingested, and it is found in the œsophagus, in the sucking stomach, in the digestive stomach, and in the alimentary canal in varying conditions, which indicate plainly the progress of digestion.

In view of the inapplicability of the suctorial parts of the proboscis to the ingestion of a solid-like pollen, this organ becomes a fresh object of interest; and I have endeavoured, by means of the greatly amplified figure on Pl. VIII., to set before the reader the details of its structure. It will be seen that the split tubes, or " pseudo-tracheæ," take their rise on either side from a curved chitinous channel of similar structure, into which any fluid absorbed by them would pass, and which in time would convey it to the anterior extremity of the hollow shown at *a*, which constitutes the floor of the mouth. In this hollow, when at rest, lie the long lanceolate tongue (*ligula*), *b*, and the horny upper lip, *c*, both of which are partly shown more highly magnified at Fig. 2. When these parts are in place, the upper lip, *c*, constitutes, with the bottom of the groove or hollow, *a*, a tubular passage, through which fluids, absorbed by the lobes, find their way to the pharynx and thence to the œsophagus. This passage presents ample space for the passage of pollen-grains also; but a comparison of the diameter of those found in the œsophagus with the width of the " pseudo-tracheæ " of the lobes at once makes it evident that it is not by their means that the pollen-grains reach the mouth.

An examination of the upper lip, as shown at Fig. 2, suggests an explanation of the manner in which the ingestion takes place, which is doubtless correct. The upper part of the extreme end is divided into teeth, which are sufficiently wide apart to admit of pollen passing between them; but between these are smaller brush-like processes, which appear especially fitted to sweep it from the flowers upon which the fly feeds into the cavity of the mouth, where it is met by the secretion issuing from the duct, which proceeds from the salivary glands, and which terminates in the tongue. This duct is shown on Fig. 1, at *s*.

The uniformity with which the pollen-grains are to be found in all parts of the alimentary system, and their great abundance in the sucking stomach, indicate pretty clearly the important share they take in nutrition. To assure myself more fully of this, I have experimented by supplying the flies, both reared and captured, with saccharine foods. The flies were kept under the most natural conditions consistent with confinement. They at first attacked the food with avidity, but although fresh leaves, etc., were given daily they soon became languid and generally died about the fifth or sixth day. On examination of a fly under these conditions, the sucking stomach is invariably found full of a clear syrupy fluid, while the proper stomach and intestines appear collapsed and empty. It seems probable that, as in the human stomach, the presence of more or less solids is an important factor in stimulating the digestive organs into action. Possibly, the effect of the cellulose cases of the pollen may be somewhat similar to that of the bran of brown bread or the husks of oatmeal when either of these are ingested by man.

Mr. J. W. Morris, F.L.S., has kindly endeavoured to identify for me the pollens found in the viscera of *Syrphus luniger, Rhingia rostrata,* and *Eristalis tenax,* and the result is, I think, a very remarkable addition to our stock of knowledge in this direction. These three species differ materially from one another in their characters and habits, and the specimens examined were taken on hedgerows or in gardens some miles apart. But two pollens were found always present, accompanied by a third in a few instances and in small numbers. The difference in appearance which may be due to the liquids of the stomach necessarily makes precise recognition difficult, if not impossible, and I think, myself, that there is room to doubt whether this apparent third variety may not be the result of partial digestion. At any rate, no more than three pollens at the most are found, two of which are, without much doubt, the pollen of some species of *Doronicum,* or "Leopard's bane," varieties of which bloom from spring to autumn, and that of some near allies of *Lathyrus,* of which the Everlasting Pea and the Bitter Vetch are examples. These are shown at Fig. 3 on Pl. VIII., together with the other variety.

As the flies visit in succession numerous flowers apparently for food, it would seem that from among them all they select the pollen of these or of nearly allied species, thus exercising a discriminatory power which I am not aware has been similarly observed among winged insects, and which, when confirmed by other observers, will add to our estimate of the high degree of intelligence already claimed by naturalists for the Insecta.

The fact that the outer shells of these minute grains pass through the fly in so perfect a condition as often to present no serious difficulty in their identification, raises some interesting questions as to the nature, the process, and the secretions by which digestion is accomplished. In order to study this, it will be desirable for a moment to consider the composition of pollen.

The grains of pollen consist of an outer case of considerable hardness, which appears to be mainly cellulose, and which contains the protoplasmic mass known as "fovilla." The hard case presents some points, which are capable of being ruptured by the expanding fovilla, which, if it has the opportunity, penetrates, by means of an extension, known as a "*pollen tube*," the stigma of a flower of its own kind, and effects the fertilisation of the ovules contained in the ovary at the base of the style.

From these circumstances we might expect that the semi-fluid contents of the pollen grains would be of a highly nitrogenous character, and would possibly contain also such elements as phosphorus and sulphur.

The following analysis of "Pollen," by Schneider, given in the *Journal of the Chem. Soc.*, Vol. X., p. 175, shows that this expectation is well founded.

Water	29·89
Ash, chiefly Phosphates ...	3·08
Albumin and Peptones ...	17·81
Sugar	25·12
Fats and Fatty Acids (colouring matter)	8·98
Cell Membrane	7·56
Pecten	7·42
	99·86

Sachs states that fovilla consists of "coarse-grained protoplasm, with grains of starch and drops of oil." Another authority states that it consists of "starch granules, oil globules, and protein compounds." From my own experience, I am inclined to think that the sugar given in the above analysis should have been shown as its equivalent of starch, from which it was probably formed in the analytical process. The proteids are mainly, if not wholly, in the form of albuminoids, and do not differ widely from the following analyses of vegetable albuminoids, given by Lintner and Szilágyi respectively :—

Carbon	..	43·33	...	44·50
Hydrogen	...	6·98	...	7·08
Nitrogen	...	9·92	...	9·49
Sulphur	...	1·07	...	1·08
Oxygen	...	32·91	...	32·95
Ash	...	4·79	...	4·9

Now, the percentage of nitrogen in beef is stated by Dr. Letheby at ten, and Professor Goodfellow tells us that it is in the form of albuminoids, which constitute fifteen per cent. of the whole bulk. And very little difference exists in the composition of albuminoids, whether they are derived from an animal or vegetable source. We have seen that pollen contains some 17 or 18 per cent. of these compounds of nitrogen. But although nitrogen is one of the most necessary constituents of animal food, and is generally ingested in the form of albuminoids, yet as such, it is incapable of assimilation, because albuminoids are incapable of undergoing diffusion through animal membrane. Hence, it is necessary for them to undergo some change which shall render them diffusable, and in man this is accomplished by the first secretion which the food meets in the stomach—that known as the "gastric juice." Under the influence of certain constituents of this secretion, the albuminoids are converted into "peptones," which, while possessing a similar composition to albuminoids, are highly diffusable, and are capable of readily passing through animal membrane. The process is somewhat as follows :—When food enters the human stomach, it sets up a nervous disturbance, which causes the walls to become suffused with blood, and under the influence of the stimulation thus applied, the gastric glands pour out their secretion to effect the necessary change.

In the Syrphus fly the similarly constituted food, on being impelled from the extension of the œsophagus, known as the sucking stomach, enters the proventriculus, which forms the entrance to the proper or digestive stomach. Here it meets with the secretion of the adjacent glands, shown on Pl. IV., Vol. V., at *n, n*, which appear to be brought into action by the presence of hard food in the proventriculus, and following on this the changed appearance of the pollen makes it clear that the digestive process has begun. For comparison a gastric gland from the human stomach is depicted on Pl. VIII. at Fig. 4, and it will be seen that it presents considerable analogy to the apparently similar gland of the fly.

On these grounds, therefore, it seems not unwarrantable to assume that the secretion of the glands (*n, n*), which is the first digestive fluid encountered, apart from the saliva, answers much the same purpose as the gastric juice in man, and I think we may fairly describe these glands as "gastric glands."

The investigation of the means by which the oil-globules of the fovilla are rendered available for the nutrition of the insect presents great difficulty, inasmuch as no organ presents itself as likely to exercise pancreatic functions, and as yet I have failed to observe the point at which the oily globules of the pollen disappear.

Another interesting set of organs, which are remarkably well seen in many of the Syrphidæ, are those which I shall call the "renal bodies." They are shown on Pl. IX. both *in situ*, in Fig. 1, and under a higher amplification the individual organs are depicted at Figs. 2 and 3, as seen on the surface and in section. These are from *Syrphus luniger*, but Fig. 4 (shown *in situ* in the wall of the rectum) is from *Eristalis tenax*, and with it are shown some of the smaller pollen-grains *(Lathyrus)* floating loose in the rectal cavity. These organs are also indicated on Pl. IV., Vol. 5, at *t, t.* These organs, which perform somewhat similar excretory functions to the kidneys of animals, are four in number, and are inserted through the walls of the rectum at a point below the valve, at the termination of the lower intestine, where the former enlarges to form a lemon-shaped pouch.

The arrangement and construction of these organs are obvious from the drawings. Each one of the conical bodies consists of a hollow cone, composed outwardly of a delicate membranous sheath, beneath which a thick layer of gland-cells are disposed. These are somewhat angular from mutual pressure, and each contains a nucleus, which, however, is generally very difficult to demonstrate.

The gland-cells are supported by a transparent and apparently structureless membrane, which forms the inner surface of the cone. A plentiful supply of oxygen is ensured by the large and direct tracheal trunks, which connect these organs with the spiracles of the abdomen. These trunks do not appear, as in the *Musca vomitoria*, to divide into distinct sets of vessels supplying respectively the inner and outer layers of the organ, but branch off in a tree-like fashion, as shown in Fig. 3, ultimately ramifying into fine tubes, which are disposed parallel to the exterior surfaces, beneath the outer membrane, through which they may be seen by careful focussing, as in Fig. 2.

The inner surface of these organs appears to be in free connection, and in fact continuous with, the cavity of the abdomen, the fluids of which are no doubt pressed into them during the alternate compression and dilation which the abdomen of the fly continually exhibits. It is probable that they contract and expel the fluid by means of muscular bands at their bases, which are not shown in the drawings. Thus, a continual passage of the abdominal fluids would give opportunity for the secretory glands to do their work. This appears to be the separation of urinary compounds, and may, under favourable circumstances, be evidenced by the demonstration of the presence of uric acid in the rectum, while it cannot be found above the valve connecting it with the intestine. The presence of uric acid in the excreta may be readily shown by acidulating it on a slide with a minute quantity of hydrochloric acid, when crystals of uric acid will be readily formed. Indeed, the first white excreta passed by the fly after its emergence from the pupa-case is almost entirely uric acid.

The only remaining feature which I propose to notice here is the three spermathecæ, with their extraordinarily long and slender seminal canal, which, if extended in a direct line, would measure about ·25 inch in *Eristalis tenax*, the species in which it can be dissected out with least difficulty. One of these receptacles is shown with the attached tube on Pl. VIII., Fig. 5. It consists of a dark brown, opaque, chitinous capsule, nearly spherical in shape when fully expanded, and formed of two or three segments, in some way jointed together. Each receptacle is surrounded by a layer of cells, which are covered by a more or less wrinkled membrane, probably containing muscular fibres. When the female is yet unmated, the spermathecæ present the appearance of two flattened capsules, placed edge to edge, from one of which proceeds the immensely extended narrow canal connecting it with the vagina. Around the opposite capsule is a considerable mass of what appears to be cellular tissue interspersed with muscular fibres, and which, when the capsules expand to a spherical form on the entrance of the spermatic fluid, appears to stretch so as to form an even covering. The earlier condition is shown at Fig. 6 on Pl. VIII.; the later state at Fig. 5, where also are shown the thin delicate tubes, which form the passage for the spermatozoa. These

tubes are encased in tissue similar to that enveloping the spermathecæ themselves, and their inner coat is continuous with that of
the interior of the capsules.   The latter are lined with epithelial
cells of an angular form and of a dark brown colour, which may
readily be seen on breaking the capsule by a little pressure on the
cover-glass, when also the junctions of the edges of the capsules
will often be evident.   The three spermathecæ are in some species
grouped together in a surrounding membranous network ; but in
some, as *Syrphus luniger*, they are arranged two on one side and
one on the other, as in the Blow-Fly.

THE CLASSIFICATION OF THE SYRPHIDÆ.

It has already been mentioned that this family has been divided
by entomologists into something like forty genera, which are
inhabitants of Britain.   But this has not been accomplished
without vast labour and an immense number of changes in the
arrangement.   Over and over again, by Fabricius, Meignen,
Walker, Fallen, and others, genera have been constituted, recon-
stituted, and abandoned, and, not uncommonly, so complete has
been the change that the generic names seem to have been transferred from one group of species to another.

I do not propose to enter upon an examination of the whole
of this extensive family.   Such a course would be beyond the
scope of the present work, but it may be useful to students of the
Aphididæ if I quote from the admirable *List of British Diptera*,
arranged by Mr. G. H. Verrall, the particulars of those genera in
which aphidivorous larvæ are most likely to occur.

The earliest writer to attempt any division of the family was
Harris, who, writing a few years after Scopoli had separated the
Syrphidæ from the Linnæan genus, *Musca*, constituted three chief
divisions based upon the distinctive wing-neuration, as shown in
Vol. IV. on Pl. XVI., Figs. 10, 11, and 12.   Although this division has not been adopted by later writers, it is convenient for the
purpose of dividing the family into sections, and I have found it
useful to group it under the heads *Syrphidinæ* (Fig. 12), *Eristalinæ*
(Fig. 11), and *Volucellidæ* (Fig. 10).

Donovan gave some illustrations of this family at the end of
the last century, and between 1825 and 1840 J. Curtis published

several very beautiful and exact coloured plates of various species, with many descriptions. His plates usually represent the rare rather than the ordinary varieties, and thus their value to the student is for special rather than everyday studies. They are, however, almost unequalled for accuracy of drawing and delicacy of colouring. Many of the generic and specific descriptions have been since abandoned.

Some fifteen years later Walker described the family among his Diptera, but added little to our information concerning it. Schiner and Meignen have written with authority on the Diptera of Europe, and the works of both, published in German, are standard. In 1864 the former proposed some extensive changes in the arrangement of this order, and a few years later Osten-Sacken produced his classification.

At present Mr. Verrall's lists of Diptera are our best authority as regards the family of *Syrphidæ*, and I take the following list of genera and species from his issue of 1888, adding to the latter, by the kindness of Mr. C. J. Watkins, a few species which have been noted since. These are italicised, while the *known* aphidivorous specimens are starred.

GENERA COMPRISED IN THE FAMILY SYRPHIDÆ,

from Mr. G. H. Verrall's list of British Diptera, published in 1888 :

Genus.	Number of Species.	Genus.	Number of Species.
1 Paragus (Lat.)	2	12 Didea (Macq.)	1
1 species added since 1888.		13 Syrphus (Fab.)	34
2 Pipizella (Rud.)	3	Several species added since 1888.	
3 Pipiza (Fln.)	5	14 Catabomba (O.-Sack)	2
4 Cnemodon (Egg.)	1	15 Telecocera (Mg.)	1
5 Orthoneura (Macq.)	3	16 Sphærophoria (St. Farg.)	6
6 Chrysogaster (Mg.)	8	17 Xanthogramma (Schin.).	2
7 Chilosia (Mg.)	26	18 Doros (Mg.)	2
1 species added since 1888.		19 Myiolepta (Newm.)	1
8 Leucozona (Schin.)	1	20 Baccha (Fabr.)	1
9 Melanostoma (Schin.)	6	21 Sphegnia (Mg.)	1
10 Pyrophœna (Schin.)	2	22 Ascia (Mg.)	2
11 Platychirus (St. Farg.)	12	23 Rhingia (Scop.)	1
1 species added since 1888.		24 Brachyopa (Mg.)	1

Genus.	Number of Species.		Genus.	Number of Species.
25 Volucella (Geoff.)	...	4	35 Brachypalpus (Macq.)	1
26 Sericomyia (Mg.)	...	2	36 Xylota (Mg.) ...	7
27 Arctophila (Schin.)	...	1	37 Syritta (St. Farg.) ...	1
28 Eristalis (Latr.)	...	10	38 Eumerus (Mg.) ...	3
29 Myatropa (Reid)	...	1	39 Chrysochlamys (Reed).	2
30 Helophilus (Mg.)	...	9	40 Spilomyia (Mg.) ...	2
31 Merodon (Mg.)	...	1	41 Chrysotoxum (Mg.) ...	7
32 Tropidia (Mg.)	...	1	42 Callicera (Fab.) ...	1
33 Criorrhina (Macq.)	...	5	43 Microdon (Mg.) ...	2
34 Pocota (St. Farg.)	...	1		

The following Genera comprise those which contain species known or believed to be aphidivorous in the larval state :

CHILOSIA (Meignen).

(Some larvæ said to be aphidivorous.)

maculata, Fln.
antiqua, Mg.
sparsa, L. W.
pubera, Ztt.
soror, Ztt.
scutellata, Fln.
longula, Ztt.
pulcripes, L. W.
præcox, Ztt.
vernalis, Fln.
nebulosa, Ver.
chloris, Mg.
grossa, Fln.
chrysocoma, Mg.
flavicornis, F.
mutabilis, Fln.
flavimana, Mg.
albitarsis, Mg.
impressa, L. W.
rostrata, Ztt.
variabilis, Pz.

CHILOSIA—*continued.*

barbata, L. W.
vulpina, Mg.
olivacea, Ztt.
intonsa, L. W.
œstracea, L.
plumulifera, L. W.

SYRPHUS (Fabr.)

Several larvæ known to be aphidivorous.

barbifrons, Fln.
lasiopthalmus, Ztt.
punctulatus, Ver.
umbellatarum, F.
compositarum, Ver.
flavifrons, Ver.
decorus, Mg.
auricollis, Mg.
cinctellus, Ztt.
cinctus, Fln.
*balteatus, De G.
bifasciatus, F.
arcuatus, Fln.
lapponicus, Ztt.

SYRPHUS—*continued*.

*luniger, Mg.

corollæ, F.

annulatus, Ztt.

vittiger, Ztt.

latifasciatus, Mag.

nigritarsis, Ztt. ?

nitens, Ztt. ?

nitidicollis, Mg.

vitripennis, Mg.

* ribesii, L.

diaphanus, Ztt.

grossulariæ, Mg.

*topiarius, Mg.

tricinctus, Fln.

Venustus, Mg.

lunulatus, Mg.

quadrilunulatus, Schum.

albostriatus, Fln.

laternarius, Müll.

glaucius, L.

guttatus, Fall.

arcticus, Ztt.

triangulifer, Ztt.

annulipes, Ztt.

lineola, Ztt.

CATABOMBA (Ost. Sack.)

*pyrastri, L.

selenitica, Mg.

SPHÆOPHORIA (Scæva).

(St. Farg.)

(Some larvæ said to be aphidivorous.)

scripta, L.

dispar, L. W.

strigata, Stæg.

SPHÆOPHORIA—*continued*.

picta, Mg.

menthastri, L.

nitidicollis, Ztt.

PELECOCERA (Meigen).

(Larva possibly aphidivorous.)

tricincta, Mg.

PLATYCHIRUS (St. Farg.)

manicatus, Mg.

melanopsis, L. W.

*albimanus, F.

latimanus, Whlbg.

peltatus, Mg.

scutatus, Mg.

fulviventris, Macq.

immarginatus, Ztt.

scambus, Stæg.

podagratus, Ztt.

clypeatus, Mg.

angustatus, Ztt.

spathulatus, Rud.

PYROPHÆNA (Schiner.)

(Larvæ possibly aphidivorous.)

ocymi, F.

rosarum, F.

LEUCOZONA (Schiner).

(Larvæ possibly aphidivorous,)

lucorum, L.

MELANOSTOMA (Schiner).

Quadrimaculatum, Ver.

ambiguum, Fln.

dubium, Ztt.

scalare, F.

mellinum, L.

hyalinatum, Fln.

EXPLANATION OF PLATES VIII. AND IX.

PLATE VIII.

Fig. 1.—Proboscis of *Syrphus luniger* seen obliquely, and having the near labial palpus removed :—*a*, The bottom of the mouth ; *b*, the tongue (ligula) ; *c*, the horny upper lip ; *k*, *k*, the lobes of the under lip ; *s*, the salivary duct.

,, 2.—Enlarged view of the extremity of the upper lip, showing the teeth and spiny processes. Below is shown the extremity of the tongue.

,, 3.—Pollens from the œsophagus, stomach, and rectum of *Rhingia*, *Syrphus*, and *Eristalis*.

,, 4.—Gastric gland from human stomach.

,, 5.—Spermathecæ and spermatic canals from female *Eristalis tenax*

,, 6.—One of the spermathecæ of *Eristalis* before expansion, showing the separate capsules and the enveloping membrane.

PLATE IX.

Fig. 1.—The renal organs of *S. luniger*, *in situ*, in the expansion of the rectum, showing their insertion through the partially transparent walls of the rectum and the distribution of the tracheal trunks.

,, 2.—One of the papillæ removed and carefully focussed for the underside of the outer membrane, so as to exhibit the termination of the tracheal branches.

,, 3.—The same, seen in section.

,, 4.—One of the papillæ of *Eristalis tenax*, seen *in situ*, displaying the arrangemeut of the tracheal tubes and the internal cavity. Pollen-grains are seen floating free in the rectum.

The question whether the intensity of the radiation of heat by the sun is affected by its condition as to spots has been studied by M. R. Savelief, of Kiev, in the light of observations made in the spring and the fall of the years 1890, 1891, and 1892. The results point to an affirmative answer, the radiation being greater as the sun-spot activity augments. A variation in one series of the experiments is interpreted as indicating that the increase is dependent, not so much on the absolute number of the spots as upon the intensity of their evolution ; or it may mean that it is immediately consecutive on their diminution.—*Pop. Sci. Mon.*

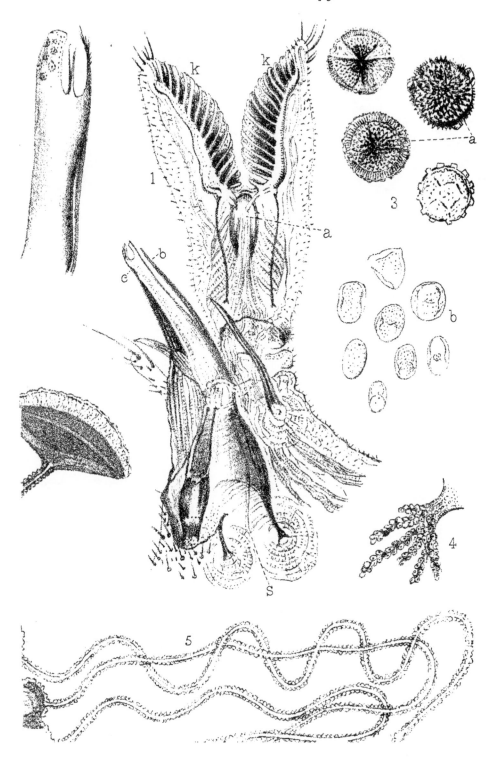

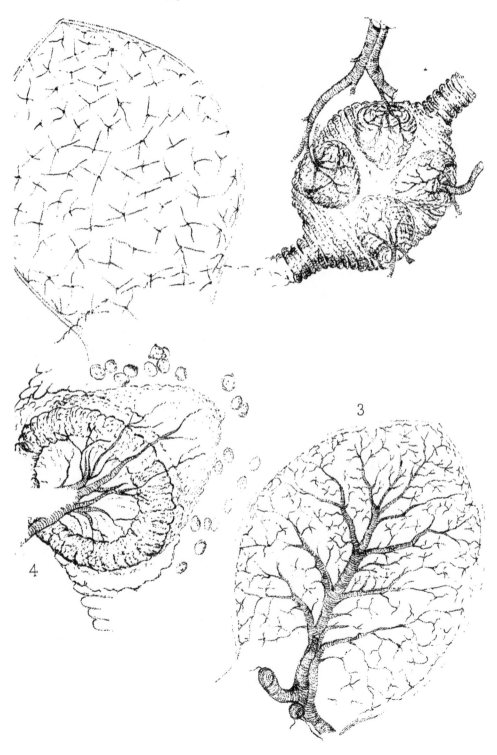

3

4

Benham's "Artificial Spectrum Top."

By G. H. Bryan, M.A.

I T is well known that if a circular disc of cardboard is painted in sectors with the different colours of the spectrum, and is made to spin rapidly as a top or "tee-to-tum," the disc appears white in consequence of the impressions produced by the various colours on the retina remaining behind for a short time and so mixing together. But that the opposite effect can be produced—namely, that of seeing colours on a rapidly revolving top where no colours really exist—is a discovery that one would hardly have expected. Yet such is the result accomplished in the latest optical novelty, invented by Mr. C. E. Benham, and sold by Messrs. Newton and Co. under the name of "The Artificial Spectrum Top."

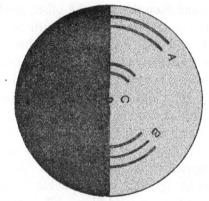

Fig. 5.

The top consists of a cardboard disc, one half of which is black, while the other is white, with a number of circular arcs of black projecting into it on either side, and subtending an angle of about 45° at the centre. The arrangement of these arcs is immaterial, provided that the arcs on one side of the centre are arranged not to clash with those on the other, and a possible arrangement is shown in Fig. 5. When such a disc is mounted on a peg as a top and spun, or carefully centred on a microscopist's turn-table and rotated in a bright light, the various arcs will produce the appearance of rings, and these rings will appear coloured. If the disc is rotated in the direction of the hands of a clock (as will be the case when the top is spun in the ordinary way between the thumb and finger of the right hand), the arcs along the portion A C will produce rings more or less blue, or at any rate fringed with blue, and those along C B will produce rings more or less red or fringed with red, and in some parts the background will be of a more or less greenish tint.   Thus, with the arrangement shown in the figure, we shall see two blue rings inside, followed by three red rings, and lastly two blue rings outside all.

If, however, the disc is spun in the opposite direction (as will be the case when it is attached to a turn-table and spun by the left hand) the colours will be reversed, the arcs along A C producing red rings, while the arcs along C B produce blue rings; the blue will, therefore, now be intermediate between the inside and outside red rings.   In the top, manufactured by Messrs. Newton and Co., there are twelve arcs, arranged in batches of three on one side of the centre and then three on the other, producing alternate groups of three red and three blue concentric rings.   As a rule, the best effects are obtained by revolving the disc in a bright light, and the colours seen depend on the rate of rotation, being usually brightest when the disc is rotating as slowly as is consistent with the markings, appearing blurred into continuous rings.

A series of investigations on these appearances has been made by Prof. G. D. Liveing, of Cambridge, who has also constructed a variety of discs with figures in black disposed on a white ground and with white figures on a black ground.   All of these, when revolved in a bright light, show remarkable bands of colour of various shades of red, green, and blue.   The general result of his observations of these discs is that if a succession of black and white objects are presented to the eye with moderate, but not too great, rapidity, then, when black is followed by white, an impression of a more or less red colour is perceived, while when white is succeeded by black a more or less blue colour is perceived.   If the succession of black and white is very rapid, the appearance presented to the eye is of a more or less neutral green or drab.

For different people the rate of rotation necessary to produce the impression of a particular tint varies somewhat.   When excentric or spiral bands of black are used, the impression on the eye is that of fringed bands, reddish on one side and greenish on the other.   Sometimes the coloured bands seem mottled or the colours appear in flashes radiating from the centre of revolution.

Prof. Liveing is not at all satisfied with the explanation published by Mr. Benham, which is to the effect that the *average* number per second of stimuli affecting the eye is increased, or diminished, according as a black portion of a band follows or precedes a white ground, in the revolution of the disc.

The only explanation Prof. Liveing can himself offer is based, firstly, on the known facts that the impression produced on the retina by a bright object remains for an appreciable time after the light from the object has been cut off, and that the duration of that impression is different for different colours (a result discovered by the late eminent physicist and physiologist, Von Helmholtz); and secondly, on a supposition (which he does not know to have been as yet verified experimentally) that the rapidity with which the eye perceives colours is greater for one end of the spectrum than for the other.   From this point of view, the explanation of the blue colour seen when white is followed by black would be that the impression of blue on the retina lasts a little longer than that of the other colours ; while the red colour seen when white succeeds black is due to the greater rapidity with which the eye perceives red light than that with which it perceives blue.

If, however, the alternations of white and black succeed each other with sufficient rapidity, the new impression of a white patch will be produced before that of its predecessor has vanished, and there will be an overlapping of impressions, and the sensation will be that of a mixture of colours of a more or less neutral tint. This is in accordance with the observation that if several circular bands, each partly black and partly white, in equal proportion, are on the same disc ; but in some of the bands the whole of the black parts is in one patch, while in others it is divided up into several patches, then, when the disc is rotated so that the band which presents only two alternations at each revolution is seen coloured, the bands which present a greater number of alternations are seen of a neutral tint.   Also, if the rotation is rapid enough, the bands *all* appear of a neutral tint.

So far as he can test the theory by his own eyes, it appears to Prof. Liveing that the residual impression, left when the light from a white object is suddenly cut off, is at first green and fades out through a more or less blue or slate colour. The object must not, however, be so bright as to dazzle the eye, as the duration and colour of the residual impression might in that case be very different. Prof. Liveing sees the colours of the discs best in bright diffuse daylight, and sees hardly any colour when the discs are in direct sunshine, but no two observers see the same appearances under the same circumstances.

The discs used in most of Prof. Liveing's experiments, and Messrs. Newton's top, are about six inches to a foot in diameter ; but the size is unimportant, and the small sketch from which the present figure was drawn, when placed on a turn-table, exhibited colours equally bright to those on the larger tops.* The black parts may either be painted on the top or made of pieces of dead-black paper, and Prof. Liveing used a kind of black paper, which he finds to be a perfect absorbent of light of all colours, thus effectually disposing of any questions as to whether the appearances may be due to colours actually present in the black or white parts of the disc.

The experiments may be varied by making the top with white arcs projecting into the black portion, and they will supply much material for recreation on a dull winter's evening, when bright gas or lamplight will answer admirably.

* If, however, the top is too small and light, it cannot be made to spin slowly enough to show the colours well, while too large and heavy a top is difficult to set spinning sufficiently rapidly.

With reference to Crolls and Ball's theories of ice ages and genial ages, Mr. Edward P. Culverwell has shown, on the basis of calculations of the daily distribution of solar heat on different latitudes at the present time and in the supposed glacial and genial ages, that the winter temperature of Great Britain in the glacial age, as dependent on sun heat, would be no lower than that from Yorkshire to the Shetlands ; and similarly that, from 40° to 80° of north latitude, the shift of the winter isothermals would be only about 4° of latitude, a result wholly inadequate to produce an ice age. The shift of isothermals in the genial age was found to be much smaller.—*Popular Science Monthly.*

A Few Points in connection with the Physiology of the Special Senses.

By Arthur J. Hall, B.A., M.B. Cantab., M.R.C.P.,

Lecturer on Physiology at the Sheffield School of Medicine.

(Being the Presidential Address delivered before the Sheffield Micro-scopical Society at the opening of the Session 1894—5.)

I HAVE selected as my subject for this evening The Special Senses, in the hope that there may be something about them which you may not all have heard before, and even if you have, may yet bear repetition. Man is usually said to be endowed with five special senses—Sight, Hearing, Taste, Smell, and Touch. Let us first enquire what that means. In the two former lectures * I have had the honour of giving you, I have insisted at some length on the fact that a living organism such as man is really an enormous colony of separate living microscopic cells, which work together in different departments for the common good. Each little mass of living substance or cell lives its own life; but in doing so depends on all the other cells for their assistance, and in turn must do its quotum for the general welfare of the whole. And as there can never be a harmonious co-operation of individuals unless there is some recognised superintendent, or master, or head, so in this colony we have a special group of cells to which this duty is assigned—namely, the Nerve-cells, making up the brain and spinal cord, or what we call the Central Nervous System. These cells, which compose our Central Nervous System, alone have the property of consciousness and the power of producing conscious effects when stimulated in various ways. It is of the highest importance that such valuable cells as these should not be exposed to injury from without, just as it is important that an army in battle should not lose its commander. And so we find the Nerve-cells forming the brain and its accessories enclosed in a hard, bony case—namely, the skull and vertebral column. But this very enclosure of our only conscious cells removes them at once from the very position they should hold if they are to be conscious of what is going on around—namely, the surface of the

* See this Journal, Vol. III., p. 257 ; and Vol. IV., p. 225.

body.   How, then, can they be of any use to the organism as
regards obtaining food for it, directing its movements, avoiding
dangers, etc., situated, as they are, in a strong, dark, bony case, in
the most protected of all parts of the body?   To understand this,
we must revert for a few moments to the study of embryology or
the development of the body.   Every human being begins life as
a single, round, nucleated cell (Fig. 6, A), measuring not 2 mm. in
diameter ; and this cell must be endowed with all the living properties of the future adult body in a primitive condition, for from

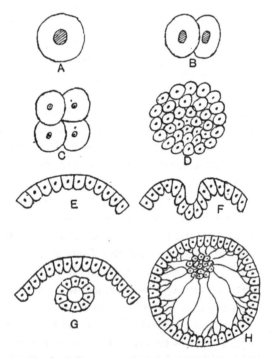

Fig. 6.—Embryological changes in Epiblast Cells, showing development
   of Central Nervous System from surface cells (diagrammatic).

   A, Ovum.   B, C, D, Segmentation or multiplication of cells.
E, Surface layer or Epiblast.   F, Dipping down of E to form Central
Nervous System.   G, Tube of Epiblast, cut off.   H, Diagram of con-
nection in adult between brain-cells and surface-cells by nerve-fibres.

it all the other cells are formed.   It soon divides and multiplies
(Fig. 6, B and C) till the single cell becomes a mass of cells like
a mulberry (Fig. 6, D) in appearance.

A little later these cells begin to arrange themselves in groups, and, neglecting the rest for our present purpose, we will confine our attention to those on the surface of the sphere. They are called the epiblastic cells, and form the surface of the organism (Fig. 6, E). A certain number of these, having a definite position, now begin to fold themselves inwards (Fig. 6, F), and finally to become detached from the rest of the epiblastic cells (Fig. 6, G), which close over them, leaving them inside as a ring of epiblast cells. This ring of epiblast cells (which were once surface cells) represents the cells from which the future nerve-cells will be formed, whilst the rest of the epiblast gives rise to the surface-cells of the adult body, which we call the epidermis, or skin. So, you see, the antecedents of the brain and spinal cord cells were originally surface cells, and occupied the same position that the cells forming the skin do now. By means of long filamentous processes, which we call nerves, proceeding from the cells of the Central Nervous System, these brain-cells join themselves to the surface or epithelial cells (Fig. 6, H). In this way the central cells can be informed, by a sort of telegraphic system, with what is going on at the surface of the body. Thus, what we call our special senses are nothing more than conscious impressions produced in the cells of our brains by the various external stimuli acting upon the non-conscious cells of our surface and conveyed up to the brain along the nerve-fibres.

And this is true of all our different so-called Special Senses. They all require for their carrying out efficiently a surface cell or cells, a nerve-fibre, and a brain-cell intact. If any one of these is destroyed, the power of receiving outside impressions accurately is lost. The eye, the ear, the inside of the nose, and the mouth are merely modifications and elaborations of these elementary nerve-epithelium continuations, adapted to receiving particular kinds of impressions or stimuli only, whereas the whole of the rest of the body surface receives simpler and less elaborate stimuli. The waves of radiant heat from the sun stimulate the epithelial cells of the whole surface of the body, producing a sensation of heat; the same rays, or very closely allied ones, stimulate the epithelial cells of the eye and produce a sensation of light.

The waves of sound produced by a heavy cart passing in the

street produce in the epithelial cells of the ear a sense of noise; but they also produce on the body surface a sense of vibration, transmitted through the earth.   Thus, the sense of sight, hearing, smell, taste, and touch are merely modifications of consciousness propagated in different ways around us, capable of being received only by certain portions of our surface, and each producing a particular impression when the stimulus is carried up by the nerves to our brain-cells.   And the converse also holds good—namely, that in whatever way those surface-cells are stimulated, the cerebral impression called forth is the one that would be called forth by the usual stimulus affecting them.   For example, a blow on the eye produces the sensation in the brain-cells of a flash of light, because the central cells merely receive a stimulus from the eye epithelium, and that is usually due to waves of light.   So also in disease, when the surface epithelium is being constantly irritated, there may be, if it is in the ear, noises; if in the eyes, flashes of light; or if in the skin, creeping sensations; and so forth.

I would next draw your attention to the accuracy of the localisation of stimuli on the body surface.   Although the seat of the conscious impression is enclosed in the bony skull (and we know that it is so because during sleep, or brain rest, the external stimuli produce no effect); yet when an external stimulus acts upon any part of our surface, the brain-cell not only receives an impression of surface irritation, but we know exactly where on the body surface the irritation has occurred.   Thus, if something touches my little finger, I immediately know the exact spot where my finger has been touched, so that I am able at once, if necessary, to move my muscles so as to remove the irritant if it be harmful, or to seize it if it is wanted.   There must, therefore, be somewhere in the brain a separate set of nerve-cells representing each square inch of the body surface; and whenever those are irritated the consciousness projects the impression to its corresponding portion of surface; something in the way, that if the little brass plate opposite my number in the switch-room of the Central Telephone Office drops, the attendant thinks I have rung, and calls out to know if I am there.   Usually I have rung; but if there is some fault in the little brass plate and it falls of itself, the resulting impression on the attendant is the same.

So in our bodies. If the central nerve-cells are themselves diseased and irritated, the impression on the brain is projected on to some external surface change ; and in such a way people who are insane see, hear, and feel things around them which do not exist except to themselves, but they see and hear them very really.

Now, this perfect localisation of sensations is not equally present in all of us at all periods of our lives. It is partly the result of past impressions, memory, and partly the combined result of using all our senses together. By education or training, the senses can be made much more acute ; that is, the power of localisation can be made more accurate, and not only that, but a slight stimulus applied to the surface can be much more fully interpreted by comparing it with previous stimuli like it.

Let us take, for example, the neuro-epithelial mechanism of the skin, which, we say, is endowed with the sense of Touch. This sense of Touch is in reality at least threefold. There is the Tactile sense, or sense of Touch proper, and the sense of Temperature, together with something quite different—namely, the sense of Pain, though this sense of pain is unfortunately not limited to the surface of the skin. That these three are quite distinct from one another, and travel by different nerve-fibres to the brain, is undoubted, for in certain not uncommon diseases of the nervous system the sense of pain may be altogether absent, whilst the senses of touch and temperature remain. Thus, the person can feel anything touching him, can tell a warm body from a cold ; but if pinched or pricked with a pin, feels no pain whatever, but merely feels the contact of the pin touching him. So also in other cases, the person can tell hot from cold, but cannot feel the contact of the hot or cold body.

The different parts of the surface of the body differ very greatly in their power of delicately appreciating touch and temperature, and those parts which are most sensitive to touch are not the most sensitive to temperature. There is a simple instrument called an Æsthesiometer, by which the tactile sensibility of different parts of the surface may be readily estimated. It consists of two fine points, which can be brought close together or separated by a sliding scale, so that the distance between them can easily be read off. The points are then laid on the skin and placed closer and closer to one another until they can no longer be distinguished as two distinct points. The seat of most acute tactile sensation is the tip of the tongue, where two points, only 1 mm. apart, may be distinguished. Next comes the inside of the finger-tips, where, at 2 mm. apart, they are felt distinct; whilst on the back of the hand they are only distinguished as two when 30 mm. apart; and on the neck and back of the trunk when from 50—70 mm. distant.

The Temperature sense is somewhat more complicated to investigate, inasmuch as some parts of the skin seem to feel low temperature more readily, and some to feel high temperature more acutely; but, speaking broadly, the seats of most acute tactile sensibility do not correspond to those of most acute temperature sense. For instance, the shoulders and back are very sensitive to changes of temperature, not of touch. I stated a short time ago that our knowledge of our surroundings is the combined result of the manner in which they stimulate our various special senses, coupled with our memory of past stimuli.

Thus, we find that whenever we see, hear, smell, taste, or touch anything, a mental picture is produced which is far more elaborate than the stimulus alone really warrants us in forming; so that in analysing our impressions we must very carefully distinguish what we actually see, hear, smell, taste, or touch from what we think we do. Familiar examples of this occur every moment of our lives. We pass a cookshop and smell certain odours, and immediately infer that certain things are being cooked, from our previous experience. We see the rays of light from this table with certain shadowed and bright lines, and infer that the table is made of wood covered with leather and solid. But the actual rays which are stimulating our eyes do not carry any such information themselves, so that a clever drawing of a table might entirely deceive us.

Now, this distinction between what we really see, to take this sense for our example, and what we think we see, which is a process of cerebration, comes home very strongly to all who take up microscopy; for in microscopy the size of the objects with which we are dealing prevents us from making any use of our sense of touch, and we are in consequence unable to correct the impressions of one sense by those of another.   Hence, if we are to do good work, we must be extremely careful to distinguish between what we see and what we think we see.   And it is of the greatest value, when looking at an object under the microscope, to write down a description of it without using any technical words whatever, merely describing the objects as dark and light patches, or areas, giving their size with the micrometer, and describing the shape of their outline, etc., together with any other physical details; and, afterwards, to write down what we infer from these details as to their possible structure and nature.   For it is surprising how often we are led unconsciously to see what we want to see, and how grossly we can voluntarily deceive our senses.

I will endeavour to prove my statement in the case of all of you here to-night by a very simple experiment.   If I take this tumbler standing on the table, you see it is upright; if I now tilt it slightly, the rim at the top is in the reverse position to you. Now, if you will close one eye or half-shut both, you can see the rim in the tilted position—that is, the proximal edge of the rim is now the distant edge, although you know perfectly well that it could not stand in such a position.*   With a little practice you can do this with the eyes wide open.   Seeing, then, how easily we can deceive ourselves voluntarily, and how much more easily we may be deceived unconsciously, it behoves us to proceed very cautiously in using an instrument like the microscope, and to make as certain as possible that our data are accurate before we proceed to draw our conclusions.

The association between stimuli and mental pictures is so close and remains so long, that often a sudden stimulus, similar to one received many years before, may bring back the whole mental picture with extreme vividness.   This is especially the case with the sense of smell.   The slightest stimulus of the organ of smell, not even sufficient to be noticeable to us as distinct, may bring back some long-forgotten scene, whilst we puzzle over what could have made us think of it.   And I may here note by the way that in man the sense of smell is apparently to a great extent a lost sense.   It is not nearly so highly developed as in the lower

* I am indebted for this experiment to Mr. Waller; a full description can be found in his *Human Physiology*, p. 542, published by Longmans and Co.

animals, because it is not so much required, in our present condition, for hunting out our prey or scenting danger at a distance. But the sense is occasionally very highly developed in individuals, and this, or something allied, may be the case in the persons who are reputed to be able to state where water will be found on boring, without any previous knowledge of the locality.

That any persons should be thus endowed by Providence with such a special power apart from the rest of mankind, for the sole purpose of making a livelihood, may be sufficient explanation for the uneducated ; but if the power be there, it seems possible that it is the epithelium of their organ of smell which is capable of appreciating some atmospheric condition insufficient to affect most of us.   At least, this is a possible explanation.

The sense of taste, so-called, is very largely the sense of smell, and not taste at all.   Thus, the ordinary aromatic flavouring substances used in cookery do not affect the tongue and palate, but the smell organ in the nose.   In fact, the only things we really taste are bitter, sweet, and acid substances.   The senses of taste and smell are, when both put together, very small in the information they give us of our surroundings.   A smell gives us no idea of the shape, size, or quality of a substance ; nor is it of any use except to warn us of dangerous materials in the atmosphere or food, or to stimulate our desire for food.

Of the other senses, time compels me to leave out very much that might be said of extreme interest to us all, and I must content myself, in conclusion, with a few words as to the sense of Pain. Why do we suffer pain ?   What good is it?   What is it ?   These questions must occur very frequently to all who are themselves sufferers, or have the sad opportunities of seeing pain in others.

To take the last question first, pain is a conscious impression of excessive or harmful peripheral stimulation.   It is to a certain extent possible to localise it, but we get little or no information as to the cause of the pain.   The whole sensation begins and ends in the consciousness of pain and the attempt to get rid of it.   Pain varies in many ways, according to the seat of the irritant.   Thus, certain noises, such as the squeak of a slate pencil, produce on certain persons not only the sense of sound, but also of actual pain.   So also with vision, though the pain differs from that produced on the body at large, Pain does not require peripheral terminal cells, like the special senses ; for if you knock the skin off your knuckles, the spot will be painful when touched, at the same time as it will have lost the delicate sense of touch. Moreover, painful impressions may arise in the deeply-seated organs of the body. What is the good of pain ? It warns the Central Nervous System of something locally wrong and ensures rest for the painful part, thus giving the first and best chance for repair to take place. When our first ancestor first injured himself, we can picture his astonishment at feeling this new sensation, his quickly discovering that movement increased his pain, and his consequent resting of the wounded part till it healed. Pain, then, has some value in this colony we call man ; it is the message from some outlying units that they are out of working order and must be given rest, whatever their rest may mean.

Notes for Beginners in Microscopy.

By R. N. REYNOLDS, M.D.

Micro-Organisms of the Teeth.

THOUSANDS, if not tens of thousands, of bacteria of various kinds and motions may be plainly seen swimming about in each field of the microscope by employing the following method :—

Place on any good microscope a Wenham paraboloid and a good half-inch objective.

The specimen may then be taken from the natural teeth of any human mouth by placing the points of curved forceps or other instrument near the gum between two of the lower back teeth, then giving one scrape upward. A pin-head size of tartar will be found on the forceps, and should be broken up in a drop of water on a glass slip. On this lay a cover-glass and place on the stage of the microscope. When fully illuminated by the paraboloid, the pieces of tartar will appear as white as snow, but we shall see no bacteria ; next turn the mirror until the light is beginning to lessen, and we shall soon reach a point at which we shall be able to plainly see that the dark water between the pieces of tartar is teaming with thousands of bright points moving about. On closely inspecting, we shall see the long *Leptothrix*, many kinds of shorter bacteria, and round micrococci, all apparently having a grand outing in a drop of water.

The half-inch objective is too low a power for the study of bacteria; but in the manner stated will show ten thousand at a glance, while a high power will show only a very few at a time.

In using the half-inch and paraboloid on the bacteria, we must not forget that after we have the bacteria in plain view, a very slight movement of the mirror will leave us unable to see them. Hence, the mirror should be moved very slowly while we watch closely in the dark water between the bits of tartar.

Vinegar Eels.—These little creatures are very popular with beginners in microscopy, and by the following method they can easily obtain a grand view of them, and one which will remain in focus for months if desired.

Take a one-dram clear glass vial; fill it about one quarter full of cider or other vinegar well supplied with eels. Into this drop a silk thread, so that one end rests on the bottom, the other end hanging out of the vial. The thread will cling to the side of the vial from the top to the vinegar, and by capillary attraction keep a film of vinegar along its side.

To hold the vial on the microscope stage, take a bit of cigar-box or other thin wood, about one and a-half inches wide by two and a-half inches long; across its centre draw two lines three-eighths of an inch apart. The space between these lines is to be hollowed out to fit the side of the vial; the hollow space is to be blacked with ink.

Next to the hollow cut a hole, about three-eighths of an inch wide and five-eighths of an inch long, through the block, the length of the hole to run crosswise of the block. This hole is to allow the light from the sub-stage illuminator to pass through. Then cut a quarter-inch square out of each corner of the block; now pass a small rubber band under one end of the block, up through the cut corners, thence across the vial groove, and under the opposite end.

The vial may now be slipped under the rubber band into the groove ; the rubber will then hold the vial in place. Turn the vial until the thread in it is on the upper side ; then place the block on the stage of the microscope ; use about ¾ or 1 inch objective ; naturally, sub-stage illumination is used. The light may now be thrown on the thread ; adjust for the thread above the vinegar ; the eels will then be seen going up and down the thread, but cannot get out of focus, although they may, of course, move out of the field. With vinegar well supplied with eels, large numbers of the eels will be in the field at any time of day or night for months. The cork of the vial must have a notch or slot cut in it to admit air to the eels.

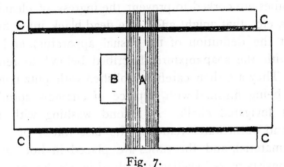

Fig. 7.

With the mirror swung well to one side and the ink-black background, the eels will appear nearly white. They will often gather in bunches on the thread. This arrangement is very satisfactory, and is very much better than that of placing a drop of vinegar on a slide. If this is done, a 1-inch or 2/3rd inch objective is best for their purpose.

A little more trouble will considerably improve the fixture, but makes it more delicate. The improvement is in cementing with Canada balsam a half-inch or five-eighth inch cover-glass on to the vial, to give a flat instead of a round surface to look through.

The cover-glass may be protected and the vial held by a thin metal holder instead of the wooden one, the metal holder passing over the vial and cover-glass.

Lacquering Microscope Tubes and Stands.*

By Dr. Frank L. James, St. Louis. Mo.

THE art of lacquering, like every other, requires some experience to get good results. The writer has been in the habit of closing the tubes with close-fitting corks and placing them in hot water. This ensures an even temperature throughout, and the corks afford a good handle to hold them with during the application of the lacquer. Another precaution, too often neglected, is the removal of all old lacquer before the application of fresh. The following is the method employed by the writer :—

The tubes are corked to prevent the ingress of cleaning material, water, etc., that might affect the dead-black inner finish (and thus affect the definition of the visual apparatus), and are then washed with the soap-mixture described below† to remove the lacquer. They are then carefully polished with putz pomade, the operation being finished with a piece of chamois, sprinkled with the finest levigated chalk. A second washing with the soap mixture and warm water removes the last traces of grease and polishing material, and the tubes are placed in hot water. The lacquer consists of red shellac, dissolved in alcohol of 95°, about 1 ounce of lac to 10 or 12 ounces of spirit. This gives a very pale lacquer, which may be darkened by the use of dragon's blood or given a yellow hue by gamboge, sandarac, anatto, etc. Appended hereto will be found some formulæ that we have tested and found reliable.

The lacquer is poured into a saucer, or small open dish, and

*From the *National Druggist.*

† This mixture is usually made out of the refuse alcohol, and benzol used in washing, clearing, etc., in mounting (small portions that cannot be returned to the container without redistillation). The bottle is kept handy, and fresh soap is added from time to time, no particular regard being paid as to proportions. A good formula would be as follows :—

Shaved Castile Soap	...	...	...	1 part.
Alcohol, 94° to 95°	...	...	...	3 ,,
Benzol ...	...	...	...	3—4 ;
Liquor Potassæ	...	...	...	1 ,,

Mix and shake from time to time.

the brush should be very soft, and for large articles flat.   For smaller articles, a round camel hair or sable pencil will answer. The tube is removed from the water, wiped, and held for a moment to ensure dryness, and the lacquer then applied as rapidly and smoothly as possible.   The brush should be drained on the side of the dish, in order to avoid getting too much of the liquid on the object at once.   If a second coat is deemed necessary, let the first become quite dry before applying it.   If the tube is not warm enough, return it to the hot water for a few moments and dry carefully.   The water should not be too hot, or there is danger of blistering the lacquer.   From 150° to 160° F. is a good temperature.   In applying the lacquer, care should be taken not to rub the object with the brush.   Apply it with light strokes, *all* in the same direction.

 1.—*Pale Gold.*

   Shellac, quite clean ... ... 1 part.
   Ground Turmeric   1 ,,
   Alcohol ... ... 4 ,,

Mix and set aside in a warm place and let stand, with frequent agitation, until the lac is dissolved.   Decant the clear liquid and preserve for use.

 2.—*Yellow Gold.*

   Shellac   ... 25 parts.
   Turmeric ... ... 4 ,,
   Dragon's Blood   1
   Alcohol, 95° ... ... 70 ,,

Digest for eight days, with frequent agitation ; decant and filter.

 3.—*Deep Red Gold.*

   Spanish Anatto ... 2 parts.
   Turmeric, powdered ... 30 ,,
   Red, Saunders' .. 3
   Alcohol, 95° ... ... 480 ,,

Infuse in the cold for twenty-four to thirty-six hours, shaking occasionally.   Let stand until settled ; then add

   Shellac ... ... 60 parts.
   Sandarac ...   ... 15 ,,
   Mastic ... ... 15 ,,
   Canada Balsam ... ... 15 ,,

When dissolved, add 10 parts oil of turpentine. Upon highly polished brass this gives a deep red gold colour that is lasting and very handsome.

For a lacquer to use on cold metal we give the following, but do not recommend it, or, in fact, any cold lacquer :—

Shellac	16 parts.
Dragon's Blood ...	1 „
Anatto ...	... 8 „
Alcohol	... 256 „

Digest together until the shellac is dissolved ; strain, etc., as No. 1.

British Hydrachnidæ.

By C. D. Soar. Pl. X.

WHEN we consider the great number of natural history societies and microscopical clubs that are in existence in Great Britain, it makes one wonder why that interesting family of the Acarina—viz., *Hydrachnidæ*—have been so long neglected. Perhaps the want of good text-books on the subject is the reason. There is no work on the Hydrachna in English. Andrew Murray, in his " Economic Entomology—*Aptera*," mentions a few species, but some of them are wrongly named and the drawings very bad. Dr. F. C. George, in the pages of *Science Gossip*, gave a few papers and a few good illustrations, but did not get further than a few species of one genus. The last edition of Carpenter dismisses the subject in about six lines. There are a few isolated papers in Microscopical Journals, but I do not think any are much more than a mere reference to the subject. On the Continent a great deal has been written from Müller to Neumann. I propose, with the editor's permission, to give a paper from time to time on the British Hydrachnidæ, each paper to illustrate a genus, with one species merely as an example ; the letterpress to be as brief as the subject will allow.

This, I hope, will lay a foundation for anyone about to take up the study of these interesting mites. The illustrations will all be from my own drawings taken from the life and all measurements in English.

Genus, AXONA (Kramer).[*]

Body chitinous, with a granulated surface and a depressed line round margin of body. Legs short, not very hairy, but adapted for swimming; all tarsi have claws. Epimera fused together into one group. Palpi, long. Fourth joint, spoon-shaped; second joint has a small, peg-like process. Eyes widely separated and near margin of body. On each side of the operculum is a special plate, containing three copulative pores. Mandibles in two distinct portions.

To illustrate above genus, I give species, *A. versicolor.*

AXONA VERSICOLOR (Müller, Kramer).

1776.—*Hydrachna versicolor*, Müller, *Zool. Dan. Prodr.*, p. 191, N. 2285.

1781.—*Hydrachna versicolor*, *Ibid., Hyrachnæ*, etc., p. 77, Tab. VI., Fig. 6.

1793.—*Trombidium versicolor*, J. C. Fabricius, *Ent. Syst.*, Tom. II., p. 400, N. 9.

1805.—*Atax versicolor, Ibid., Syst. Antliatorum*, p. 367.

1835—41.—*Arrenurus versicolor*, C. L. Koch, *Deutschlands Crust.*, etc., N. 13, Figs. 16, 17.

1854.—*Arrenurus versicolor*, Bruzelius, *Berkr. ö. Hydrachn. som. förck.*, I. Skáne, p. 33.

1875.—*Axona viridis*, Kramer, *loc. cit.*, p. 311, Tab. IX., Fig. 19.

1879.—*Axona versicolor*, Neuman, *Sveriges Hydrachnider*, p 74, Tab. XI., Fig. 2.

This very beautiful mite has been shifted about from one genus to another several times; but I hope it has at last found a resting place where placed by P. Kramer in genus *Axona*. The mites, both male and female, from which the drawings are taken were found in the river at Woking on July 14, 1894. The female is about 1/50th of an inch long; the male a little longer. In colour it predominates in a blue-green, with patches of red and white. The epimera is nearly all (but the actual centre) a very vivid green. The centre is a beautiful pink. Koch gives a figure of both male and female, but different in colour. In the specimens from which

[*] P. Kramer, *Beitr. zur Naturgeschich. der Hydrachniden*, p. 310, "Axona," 1875.

I drew my figures, the colour is identical in both. Müller gives a figure, but not coloured. Neuman gives a beautiful drawing in colour, but the colour does not quite correspond with those found by me; neither does the shape, but this may be only a local variation of the same species.

The body of the female is nearly round, flattened in the front part. The body of the male is pear-shaped. On the fourth joint of the fourth leg of the male will be seen a peculiar process, which is also found on males in other genera. The legs of the male are also stronger and better developed than they are in the female.

EXPLANATION OF PLATE X.

> Fig. 1.—Dorsal view of female.
> ,, 2.—Ventral view of ditto.
> ,, 3.—Side view of palpi of female.
> ,, 4.—Fourth leg of female.
> ,, 5.—Dorsal view of male.
> ,, 6.—Ventral view of male.

PAIN IN THE EYES.—Pain in the eyes after reading or minute work of any kind is often due to spasm of the muscle of accommodation. Engravers and workers with the microscope frequently suffer severely. The pain comes on after prolonged application, and is usually of a dull aching character. Not infrequently it is attended with a little feeling of sickness and considerable depression of spirits. The best thing is to lie down in a dimly-lighted room when the pain comes on, and place over the eyes and eyebrows a pad of lint dipped in cold water. A small piece of mustard-leaf to the temples or behind the ears will ease the pain. We have found relief from bathing the eyes with a very weak solution of atropine, prepared by putting a drop or two of the solution of atropine (*liquor atropiæ*) in a tumbler full of water. It is to be used occasionally as a lotion, but must not be taken internally. Arnica often does good; it should be taken according to Pr. 42, and also applied locally in the form of the arnica lotion (Pr. 94). When the pain is the result of prolonged work by gaslight, nux vomica (Pr. 44) may be used.

—From "*The Family Physician*" for March.

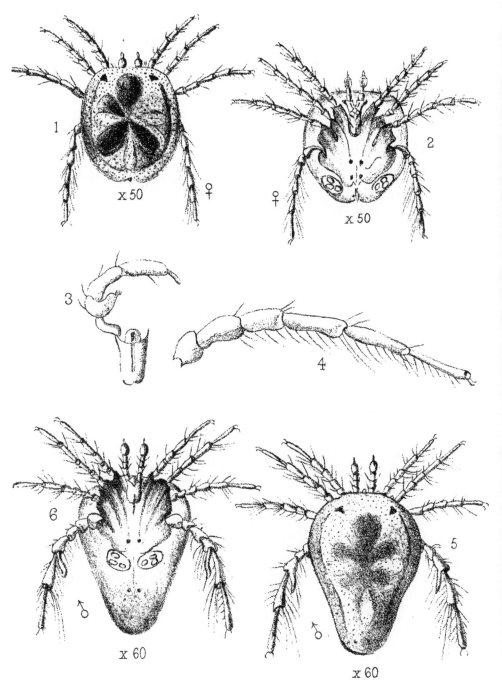

1 x 50 ♀

♀ 2 x 50

3

4

6 ♂ x 60

♂ 5 x 60

Leaves from my Note=Book.

By Mrs. Alice Bodington, British Columbia.

SOME INSTANCES OF THE INFLUENCE OF THE ENVIRONMENT IN PRODUCING RACE PECULIARITIES.

1.—Cats and Rats in Cold Storage Warehouses.

2.—The Paradise Fish of Japan.

3.—The Indian Ghost Flower. (Plate XI.)

MR. Herbert Spencer, in his chapter on Heredity, says that, "excluding those inductions that have been so fully verified as to rank with exact science, there are no inductions so trustworthy as those which have undergone the mercantile test. When we have thousands of men whose profit or loss depends on the truth of the inferences they draw from simple and perpetually repeated observations; and, when we find that the inference arrived at and handed down from generation to generation of these deeply interested observers has become an unshakeable conviction, we may accept it without hesitation. In breeders of animals we have such a class led by such experiences, and entertaining such a conviction—the conviction that minor peculiarities of organisation are inherited as well as major peculiarities." *

With the remembrance of this chapter in my mind, I was particularly interested in an account of the development of a peculiar breed of cats in the "cold storage" warehouses of Pittsburg, Pennsylvania† ; these warehouses being maintained constantly at a temperature below freezing point. At first, no rats could maintain existence under such Arctic conditions, a most convenient state of things where toothsome eatables are kept. But in a few months nature proved herself equal to the occasion, and a breed of rats appeared which could withstand the low temperature. Not only were their bodies clothed, as might be expected, with remarkably long and thick fur, but even their tails were covered with a thick growth of hair. Dr. Manley Miles, in a paper read at the Brooklyn Meeting of the A.A.A.S., August,

---

* *Principles of Biology*, Vol. I., p. 241.

† *Pittsburg Despatch*, quoted in *Public Opinion*.

1894,* says that the conditions of artificial selection differ so much from those of natural selection that it is difficult to carry on experiments in artificial selection, which are biologically valuable ; but in the case of the rats and cats in the " cold storage " warehouses of Pittsburg, *natural and artificial selection have existed side by side* with results equally interesting, as showing the action of the environment in modifying organisms.

The thickly-furred rats above mentioned made themselves at home in all the warehouses, and efforts were naturally made to introduce cats, but at first in vain. The first cats turned into the icy cold room pined and died. But advantage was taken, as is usual in artificial selection, of a slight natural difference, and one cat, which possessed unusually thick fur, proved not only able to withstand the cold, but actually throve and grew fat. By careful nursing a brood of seven kittens was raised in the rooms of the Pennsylvania Storage Company, and developed into sturdy, thick-furred cats, suited to an Icelandic clime. These original kittens were distributed amongst the other " cold storage " warehouses in Pittsburgh, and have been the progenitors of a peculiar breed of cats adapted to the conditions in which they must find their prey. Their tails are much shortened, and their hair is as thick and as full of under fur as that of the wild cats of the Canadian woods. But one of the most striking features in these cats, from a biological point of view, is the development of the sensitive hairs on the face, popularly called " feelers." In the ordinary cat these sensitive hairs are about three inches long, but in the cold warehouse cats they grow to a length of five and six inches. This is a modification which has arisen *de novo* ; probably from that condition of their environment, which obliges the animals to find their prey in semi-darkness. There was, perhaps, a " predisposition on the part of the germ " in the mother cat, as Professor Weismann would say, to grow thick fur ; but the germ does not seem to have indicated any change in the feelers of the ancestral pussy.

The storage people state that if one of these furry cats is taken into the open air, especially in summer time, it will die from convulsions in a few hours.

* " Limits of Biological Experiment," *American Naturalist*, Oct., '94, p. 845.

PARADISE FISH.—I have read an account in an American Magazine of the so-called " Paradise Fish " of Japan, but I can find no account of it in any work I possess. It seems from its habits to be a species of Stickleback (*Gasterosteus*). At the mating season, the male, which is ordinarily of a dull silver hue, exhibits brilliant stripes of red, blue, and green, its ventral fins showing streaks of brightest orange ; hence, doubtless, the Paradise Fish has acquired its name. It breeds freely in confinement, and requires no change of water in the globe or aquarium except enough to make up for evaporation. It is very tame and " surprisingly intelligent," and its habits at the breeding season render it a peculiarly amusing pet.

When the female is about to lay her eggs, the male fish makes a nest at the bottom of the water composed entirely of air-bubbles. For this purpose he swallows air and ejects it in the shape of bubbles, which are held and made permanent by glutinous capsules from a secretion in the fish's mouth. When the female lays her eggs she would devour them immediately, but the male will not permit her to do so. He takes them in his mouth, and going beneath the nest ejects them, when they rise and find a resting-place amongst the bubbles. Or, sometimes, he will conduct his mate under the nest, so that the eggs as she lays them will ascend and lodge in the nest. The laying being concluded, and all the eggs disposed of in this manner, he keeps guard at the nest, not allowing the female to approach it, and even attacking her fiercely if she makes the attempt. At the same time, he occupies himself continuously in making fresh bubbles of air in the place of those that burst, and at intervals he spouts streams of air all round the nest, apparently for the purpose of keeping it thoroughly aërated. After about five days, during which this performance is kept up, the young are hatched out. *They cannot swim, but cling like little tadpoles to the bubbles.* If one falls to the bottom, as happens every now and then, the father fish darts after it, takes it in his mouth, and conveys it beneath the nest, where he disgorges it among the bubbles again. Thus the Paradise Fish continues to take care of its offspring for several weeks.

Everyone knows of the Water Spider, which makes a little diving-bell of bubbles of air, in which she lives quite dry and spins her silken cocoon with its hundred eggs; only coming to the surface to secure water-flies and other small insects and to provide herself with fresh bubbles of air, which adhere to the hairs on her body and keep her perfectly dry. This is a sufficiently strange instance of the adaptation of an air-breathing animal to life at the bottom of the water; but it would be impossible to say for how many thousands of years the Water-Spider has made her diving-bell; whereas the Paradise Fish must have learned the art of bringing up its young in insufficiently aërated water during that comparatively brief time since human civilisation has begun, and men have had leisure to enjoy the keeping of fish as pets and the art to construct receptacles in which to keep them.

It would be particularly interesting to know in what manner the fish supplies itself with fresh air on ordinary occasions, since, as it never requires change of water, it must possess some means of oxygenating the water in which it lives.

INDIAN GHOST-FLOWER (*Monotropa unifolia*. Pl. XI).—This poetical name has been given to an aberrant member of the genus *Ericaceæ*, which, from its fungus-like manner of obtaining its nourishment, has taken on the superficial appearance of a fungus. Walking in a trail through our woods one afternoon last autumn, I saw, as I imagined, a group of very pretty graceful fungi amidst the bracken. On picking a bunch of these apparent fungi, I found the wax-like, colourless stem supporting an equally wax-like, colourless flower, the whole being in shape very like a minute tobacco-pipe. Further examination showed a flower which placed the plant amongst the most highly developed of the dicotyledons. No amount of search showed any trace of a root. The pointed, colourless end of the Ghost-Flower simply penetrated deeply into a deposit of completely decayed wood. A few small, colourless scales represented all it possessed of leaves, the whole plant being absolutely destitute of chlorophyll, and having the appearance of being made of white wax, resembling in this respect a true fungus which grows in the same woods. No mere description, however, can give a correct idea of the peculiar stiffness of these groups of little pipes. In decaying, they also resemble fungi in turning of a black colour. I find no mention of this peculiar plant in Balfour's *Botany* nor the *Encyclopædia Britannica*, but it is known on the

s. ad nat del

F Phillips Sc

Monotropa uniflora

eastern as well as on the western side of the Rocky Mountains. It seems a most interesting example both of adaptation and of heredity, the flower being undegenerate, and like that of any other heaths, whilst the rest of the plant is changed almost beyond recognition, owing to its saprophytic habit of life. This year I have, as yet, been unable to find the plant, or I would give a more exact description of the flower. A lady who has made a large collection of British Columbia plants says that she also has been unable to find any semblance of a root in the Ghost-Flower.

I do not suppose that any of the evidence of adaptation to the environment given in the foregoing instances will have any effect on those who deny this power of adaptation ; but the facts themselves must be of interest both for Neo-Lamarckians and for followers of Professor Weismann.

[After sending the preceding paper, Mrs. Bodington wrote to the Union Storage Co., of Pittsburg, Pa., and in forwarding us their reply she very pathetically says :—" It turns out that there are no rats with furry tails in the business ; no cats with abnormal whiskers ; no convulsions on being taken into the open air ; no constant cold and darkness, and nothing to demolish Prof. Weismann." The following is the letter referred to.—ED.]

" Dear Madam,—

Your favour of the 3rd inst. was handed to the writer a few days ago. While there is some foundation for the newspaper article, a copy of which you enclosed, it is somewhat exaggerated. We are engaged in the storage business, having several warehouses for general storage, and one large house for cold storage. The cold storage house is separated into rooms of various sizes, varying from 10 degrees to 40 degrees above zero. Our experience has taught us that the best way to destroy mice that are likely to come in goods sent to us for storage is by the use of cats.

About a year ago, for the first time, we discovered some mice in one of the rooms in our cold storage house. We removed one of the cats from our general storage warehouse into the room referred to in our cold storage-house. While there, she had a litter of seven kittens. Four of these we transferred into one of the other warehouses. leaving three in the cold storage house, where they now are. After the kittens were old enough to take care of themselves, we put the old cat back into the house we had taken her from. The change of climate or temperature seemed to affect her almost immediately. She got very weak and languid. We placed her again in the cold storage-house, when she immediately revived, and seemed to be filled with new life and energy.

While the feelers of the cats in the cold storage-rooms are of the usual length, the fur is very thick, and the cats are larger, stronger, and healthier than the cats in any of the other warehouses. They have been transferred from room to room, the temperature, as above stated, varying from 15 to 40 degrees. The building is lighted by electric lamps, but most of the time there is no light. Until last Saturday the cats had never seen natural light. Saturday afternoon we had two of them carried to the shipping platform on the first floor. When the men got on the elevator with them, the cats looked surprised and startled, and struggled to get away, although always very playful and happy in the rooms above. They were evidently so badly frightened that I had them immediately taken back to the rooms. The original cats are common, every day pussies.

We paid no particular attention to their diet, giving them milk once a day and meat once a week. Sometimes the workmen will give them some scraps from their dinner-pails.

Will be glad to give you any further information you may want regarding the matter, and follow any suggestions you may care to make.

One of our warehouses for general storage is a six-storied building with cellar, the average height of the floors being 12 ft. There is no communication from one floor to the other except by elevators. On each floor in the house we have one or two cats. In cold weather, after the warehouse is closed and absolutely dark, the cats jump from the floor above to the floors below, in that way going from one floor to the other until they reach the cellar, for the purpose of spending the night close to the steam elevator engine. Each morning the cats are carried from the cellar floor to the different floors above.   Yours very truly,

S. BAILEY, Jnr.,

Sec. and Treas."

EXPERIMENTS continued through many years, by Dr. S. Rideal, show that the chemical activity of sunlight during winter on the high Alps is much greater than at lower levels, and enormously greater than in large towns at the same season. This increased activity may contribute in an important manner to the beneficial effects of health of residence in such regions.

Bacteria of the Sputa and Cryptogamic Flora of the Mouth.

By Filandro Vicentini, M.D., Chieti, Italy.

SECOND MEMOIR.

Translated by Professor E. Saieghi.

Recent Bacteriological Researches on the Sputa; the Morphology and Biology of the Microbes of the Mouth.

FURTHER REMARKS ON THE BACTERIA AND BACILLI FOUND IN THE SPUTA.

---

§ III.

OPINIONS HITHERTO HELD RESPECTING THE MICRO-ORGANISMS OF THE MOUTH, according to Miller.

HISTORY AND GENERAL FACTS.

WHEN I for the first time began to examine the fructification of *Leptothrix*, of which no mention had hitherto been made in the ordinary text-books of microscopy, I, of course, consulted several of the special publications quoted in the Bibliographical Appendix, in order to ascertain whether others had described these fructifications before; but my search was fruitless. It is true that Robin believed in their existence, as appears from the quotation at the beginning of this Memoir; but the granules mentioned by him are simply the reserve gemmules of this microphite. However, I should not have been induced to treat on this subject so early, had I not unexpectedly met with a very conspicuous specimen of such fructification in some pulmonitic sputum (see Fig. 16). I then wished to acquaint myself with the last work of Miller, already quoted, and on perusing it, I found that he had not even touched on that point; but, on the contrary, his views were, in a sense, diametrically opposed to my own.

Of the specimen found in the pulmonitic sputum I shall speak in Section 4. According to Miller, the opinions hitherto acquired on the microbes of the mouth would be as follows :—

In the 3rd chapter of his volume, " The Micro-Organisms of the Buccal Cavity : Local and General Complaints produced by them" (the most complete and most recent work on the subject), the illustrious Berlin Professor goes on to relate the ancient and recent scientific opinions on the microbes of the mouth, beginning with their discoverer, Leeuwenhoek, who, in 1683, discovered first on his own teeth, always kept clean, and afterwards on those of an old man, "*magna cum admiratione*," animálcules, drawn by him (Fig. 10 of Miller), some of which resemble the Comma bacilli :—"*Multa exigua admodum animalcula jucundissimo modo se moventia.*"

Lebeaume compared the tartar on the teeth to coral formations. Mandl held that the tartar proceeded from the chalky remains of the vibriones described by him, which he thought would be killed by heat, by hydrochloric acid, and alcohol. Bühlmann was the first to observe and describe the filaments of *Leptothrix*, without, however, giving any opinion as regards their vegetable or animal nature.

Henle was the first to declare that these microbes were of a vegetable nature.

Erdl treated the decayed teeth with hydrochloric acid, and from the crown he obtained a kind of delicate, enveloping membrane, composed of parasites.

Ficinus dealt diffusely with microbes of the buccal cavity, which he held to be of an animal nature, and gave them the generic name of *dental animalcules*. He described the filaments of Leeuwenhoek and of Bühlmann, the granules, the epithelia, the corpuscles of mucus, forming the *patina dentaria*, as well as certain infusoria that accidentally dwelt there. He not only held the bacteria and bacilli to be real animalcules in brisk motion, but he imagined also that they possessed a mouth. He grouped them with infusoria without cilia, probably shelled, after the types of the genera *Paramœcium* and *Colpoda*.

In 1847 Robin thought that the principal micro-organism of the mouth was an alga which he called *Leptothrix buccalis.*

Klencke, continuing the researches of Bühlmann and Ficinus, drew the dental animalcules, the filaments of Bühlmann (reproduced in Fig. 11 of M.) ; and, following the opinions of Ficinus, he even drew the mouths on their ventral side, as can be seen in the quoted figure.

Meanwhile, the ideas of Robin were accepted, and *Leptothrix* was generally held to be an alga or fungus *sui generis*, excepting by Frey, who, together with Hallier, thought it to be a form of *Penicillium glaucum*, and by Tilbury-Fox, who restricted it indiscriminately to a form of *Oidium.**

After these historical hints, Miller goes on to describe his methods of investigations, from which it appears that the author and the preceding investigators have not acted with that necessary care and discernment, which has been demonstrated, in collecting and preparing the contents of the mouth ; nor have they instituted comparisons with the sputa. They directed their best attention to the culture of the microbes of the mouth and to the study of the *patina dentaria*. They did not avail themselves of the immersion methods except in the special examination of certain fragments or isolated microbes and of their cultures. It is no wonder, therefore, that they should not have met with the fructifications and the productions of *Leptothrix* by points.

Miller speaks of cultures on agar or on calf's blood serum, and from thousands of specimens he gives the preference to agar peptonised with broth, with an additional 0·5—1 per 100 of sugar.

The cultures were made from saliva, the tartar of decayed teeth, or scrapings from the surface of the tongue or from dental ulcerations and altered pulps of teeth. The cultures from saliva in the gelatine were negative. Often the bacteria reared in agar did not succeed in gelatine, owing to the low temperature maintained.

For a general study, he considers an amplification of 20 to 300 diameters sufficient, but for the special morphology he used a homogeneous immersion lens. He speaks of fungi derived from

---

* Frey, Bibliography.   Tilbury-Fox, by Beale, Work, p. 491.

external sources, as the Bacilli of tare, potatoes, lactic acid, green pus ; and also of the *Micrococcus tetragenus*, of *Mycoderma aceti*, of *Staphylococcus pyrogenes*, *aureus*, and *albus*, to the effect of excluding them from the cultivations; But he touches again upon some of these forms in Chapter IX., as we shall see later on.

Miller classifies, afterwards, the microbes really constant, or primary, of the buccal cavity, into six different species. Of these we give a list, putting our denominations to each species of Miller.

Synopsis.

Miller :	Vicentini :
1—*Leptothrix buccalis (L. innominata).*	*Leptothrix buccalis.*
2—*Bacillus buccalis maximus.*	Fragments of stumps.
3—*Leptothrix buccalis maxima.*	Stumps.
4—*Jodococcus vaginatus.*	Special sheaths of bacteria proceeding from some reserve gemmules.
5—*Spirillum sputigenum.*	Pointed or virgulated Bacilli (copulatory organs ?)
6—*Spirochæte dentium.*	Spirillum (fragments of very slender filaments).

The above species or types are found in every mouth. The *Spirillum sputigenum* is sometimes found, in almost a pure culture, in the carious cavities ; and the *Spirochæte dentium* near the gums. The cultures from substances mixed with saliva, acids, alkalies, product of caries, dental mucus gathered out of the ebullition of decayed teeth, do not succeed ; some simple cultures, however, give productions of 15 or 20 articulations of certain species, but these cannot be reproduced a second time.

I will now sum up the characteristics which Miller attributes to these six primary species, beginning with *Leptothrix innominata*.

LEPTOTHRIX BUCCALIS (*Leptothrix innominata*).

The name of *Leptothrix buccalis* is attributed to Robin, who applied it in general to all buccal microbes. Hallier, Zopf, and all others who wrote on this subject conformed to his opinion, maintaining that the motile bacteria are spores of *Leptothrix* in activity, and cocci and bacteria not motile are wandering spores in a quiescent state.

Leber and Rottenstein found the fine violet staining with iodine and the acids to be a characteristic of buccal microbes (included by them in a single species of *Leptothrix*). Miller, however, considers such forms to be different species, although they show the same behaviour towards staining re-agents, for the reason that the filaments of *Leptothrix* are not articulated (as stated by Robin). On the contrary, the bacilli or filaments under iodine re-agents are articulated. Now, can this distinctive character be sufficient to classify a species? I shall show later that, besides articulations, bacilli may be found, or knotty little rods in continuity with small chains, and the latter with fertile filaments (not articulated), as in Figs. 9 and 10. This demonstrates the fact that such filaments or bacilli, either articulated or not, or simply in beaded forms or chains, do not constitute a difference of species. Moreover, the filaments more woody and articulated, found in the lower layer of the patina, or, owing to the friction, are left exposed, so that the remaining stems become thick and hard, and their internal gemmules give rise to enclosed bacteria. Vignal was not of a different opinion in holding *Leptothrix* to be a fungus of the mouth, which he had reared, although it showed transverse segmentations easily discernible in aniline. Likewise, Miller, in his article on *Leptothrix gigantea*, as we shall see later, entertained the same opinion.

Quite recently, the dumb-bell bacteria and other fungi of the mouth were identified with *Bacterium termo* by Stockwell, Clark, and others. In the preceding Memoir we held this identity as plausible; but Miller positively rejects it because the title of *Bacterium termo* would embrace a mass of very different forms and species. He likewise rejects the title of *Leptothrix buccalis* (given by Robin) and of dental filaments (by Bühlmann), names which Miller would have banished from microphitology. However, he would give the name of *Leptothrix innominata* to those little-known filaments which appear to constitute a group or a species differing from other fungi. It is clear that the author alludes to our fertile filaments, making of them a separate species.

The *Leptothrix innominata* of Miller (the real *Leptothrix* akin to fructifications to our knowledge) is to be found in the *patina dentaria* (*materia alba* of Leeuwenhoek) constantly, but in different quantities in every mouth. It is, of course, very scarce in well-cleaned teeth, because, lying on the top of the tufts, it is the first to be removed by friction. The *patina dentaria* exhibits large and small heaps of round granules, and, on the edges, slender filaments variously bent. The granules were considered as the matrix of the fungus, having been taken for spores of *Leptothrix*. The author thinks them to be either micrococci foreign to *Leptothrix*, or simply linking points of its filaments.

The filaments are of varied length, from 0·5 to 0·8 micro-millemetres, sometimes twisted, usually still, and without articulations (but this is not always exact). They generally have irregular contours and appear to be badly nourished or are dead, but we shall see that they contain gemmules and bear spores (ears?). In their surroundings there are also numerous shorter filaments or small rods, which the author thinks might be simply fragments of longer filaments, or cellules of the fungus not having yet reached their full development.

By using the solution of iodine, slightly acidulated with lactic acid, under a power of 350 diameters, epithelia and clusters of micrococci can be detected intermixed with them (Fig. 12 of Miller), and even various forms of little rods and filaments of *Leptothrix* lightly tinged with yellow. Other larger bacilli become tinged with deep violet, and these are called by the author *Jodococcus vaginatus* and *Bacillus buccalis maximus*.

The same author again speaks of the *Leptothrix innominata* on page 203. He declares positively that it cannot be cultivated in every medium. This is true, if he intends to speak of the ordinary culture media, but not equally so with regard to sputa, in which are to be found not only natural cultures, but even the most vigorous fructifications of this parasite, as we shall see later on.

Other Primary Micro-Organisms of the Mouth.

Bacillus buccalis maximus.—This consists of isolated bacilli or filaments, more frequently of bundles. The filaments of the length of 30—50 micro-millemetres (Fig. 13 of Miller, stained with the acidulated solution of iodine) are distinctly articulated.

The isolated bacilli or articulations are from 2 to 10 micro-mille-metres; their thickness is from 1 to 1·3 mm. Of the primary fungi of the mouth this is the greatest; all its parts cannot be coloured with iodine (Fig. 14 of Miller to be compared with our Fig. 2, *m, m, n, n, n', q, r*, and with Fig. 5, *a, b, c, d)*. It is not detected in the dental tubuli, probably because it is too big·to penetrate there.

Miller holds it to be of a different species from *Leptothrix innominata*, from its size, and through possessing segmentations or knots, and not bending in zig-zag ; and also on account of its very strong behaviour towards iodine re-agents. But these points, very feeble in themselves, are not even constant. Take, for example, the specimen drawn in Fig. 9, *d*, in which the filament of *Leptothrix* (fertile filament not segmented) springs from a portion of the chain which rose on the top of a large, knotty filament. If these large filaments appear in stumps or in separate networks it is owing to the growth of superior vegetation and the subsequent rubbing off of the latter by friction.

I shall presently show that *Leptothrix* living in the water is absolutely identical with this *Bacillus maximus* of Miller.

I have, however, to observe that very distinct segmentations exist sometimes, even in the most slender filaments. In Fig. 5, *g*, I observed an opaque segment, deeply coloured in the middle, with two clear, pale segments at the ends. This filament, compared with bacillus *d* (same figure), hardly gives the thirtieth part of its thickness. Miller himself, in his previous work of 1883 on *Leptothrix gigantea*, describes small rods and cocci within the sheaths of filaments of the said *Leptothrix* (Figs. 2, 3, and 4 of relative plate). In Fig. 5 he even drew a cumulus of small rods, sprung out of the filaments through bursting. The very slender filaments drawn in Figs. 2, 3, and 4 are also articulated.

Leptothrix buccalis maxima.—The same author is not certain whether this is a really separate species or a younger form of the same *Bacillus buccalis maximus*. The only difference consists in its resisting an iodine reaction, and in the greater distance of its articulations.

Jodococcus vaginatus (Fig. 15 of Miller, reproduced in our Fig. 7) is found rather plentifully in uncleaned mouths. Miller found it only in two children. It is formed of from four to ten cellules or nuclei, seldom more, placed obliquely in chain, and having the shape of little flat or round shields, or tetrahedrons. The chain is either bare, as in *p, p,* of Fig. 2, or sheathed. The sheath, large, 0·75 mm. (I have found some even larger) is colourless or takes a yellowish colour, with a long iodine saturation. The nuclei are tinged with dark violet. At times the sheath is broken, the place of the nucleus is empty, or it has dropped off.

I shall not repeat what I demonstrated in Section I. with regard to these chains—of their envelope tardily discovered, of the presence on the same of curved diplococci, morphologically identical with the gonococcus of Neisser. I shall only state that, as such forms are found, even of greater dimensions, in urine, and being intermixed with filaments and other forms of *Leptothrix*, their relation with this parasite, which is most widely spread, becomes still more probable. It is, perhaps, the only one which lodges simultaneously in the mouth and in the external genito-urinary mucous membrane.

But, even admitting the *Jodococcus vaginatus* of Miller to be entirely different from the small sheathed chains described by us in Fig. 4, their entity as a distinct species still remains doubtful. The same author, in the quoted work of 1883, described and drew (annexed plate, Fig. 8) some articulations divided in two and in four small cocci, arranged, as in the *Jodococcus vaginatus*, by couples or tetrahedrons. He even found in the pig *whole filaments* of *Leptothrix*, made up with a series of those cocci, as we shall see later.

Spirillum sputigenum (Figs. 16 and 17 of Miller and our Fig. 5, *h*).—It is found in every mouth, and mostly upon unclean teeth. We should note this circumstance, as it corroborates our views. Cleaning the teeth destroys, every time, the pseudo-inflorescences or productions by points, from which would rise those Comma bacilli, together with the spindle-like and snake-like bacilli. We shall see, in fact, that the simple act of mastication is apt to destroy those pseudo-inflorescences, so that we must look for them on empty stomachs. Is it to be wondered, then, if their products, and chiefly the *Spirillum sputigenum*, are so scanty in those persons who regularly clean their teeth ? Miller goes on to say that in unclean mouths they are numberless.

These microbes are in the shape of small Comma rods, endowed with very quick, screw-like movements. When linked by twos, they form themselves into small snakes, like the letter *s*.

Lewis, as we have already said, identified this microbe with the bacillus styled *Cholerigenus*. However, its presence in the mouth had been detected before, and by a few attributed to fragments of *Spirochæte*. Clark held it to be the cause of caries, for it penetrates the dental tubuli as shown in Fig. 17 of Miller. This author excludes every relationship between the *Spirillum sputigenum* and the bacillus *Cholerigenus*, judging from the culture which for the latter is positive whilst for the Virgula of the mouth it is always negative. However, this is a new confirmation of our views, for, if the Commas in question are real organs or copulative filaments, it is quite natural that, whilst they fecundate the articulations or the spores, they should be incapable of reproduction themselves.*

In addition to the *Spirillum sputigenum*, Miller classifies two other types of curved filaments : one short, massive (Fig. 32), and motile, which liquifies gelatine ; the other slender, still, and more curved (Fig. 18). In growing old, it gives rise to a small

---

* After having completed our work, we found in the *Lancet* of June, 1890, an important article by Dowdeswell, on the Comma bacillus of cholera. He touches on the widely-spread hypothesis that those Commas may be only fragments of spirilla, and then he describes three cycles in their evolution :—
A, Commas or fragments of spirilla which may or may not end in sporules ; *B*, Active cellules, with cilia and amœboid forms ; then round quiescent cellules (sporanges) ending in minute sporules ; *C*, Active filaments ending in sporules of the first generation. The author, however, has not succeeded in reproducing, *by any method*, the normal Commas, vital and reproductive for themselves. He affirms that cholera bacilli are to be found, normally, in the large intestine of the guinea pig, and that the subcutaneous inoculation of a large number of such bacilli really produces choleriform symptoms in that animal ; but it is not a true cholera infection. Considering the high temperature that accompanies it, it is a true septicæmia. The fact, then, that the author has found forms of *Leptothrix* in the Comma bacillus cultures, goes still further to support our views. Dowdeswell, Note on the " Morphology of the Cholera Comma Bacillus," in the *Lancet*, 1890, Vol. I., page 1419—23.

chain of cocci.   The author has not detected any spores.   Although this microbe grows in the gelatine, it does so in a different manner from the cholera bacillus.

Spirochæte dentium (Fig. 20 of Miller and our Fig. 6, *c*) is not found in the decayed teeth, but on the edge of the gums, together with the *Spirillum sputigenum* (affection of the gums).   It exhibits spires 8·25 mm. long, of various thicknesses.   The more slender ones hardly become coloured.   The author himself doubts whether the slender spires constitute a separate species, because he considers the largest amongst them to be akin to the *Spirillum sputigenum*.   Their development is probably unknown.   " We know little or nothing (he says) about the vital conditions and their manifestations, such as fermentation, pathogenic action, etc."

I shall observe that, amongst the materials which came under my examination, I have constantly found these *spirilla* or *spirochæta* with the *Leptothrix*.   In the former Memoir on relapsing fever, I spoke of Spirilla and Spirochæta, which, together with filaments, small rods, and other portions of *Leptothrix*, are produced in infusions of potatoes.   In a later paper upon a Diplococcus identical with the Gonococcus of Neisser, in a case of Carcinoma of the bladder, I mentioned the striking *Spirochæta* that are found in the sediment of bottles filled with the stale water. Together with Zopf, I have retained the name *Spirochæta* for those spirilla which bend in the middle, and thus mark a sort of transition between the so-called *Spirillum* and the *Spirulina*.   I may add that, in the sediment of the bottles, the *Spirochæta* surround the tiny islands or lumps of a stringy microphite, which, in its natural state, resembles the filaments of the *Leptothrix gigantea* from the dog, especially the more slender ones, drawn in Figures 2, 3, 4 in the already mentioned article of Miller.   On the other hand, the intertwined filaments of that sediment, being coloured with aniline, resemble entirely the lumps of the *Bacillus buccalis maximus* of Miller, after Fig. 13 in the last work of this author.   According to Zopf, *Leptothrix* is a microphite that lives in the water.*

* Zopf, *Die Spaltpilze*, Breslau, 1883, page 80.   Vicentini, On a diplococcus analogous to the gonococcus of Neisser, found in urine, in a case of Carcinoma of the bladder—Vol. XLIII., *Degli Atti Della R. Accademia di Napoli*, 1889.

I have already stated that, in the midst of those masses of filaments of the sediment in the bottles, there are preserves of Vorticellas.

Now, it is not improbable that the *Spirochæta* rise there from the more slender filaments of the *Leptothrix gigantea*, and in the mouth from those of *Leptothrix buccalis*, as I shall show later; and that the *Leptothrix gigantea* and the *Bacillus buccalis maximus*, or *Leptothrix buccalis maxima* of Miller, may be the same thing. I have remarked that, in the sediment from the bottles, the vorticellas and filaments are never wanting; whilst the *Spiroehæta* are found there only in summer, and then in brisk motion. Hence, we may infer that the temperature may greatly influence the production of the Spirilla and the Spirochæta, as Miller had argued from the cultures, and as their abundance in the mouth would prove.

I shall refer later to the Spirilla that, in certain cases, I have seen accompanying the *Leptothrix* of the prepuce and the urethra.

Secondary Micro-Organisms.

Leptothrix gigantea (Fig. 21 of Miller).—This form was found on the teeth of a dog affected with alveolar pyorrhæa. It was also found on other carnivorous and herbivorous animals. It exhibits a vigorous development ; has a bushy shape, like the *Crenothrix*, the stems of which deflect at the top, with cocci, small rods and filaments sometimes alternated on the same line. The stems are polymorphic of varying thickness. Some increase in size towards the top ; some are partially or wholly spiral ; and others double and twist as in *Spirulina* (Fig. 12 of the plate annexed to the article of 1883).

In this article, published in the *Berichte der Deutschen Botanischen Gesellschaft*, are to be found more particulars upon the microphite in question. The oldest filaments are articulated ; some are spiral and mutually twisted at the base (Fig. 1, *E, a*). The articulations of the more slender filaments cannot be detected without colourisation. There are sheathed filaments whose internal articulations sometimes evacuate and form cumuli of little rods of different sizes (Fig. 5) ; others become compressed or swollen, or deviate from the axis (Fig. 1, *A, B, C*).   By staining process we find, particularly in the pig, articulations subdivided into two or four cocci; or groups of two or four, which at times occupy the whole length of the filament (Fig. 8).   In the tufts can be seen straight and tortuous filaments.   The bending is graceful and extended.   These are styled by the author, *Vibrionic* or *Spirochætic* forms (reproduced in our Fig. 8).   In the more slender filaments we can observe the transition from the vibrionic to the spirillic form upon a single filament (Figs. 10 and 11).   In the spirillic forms of these slender filaments, by means of a proper colouration, the articulations can be detected.

In 1883, Miller suspected that the same slender spires of the *Patina Dentaria* would rise from the fragments of long and slender tortuous filaments (Figs. 13 and 17).   In the dental Spirochæta there is no trace of articulations, but in the marshy Spirochæta, through the weak colouring of Zopf, are detected several articulations in every filament (Fig. 20).   Hence, then, the author inferred that relations existed between the Spirochæta and other microbes of the mouth.   But now, in his work of 1889, he seems to have discarded his former theory of unity of forms.

Miller is unable to say whether the Leptothrix, detected by him and by others in the tame herbivora and in the pig, is identical with that which lodges in the mouth of man and of the carnivora. He says that it would be necessary to institute pure cultures. Anyhow, it is important to admit the existence of *Leptothrix*, even in the herbivora.   Whilst before it was held that *Leptothrix* lodged exclusively in the mouth of carnivora, Zopf had already admitted its existence in the herbivora.*   It is true that it is less frequent in the herbivora; but, if we are not mistaken, it must be taken into consideration that these, and especially the ruminants, are continually using their teeth, so that the parasite cannot thrive there at ease.†

---

* Cornhill and Babes' quoted work, page 135 ; Zorpf already quoted. This author thinks it probable that *Leptothrix buccalis* originated from the external world, especially through water and food.   We have already mentioned the *Leptothrix* found by us in the sediment of a bottle of water ; but how can we explain the *Leptothrix* in genito-urinary passages ?

† Miller, *Uebereinen Zahn Spaltpilz, Leptothrix (Berichte der gigantea Deutschen Botanischen Gesellschaft*, Heft 5, 1883, p. 221).

Fungi of the Mouth which may be stained blue and violet with Iodine Colours.—Besides the *Bacillus buccalis maximus* and the *Jodococcus vaginatus*, Miller points out three others with a striking iodine reaction, viz. :—(*a*) *Jodococcus magnus* (Fig. 22), which colours violet ; (*b*) *Jodococcus parvus* also stains violet ; and (*c*) another *Jodococcus*, which takes the rose colour with iodine, and can be cultivated a first time, but it does not reproduce in a second culture.

Fungi of the Mouth which can be cultivated.—Some of these are not pathogenic ; some of a doubtful pathogenic efficiency. In Fig. 23 of the author are depicted several various types of these bacteria of the mouth ; in *a, c, g,* screw forms ; in *b,* cocci ; in *d,* small rods ; in *e,* a sheathed chain of cocci ; in *i,* a long chain (but these are numberless in the patina of the tongue, so that we believe it to be impossible to separate them from *Leptothrix*) ; in *f,* chains of small rods ; and in *k,* thread forms (to which the same remark applies).

Miller himself considers it a confusion, difficult to clear up without immense labour, as all the bacteria of the external world can form a nidus in the mouth. The author depicts 19 out of the 22 species which he separated in 1885, viz., from Fig. 24 to 27, cocci ; 28 to 29, small rods ; and 30 to 31, still longer rods ; in Fig. 32, a *Vibrio viridans* that liquefied the gelatine and produced a green colouring matter ; a spirillum (Fig. 33) ; long threads (Fig. 35) ; and small chains, partially sheathed (Fig. 34).

We shall omit seven other types obtained from simple cultures.

Miller afterwards points out the investigations of Vignal, which he finds fit and accurate. Such investigations led to the cultivation of 17 species of fungi in pure cultures ; some identical with known fungi; others unknown. The more frequent was *Bacterium termo* ; then come *Bacillus ulna* ; then the Bacillus of potatoes ; three other unknown Bacilli ; the *Bacillis subptilis* ; then the *Staphylococcus pyogenus albus*; then the *aureus*; five other unknown Bacilli ; and, finally, the *Leptothrix*.

From that view, there would result for Vignal a great preponderance of bacilli; for Miller, on the other hand, of cocci. Miller attributes this divergence of results from Vignal having used the

tothrix buccalis of Robin ; of which, in the following section, we will give our description in its principal biological phases.   At most, some doubt might be entertained in regard to the Spirilla or Spirochæta, whether to hóld them as a species, or belonging to a distinct species ; but their relationship with *Leptothrix* was recognised even by Zopf, who, on this point, wrote : " In certain cases, the filaments (of *Leptothrix*), either on a given part or in all their length, take a spiral shape, and the fragments of those fibres form the *Spirillæ, Vibriones*, and *Spirochætæ*." *

Miller himself in mentioning his studies, made in 1883 upon those of Zopf, also admitted the derivation of the *Spirochætæ* of the human mouth from the twisted threads of *Leptothrix*, of which he gave a representation of the fragments in the figures 13 to 20 of his plate ; and, in the description of the branching filaments of *Leptothrix gigantea* of the pig (Figs. 6 to 8) and of the sheep (Fig. 9 to 12), he represented by the side of straight filaments other vibrionic and spirillic ones.†

We have besides, by our observations, pointed out that the *Spirilla* or *Spirochæta* are constantly united with *Leptothrix*, even in the infusion of potatoes and in stale water (in a warm atmosphere).   Later on, we shall see that they are also found in certain urinary mucous flakes, always accompanied by vigorous fructifications of *Leptothrix*.

PATHOGENIC FUNGI OF THE MOUTH.

We conclude this short extract by summing up the views given by Miller in Chap. IX. of his work upon the experimental infections produced through the inoculation from certain microbes of the mouth, or accidentally lodged in it.

The author begins by reminding us that the venomous property of saliva was long before known when, in 1873, Wright and Senator used it for the purpose of inoculation with deadly result. It was suspected that the poisonous effect depended upon a chemical cause.   Moriggia and Marchiafava, in 1878, inoculated with the saliva of children dead from rabies.   The first who

---

* Zopf, *loc. cit.*, page 80.

† Miller, Article *cit.* 1883, pp. 223—24.

gelatine, on which not all of the above mentioned microbes thrive.

Black, through his own researches, has found in the contents of the mouth, *Streptococcus continuosus, Straphilococcus medius, Staphylococcus magnus, Coccus cumulus minor,* and *Bacillus gelatogenes.**

We have not been able indeed to convince ourselves that, from those experiments of cultures, a basis may be established for the identification of the species of bacilli or bacteria, owing to their unlimited polymorphism, according to the various nourishing media and the different conditions of their surroundings ; so that, wherever the decisive proof of the fructification is wanting, the results of the cultures for creeping or immersed vegetation cannot be freely accepted as the same.   In doing so we might run the risk of multiplying the species and the true or supposed pathogenic properties of bacteria, *ad infinitum.*

Different particles or cellules of the same organism may, in fact, need various nutrient media ; they may liquefy certain substances, and separate others of a very different nature ; as, for instance, the bony corpuscle, the hepatic cellule, the blood corpuscle, the Renal Epithelium, etc.   The poison of insects or snakes is not diffused throughout their organism, but it is secreted by special glands.   Likewise, in the higher vegetation, the leaves, the roots, the buds, the seed, etc., often contain principles or alkaloids totally different from the other parts.   And the same particle or cellule can assume various forms according to its stage of development, its surroundings, or nourishing pabulum.

Furthermore, the virulence of bacteria may decrease or disappear in certain cultures ; increase or reappear in others.   Now, if the virulence constitutes for itself a truly specific character at every step, we shall have to admit, in those microbes, changes of species; and thus, in order to defend the true and well-thought pleomorphism, we must necessarily fall into an illimited and unlikely one.

However, referring to the six species of micro-organisms admitted by Miller in the cavity of the mouth, we, with due respect for the author, cannot abandon the opinion (which, although old, is not ours) of the fundamental unity of all these forms, and their derivation from a single species, the *Lep-*

---

* Miller, *die Mikroorganismen der Mundhöhl,* etc., 1889, pp. 43—75.

tothrix buccalis of Robin ; of which, in the following section, we will give our description in its principal biological phases. At most, some doubt might be entertained in regard to the Spirilla or Spirochæta, whether to hold them as a species, or belonging to a distinct species ; but their relationship with *Leptothrix* was recognised even by Zopf, who, on this point, wrote : " In certain cases, the filaments (of *Leptothrix*), either on a given part or in all their length, take a spiral shape, and the fragments of those fibres form the *Spirillæ, Vibriones,* and *Spirochætæ.*" *

Miller himself in mentioning his studies, made in 1883 upon those of Zopf, also admitted the derivation of the *Spirochætæ* of the human mouth from the twisted threads of *Leptothrix*, of which he gave a representation of the fragments in the figures 13 to 20 of his plate; and, in the description of the branching filaments of *Leptothrix gigantea* of the pig (Figs. 6 to 8) and of the sheep (Fig. 9 to 12), he represented by the side of straight filaments other vibrionic and spirillic ones.†

We have besides, by our observations, pointed out that the *Spirilla* or *Spirochæta* are constantly united with *Leptothrix*, even in the infusion of potatoes and in stale water (in a warm atmosphere). Later on, we shall see that they are also found in certain urinary mucous flakes, always accompanied by vigorous fructifications of *Leptothrix*.

PATHOGENIC FUNGI OF THE MOUTH.

We conclude this short extract by summing up the views given by Miller in Chap. IX. of his work upon the experimental infections produced through the inoculation from certain microbes of the mouth, or accidentally lodged in it.

The author begins by reminding us that the venomous property of saliva was long before known when, in 1873, Wright and Senator used it for the purpose of inoculation with deadly result. It was suspected that the poisonous effect depended upon a chemical cause. Moriggia and Marchiafava, in 1878, inoculated with the saliva of children dead from rabies. The first who

---

* Zopf, *loc. cit.*, page 80.

† Miller, Article *cit.* 1883, pp. 223—24.

attributed to bacteria the venomous action of the saliva or sputa were Raynaud and Lannelongue, in 1891. The rabbits that were inoculated with the saliva of a boy who died of rabies, died; whilst the inoculations with the blood or the buccal mucus were harmless. Pasteur saw two rabbits die after inoculation with the same saliva; and the cultivated microbes exhibited the shape of dumb-bell bacteria in a gelatinous capsule. He then believed he had discovered the cause of hydrophobia; but Cohnheim, having noticed the rapidity of the infection, qualified it for septicæmia.

After these first experiments, Vulpian produced the same infection by inoculating with the normal saliva; and, in the blood of the inoculated animals, Rochefontaine and Arthaud detected microbes quite identical with the above-named. Sternberg and Claxton confirmed the observations of Vulpian; but Griffin demonstrated that the saliva of the parotides, gathered separately, is harmless; so that the infection is to be attributed to secondary products of decomposition, formed in the buccal cavity. Gaglio and De Mattei held the same view, and noticed that the saliva became harmless after boiling.

Fraenkel in 1884 confirmed anew the observations of Vulpian. He found that dogs were more receptive, the mice less so, and still less guinea-pigs. Dogs, pigeons, and fowls were refractory.

Miller obtained identical inoculations with the excretion of a woman affected with micosis of the tonsils. He separated afterwards from a decayed root five fungi, of which one was distinctly pathogenous.

Klein and others studied the venomous action of groups of microbes of the human sputa, so that in late years pathogenic properties were gradually attributed to a considerable number of bacteria.

Pathogenic Fungi not Cultivable.—The author repeats here what has already been said of the primary microbes of the mouth, to demonstrate that *Spirillum sputigenum* and *Spirochæte dentium* are especially not cultivable. Kreibohm found in the patina of the tongue two pathogenic bacteria unfit for cultivation in any artificial medium whatever. Miller, with fungi from a gangrenous pulp of the teeth, obtained a gangrenous product for re-inoculation.

Pathogenic Fungi Cultivable.—(*a*) *Micrococcus of the sputum, Septicæmia* (Figure 96 of Miller).—This is the same bacterium that Pasteur and others obtained from healthy and pathological sputa. Klein was the first to cultivate it. It results in oval cocci, diplococci, and capsulated chains identical with *Pneumococcus.* Subcutaneous injections with its cultures produce septicæmia and death in 24—36 hours. Even the blood of animals inoculated is totally infected; but the infection, once overcome, is contracted no more. Its virulence is attenuated by cultures in milk. This bacterium (continues Miller), *lodging constantly in the human mouth,* passes from there into the pulmonary tubes; so that the uncleanliness of the mouth would be the real predisposing condition to pneumonia, in so far as it allows the diplococci in question, deposited there from the air, to multiply. From those views to ours, to the entirely secondary propagation of such bacteria in the pulmonary products, there is but one step. (See Section 4 of Memoir on Whooping-cough.)

(*b*) *Bacillus crassus sputigenus* (Fig. 97 of Miller, from the blood of a mouse).—This bacterium was found by Kreibohm once on the patina of the tongue and twice in the liquids of the mouth; it is longer than it is broad, but often shorter in its younger articulations; resists Gram's method; sporifies through heat. Mice, inoculated even with small quantities, die in forty-eight hours, exhibiting myriads of bacteria in their liver and blood. A copious injection of it into the veins kills rabbits and dogs in from three to ten hours from gastro-enteritis.

(*c*) *Staphylococcus pyogenus albus* and *aureus,* and *Streptococcus pyogenus.*—These bacteria, held as promoters of suppuration, are brought from outside into the mouth, where they seem to find a fit soil. Black, upon ten specimens of buccal liquids and scrapings from the tongue, found seven times *Staphylococcus pyogenus aureus*—four times the *albus* and three the *Streptococcus,* which he believes with a careful search can always be found. Vignal never came across the true streptococcus; seldom found the two staphilococci. Netter, upon 127 specimens, found only seven times the *Staphilococcus pyogenus aureus.* Miller also rarely found these micrococci.

(*d*) *Micrococcus tetragenus* (Fig. 98 of Miller).—This micrococcus, relatively frequent, was first found by Koch and Gaffky in the cavities of a consumptive subject; it is more frequent in tuberculosis. Biondi found it in three cases out of five. It is in the shape of cocci disposed two by two or four by four; with gelatinous matter interposed, it does not liquefy the gelatine. White mice and guinea pigs inoculated with its cultures die in from three to ten days; those inoculated with sputa perish in from four to eight days. Their blood and their organs are infested with micrococci.

Fungi of Biondi.—This investigator has isolated from the mouth five different pathogenic bacteria, including the *Micrococcus tetragenus* above-mentioned. The four others are :—(*e*) a *Bacillus salivarius septicus*, rather frequent; (*f*) a *Coccus salivarius septicus*, obtained in a single case only from a woman affected with a serious septicæmia in her confinement; (*g*) a *Streptococcus septicus*, similar to that of erysipelas *phlegmon* and puerperal *Metritis;* (*h*) a *Staphylococcus salivarius pyogenes*, found once in the sputum of a scarlatinous quinsy.

The Fungi of Miller.—After many cultures and inoculations from buccal liquids and dental gangrenous pulp, Miller observed more serious effects from the inoculations of the pulp, even in comparison with inoculations of cultures of the same pulp. The effects were varied in different animals and according as the inoculation was made under the skin or the peritoneum cavity. The pathogenic fungi isolated by the author were :—(1) a *Micrococcus Gengivæ pyogenes* (Fig. 99 of Miller), obtained from alveolar pyorrhœa cocci that did not liquefy the gelatine ; (*k*) a *Bacterium Gengivæ pyogenes* (Fig. 100 of Miller), found with the former, in the shape of bacteria four times longer than broad, apt to liquefy the gelatine ; (*l*) a *Bacillus Dentalis viridans* (Fig. 101 of Miller), obtained from the decayed roots of teeth, in bacteria of irregular forms ; (*m*) a *Bacillus Pulpæ pyogenes* (Fig. 102 of Miller), obtained from the gangrenous pulps, in bacilli, frequently irregular, isolated, linked together, or in small chains, even with eight articulations, apt to liquefy the gelatine.*

We cannot further describe these culture experiments. We

* Miller, *loc. cit.*, pp. 199—220.

shall only remark that the pneumococci, or the exciters of salivary septicæmia, being constantly found in the contents of the mouth, are to be held, in our opinion, as one of the forms or phases of the microphite or microphites of the mouth (zoögleic phases of Billet). With regard to all the other bacteria we have just mentioned, it remains to be observed whether they are accidental— *i.e.*, proceeding from without—or normal, although modified in their properties by variable conditions of pabulum or surroundings ; and whether their numerous varieties are perhaps more apparent than specific.

As regards the experimental inoculations, we shall simply notice the fact that they introduce, under the skin, in the cavity, or in the circulation, all at once, myriads of bacteria. A single centigramme of culture or of polluted material may contain, on an average, twenty-five billions of bacteria, as we have elsewhere demonstrated ; and twenty-five milliards inoculated in a mouse are as fifty or a hundred trillions (Italian measure) inoculated in a man. Is it wonderful if local or general effects may be produced from the sudden irruption of such a mass of parasitical germs, even though in themselves innocuous ? Nevertheless, that is not the real question ; the real question is whether the different kinds of bacteria (probably belonging to the same species) have the power of attacking, *sponte* and on their own account, the human economy ; and whether, even penetrating it from their media, in an infinitesimal number (as is the case in the ordinary infectious diseases), can they develop in the human species the same effects following upon the inoculation of their elements on the brutes?

In other words, one is the receptivity through the natural passages and another through the continuous solutions ; one is the slow, successive, and external action of a few germs, another that of the sudden and internal mass of germs. Finally, one is the receptivity of this or that animal species, and another that of the human species. And all this is irrespective of the other more or less questionable points, and especially of the query : If the morbific action of the inoculations depends upon the bacteria, as such, or upon the polluted matter in which they are suspended ; or, it may be, upon their products of secretion and decomposition, as is at the present time generally argued.

But, leaving for the present those complex questions, I will proceed to speak of the morphology and biology of *Leptothrix*.*

---

§ IV.

REMARKS ON THE MORPHOLOGY AND BIOLOGY OF THE LEPTOTHRIX BUCCALIS.

The Four Phases of Leptothrix and particularly the Fructifications by Ears (Spicæ).

First Phase.—The first phase of the *Leptothrix* is that of a Bacterium or Bacillus, already described in the Memoir on Whooping-cough and in the first paragraph of the present one. This phase is common to all the other species of bacteria, but which (at least in *Leptothrix*) does not represent its whole cycle of life, but only its primordial stage of immersed vegetation, or a vegetation destined to propagate in liquid or semi-solid substrata (media).

Second Phase.—Examining the cuticle of the tongue, stained with gentian violet, we find there by preference the forms of the second stage of life of this parasite—namely, that of chains, bundles, and masses of intertwined filaments. We have already said that in these preparations the buccal epithelia predominate ; isolated filaments are found there, as well as many specimens of large dumb-bell bacteria of the type *p.p.* and *p'* (Fig. 2), and conspicuous masses of diplococci. These preparations show that the large dumb-bell bacteria are derived from the diplococci, the two original cocci linking together. The degrees of transition to be observed there are many. The same chains are often surrounded

---

* Two fresh works were lately published on this argument : one by David *(Les Microbes de la bouche, precédé d'une lettre-préface de M. L. Pasteur)*, a simple compilation ; the other by Billet *(Contribution à l'étude de la Morphologie et du développement des bactériacées)*, the result of original researches on the natural history of four species of micro-organisms (*Cladothrix dichotoma, Bacterium Balbiani, Bacterium osteophilum*, and *Leptothrix parasitica*, a parasite of sea-weeds). We shall notice these works later on ; we may at once say that not even these authors hint in the least to the SUPERIOR PHASES of *Leptothrix buccalis*, after our type or other's.

by similar masses of diplococci; even the entanglements of the type *u, u,* are not wanting (Fig. 2).

The chains belong to types *c* and *d* (Fig. 2), but are generally longer, thick, branched off in bundles of 10 or 15, which take up to three visual fields, and exactly represent the disposition of the bundles of filaments. In the same chain small diplococci may alternate with elliptical bacteria and medium-sized dumb-bells—all perfectly equal, which form the largest number.

In these we have been unable to find fructifications, so that this second phase may be held as analogous to mycelium, or the vegetation, of fungi, partly immersed, partly creeping.

THIRD PHASE (*Incomplete aërial vegetation*).—This is generally met with in the tartar of the teeth. Its predominant elements are the large filaments, often very long, bent, and re-united in bundles, taking a brilliant colour with iodine; and the stumps. I shall not describe these well-known forms, having already done so in speaking of the primary and secondary species of Miller.

I would rather point out some features, noticeable in a certain number of these filaments, when I have the opportunity of finding them isolated (especially in the unstained preparations), such features not having been considered by previous authors.

The first is a division into two or three branches (Fig. 9), which I think must be taken for a kind of radical system of this microphite, for divisions are never detected towards its apex; and, at its opposite end, a swelling or head is noticed, which is a proper form of points twined upwards (as in *b*). In *a* the filament is broken towards the top and exhibits two long roots; in *b*, there are three hood-shaped roots. Both appear to be swollen at their ends, like *haustoria*.

What can be the use of these barbs or roots, unless to obtain a firm foundation in the ground? And to what purpose is this firm foundation, unless to support higher forms of vegetation?

The other feature is that of the swelling, which we will call *apices* or heads, being found at the top of the respective filaments, as in *b* and *c*, and these are of varied form, at times containing a kind of nucleus. There are others which might be styled knotty, being found along the filament; of these we give no drawing, as they are similar to the preceding ones.

With the above-mentioned stumps are intermixed almost all the forms of bacteria, bacilli, spirilla, etc., besides a large number of very varied small chains, after the types c, d, k, i, x, and x' (Fig. 2).

The larger size and the facility to colour of these masses and stumps, with regard to fertile filaments, would depend, I think, upon a very natural cause. The friction, removing the points of the filaments and the feitile filaments, seems to impart a greater development to the remaining stems in the sense of thickness; as, for instance, it happens in the pruning of trees, through the retrocession of the ascending sap.

In the present case, the retrocession of the germinal matter would be the cause of the consecutive increase of the sheath and of the gemmules contained in it. Thus might be easily explained the appearances, so various, of the stumps and their fragments, compared with the fertile filaments; appearances which induced Miller to make of residual filaments two distinct species, under the name of *Leptothrix buccalis maxima* and *Bacillus buccalis maximus*, in opposition to our fertile filaments, which for Miller constitutes a third species, the *Leptothrix innominata*.

It remains, then, to be seen whether the longer filaments, sunk down and intertwined, of *Bacillus buccalis maximus* (Fig. 13 of Miller) may not concur, with the above described chains, to form that kind of mycelium or creepir.g vegetation, upon which is based the aërial vegetation of this microphite.

From the tiny islands of the stumps in question, prepared in the manner already described in Section 2, spring at last the fructifications. The bigger filaments, that we would call woody, gradually become thinner and pale, showing in their interior countless granules or parietal gemmules. These are the fertile filaments. They may spring either from the proper filaments, with continuous contour, or from the little chains after type d (Fig. 9), which are seen occasionally on the top of those filaments; and in this last case, instead of a gradual thinning, may abruptly pass from the chain to the fertile filament, as shown in the figure.

It is by no means uncommon to find parietal gemmules, as Robin mentions the circumstance, and gives a first drawing in Fig. 2h of his volume of 1853, magnified to 800 diameters.*

* Robin, *Hist. Nat. des Végétaux*, etc., 1853. (See Bibliography.)

FOURTH PHASE (*Completed Aërial Vegetation: Fructification*).—
In Figs. 10—13 are reproduced four specimens of fructifications,
taken from the *patina dentaria*. In Fig. 10 are seen three isolated
"ears" *(spicæ)*, stained with gentian violet; in Fig. 11, a tuft,
caught as the bird flies, stained with weak methyl violet; in Fig.
12, a part of a more vigorous tuft, seen in profile, stained with
weak fuchsine; and in Fig. 13, two fructifications, set on the same
tiny island of *materia alba*, stained with solution of iodine. The
Figures 10, 11, and 13 are drawn magnified to 1,250 diameters,
and the Fig. 12 to 850 diameters.

The fertile filaments are at times partially straight, as in Fig.
10; at times bent (even occasionally tortuous, as in Fig. 13), and
are sometimes wanting, because the fructifications have been
carried away by mechanical force, although grouping together, as
in Fig. 12. In Fig. 10 are seen, in *a, a*, the gemmules of reserve
adhering to the walls; in *b* are found the little spores properly
lodged; but in *b'* we notice only five, the others having dropped.
In *c* the penultimate articulations of the stalk appear older and
woody; the last is granular, like the two articulations on the apex
of the younger filament, *d*.

The sporules, very small, round, brilliantly coloured, show
themselves in three vertical rows in all the specimens without
exception; *i.e.*, they never appear in two rows, whence we may
infer that they are, in fact, disposed in six longitudinal lines (or
series), for, if they were only in four or five, they should, at times,
exhibit only two visible rows.

In turning the micrometer screw, the cylindrical relief of the
"ears" appear manifest, for at a higher focus the sporules of the
middle row are better seen, and at a low focus those of the lateral
rows. In several specimens, especially when coloured with picric
acid, we can detect the prolongation of the stem into the interior
of the ear; but it is not so with aniline colours, which render the
viscid substance (in which the sporules are suspended) opaque,
staining it with a weaker tint. But if, in the preparation, we sub-
stitute glycerine for water, the stalk, a long time afterwards,
reappears distinctly in all its length.

At any rate, the movements given to the preparation with the
thumbs show that the stalk and the "ear" form a continuous

whole; then we observe the "ears" on the top vibrating like the ears in a field of wheat.

It may be objected that the sporules are not immediately grafted on the stalk or fertile filament, but are distant from it ; it must, however, be taken into consideration that the distance between the sporules and the internal filament is not, in fact, greater than that which is constantly seen between the articulations of the single chains of bacteria, which, as it is scientifically ascertained, are generated one from another. Perhaps a connection exists between them, through threads or ligaments, not discernible by the optical power at our command.*

We cannot even suppose that the sporules are simply adhering to the filament through external action (due to the lowering or sinking, so to speak, of the filaments themselves into the granular masses), because a simple mechanical incrustation would not explain the regular straight line of the sporules in six longitudinal rows, nor their perfect similarity of volume and shape.

The incrustation would bring a medley of cocci and bacteria, of varied type and size, irregularly mixed up. Besides, we shall observe productions of these "ears" (and even more conspicuous) in the sputa, without any trace of bed or granular mass on which the filaments might have become incrusted from the outside.

The length of ears is frequently considerable, especially in those detached from the matrix, which at times, with their stem, occupy all the visual field. The more conspicuous specimens (Fig. 16) have been found in the pulmonary sputum, where the

---

* When we presented this Memoir, we promised in a note to resume the study of "ears" in detail, with more powerful objectives. Now Messrs. Bézu, Hausser, and Co. have succeeded in constructing a new objective with homogeneous immersion, of 1/25th inch. With this power we have been able to detect, in the most evident manner, the peduncles or engrafted threads of the sporules on the principal stalk. With the new grand model (No. VII.) of the same opticians, furnished with an Abbé condenser of the numerical aperture, 1·40, by using oblique light, and by turning the stage of the microscope, or substituting a dark diaphragm for the ordinary one, the disposition of the sporules in six longitudinal lines is fully confirmed. We shall treat this point again later ; but from this moment we think it is better to give the name of " bunches of grapes" instead of that of "ears" (to indicate the fructifications of *Leptothrix*), as they really are such.

longest were one-sixth of a millemetre high, equivalent to the thickness of a thin cover-glass, as in the Figs. *f, g* of the figure.

The colouration of the ears in question with carmine, protracted even for weeks in a watch-glass, gave no satisfactory results. With methyl violet the specimens appear less transparent; in other respects, they are like those treated with gentian violet. In Fig. 11 one may be observed, in which the *ears* are in the direction of the visual axis. With fuchsine the most fine and brilliant colourations are obtained, as shown in the specimen caught in profile, represented in Fig. 12, after being immersed for a few minutes.

In the uncoloured preparations, the *ears* are hardly or not at all discernible, so that they could not be detected by him who should ignore their existence; the sporules being, in their natural state, transparent and nearly invisible; but these preparations are helpful to him who knows them, in so far as they aid to detect, on the outline of the mass, a larger number of ears which have not been spoiled or carried away by the manipulations, or hardened and consequently become brittle.

Anyhow, the flexibility of the ears is preserved even in the colourings with the solution of iodine, not acidulated, or with picric acid. The sporules appear distinctly with the solution of iodine (Fig. 13); but with picric acid the colouring is diffused and pale, although the pale tint is useful in its turn in disclosing at one time a larger number of fructifications, without rendering any of them opaque.

Here we shall stop awhile to consider the great affinity of those sporules through the aniline colours; which affinity brings them so closely to the tubercular bacilli as to make us suspect that the single articulations of such bacilli proceed from the sporules, which, having dropped, transplant themselves into the buccal epithelia and the air-passages, and grow like beaded forms. We were led to that supposition from the following fact :—In the colourings with fuchsine of the *patina dentaria*, which last a very short time, the central portion of many masses remains uncoloured. How great was our surprise in seeing, some time afterwards, in those uncoloured areas, minute red lines, perfectly similar to the tubercular bacilli! Then those coloured lines gradually increased, and we saw the summits of ears developing, as mounds, brilliantly coloured, in a clear area. This fact is to be attributed to the subsequent action of the colouring matter.

Besides the bacilli of Koch in rosaries, we have others in small rods or filaments containing granules or vacuoli, as in Fig. 1 of the work of Cornil and Babes, etc. These rods appear more clearly when stained by Gram's method—viz., with double colouring, first with gentian violet and then with solution of iodine. Now, the fragments and slender stems of *Leptothrix*, in the *patina dentaria*, behave likewise. If the solution of iodine should be used first, the successive colouring with gentian violet does not succeed; but if we colour with violet, as usual, mounting the preparation when fresh, and then introducing by capillarity the solution of iodine, the violet colouring becomes dark blue and the small rods of *Leptothrix* appear very clearly.

On another occasion I shall treat those questions relating to the finding and colouring of the tubercular bacilli.

Productions by Points (Male Organs?)

I have already said that such productions or pseudo inflorescences are scanty, so that we must seek them patiently in the preparation. They have, however, the advantage of showing themselves clearly, even without colouring. It is absolutely necessary to examine the *patina dentaria* of persons who have fasted and who are not in the habit of cleaning their teeth, otherwise we may not find these productions. The preparation should not be pressed too much, in order to avoid the disintegration of the specimens. They are easily recognised with an objective, No. 8 of Hartnack. The above-mentioned difficulties indicate why such forms have not been considered by previous investigators.

In Fig. 14 are shown three of these productions in the natural state, taken from the *patina dentaria*, and magnified to 690 diameters. They are formed of one or more internal stocks and a considerable number of points which surround them, like the horsehairs of a cylindrical brush. In *a* the production rises upon a multiple stalk, formed of a bundle of filaments apparently broken off. The stalk itself is crowded with small points near the trifurcation; the three small divided branches bear longer points up to the top ; but the right one is also subdivided in two second. ary tiny branches. In *b* is drawn an isolated pseudo inflorescence, formed of small and medium points, like the small branches of the preceding specimen. In *c*, on the contrary, is seen a production broken on the top, formed of small, medium, and large points; the small points similar to the preceding ones; the large ones totally similar to the pointed and spindle-like bacilli of type *e* (Fig. 5), of which, for comparison's sake, we have reproduced one in its natural state in *d* (Fig. 14), magnified to 1,250 diameters. Most of the small points resemble the Virgula bacilli of type *h* (Fig. 5); some, however, look like the spermatozoa of type *h'* (same figure), having at one end a kind of small head, according to the degree to which they are magnified.

In Fig. 15 is seen a portion of a small branch, coloured with the solution of iodine, magnified to 2,500 diameters. *Points* of various size and shape are set in it, which, like the central rod, contain clear spaces or uncoloured empty ones. If it were not too bold a supposition, I should be inclined to suggest that the three small *points* are not adhering to the central rod, but are *fulfilling* their functions upon it, and perhaps even upon the bacillus which adheres to it lower down—a *quid simile* with the act of fecundation.

I repeat that it is not my intention to pronounce on the nature of these productions by *points* or pseudo-inflorescences, that being a subject for micetologists. I have only wished to describe facts resulting from my own observations. However, I cannot help noting the very extraordinary mobility exhibited by those *points*, large and small, dropped from the stalks or branches above described.

Those movements are better detected in uncoloured preparations, mounted in simple distilled water, in which they are preserved longer (even from day to day), especially in the minute *Virgula* bacilli; so that I suggest to those who will verify this part of my observations, not to use any tint at all. The stains used, particularly the aniline colours, seem to have a poisonous effect upon those bacilli, or at least exert a lethargic action on their motility. In the largest or spindle-like bacilli of type *d* (Fig. 14) the movements are those of translation, so that they fully justify the name of *Bacillus tremulus*, given to them by Rappin, if that name does not imply a separate specific entity. We notice that those bacilli vibrate not only sidewise, like fishes, but flash here and there with a rapid translatory motion, dashing first upon one tiny island and then (after a pause) upon another, or disappearing from the visual field. It is impossible not to recognise in those motions the character of vital movements, either voluntary or teleological ; and excluding the first hypothesis, which would lead us to the *animalculæ* of Leeuwenhoek, Ficinus, and Klencke, there only remains *one external purpose* to this bacilli, which is directed to the performance of one function. That function can only be an important one, considering that those motions are the highest and most persisting manifestation of life exhibited by buccal microbes. Now, what more important function than that of fecundation ? *

The movements of the *Virgula* bacilli of type *h* (Fig. 5) are generally undulating and serpentine ; but they have also a translatory one, as we have suggested before.

In conclusion, we may suppose that those points, or so-called bacilli, perform a function similar to that of *spermatia* or *antherozoids* of various fungi and sea-weeds (ALGÆ). In such hypotheses the absolute sterility of their cultures would be wholly explained, being quite natural (as before suggested) that simple male organs should not be able to reproduce by themselves alone. It remains to be seen whether these productions *by points* have any relationship with the bacillus *n* (Fig. 2 to the left below), a large bacillus veined through its length, which contains, as in germ, slender and short lineal bacilli towards the inferior end, so that we should be inclined to compare it with a *spermogene* or *antherid.*†

We have hitherto dealt with productions by *points*, in their higher phase ; but near these we find other less developed forms,

---

* In the quoted article of Dowdeswell, to our surprise, we find that he entertains doubts upon the nature (whether vegetal or animal) of the *Cholera* bacillus. See *Lancet*, 1890, Vol. I., *loc. cit.*, p. 1422.

† The spindle-like and *Virgula* bacilli of the mouth closely resemble, in shape and size, the spermatia of the *Sphærella sentina*, of the *Fumago salicina*, of the *Apiosporium citri*, etc. The *spermatia* were considered, since 1877, as male organs, when Cornu noticed that some of them were germinating by themselves, so that their nature would be still uncertain.

supported likewise by one or more internal stems and enveloped in a mass of very minute lineal germs, quite similar to those contained in n (same figure). Such lineal germs, engrafted in square on the stalks, resemble spindle-like, serpentine, or *Virgula* bacilli in formation, so that the whole may be taken for the first stage of the pseudo inflorescences, or productions by points, already described. Owing to the difficulty of reproducing them faithfully, we have omitted their graphic representation in the Plate.

(To be continued.)

On the Magnitude of the Solar System.*

By W. Harkness (Washington).

NATURE may be studied in two widely different ways. On the one hand we may employ a powerful microscope which will render visible the minutest forms and limit our field of view to an infinitesimal fraction of an inch situated within a foot of our own noses ; or, on the other hand, we may occupy some commanding position, and from thence, aided by a telescope, we may obtain a comprehensive view of an extensive region. The first method is that of the specialist, the second is that of the philosopher ; but both are necessary for an adequate understanding of nature. The one has brought us knowledge wherewith to defend ourselves against bacteria and microbes which are among the most deadly enemies of mankind, and the other has made us acquainted with the great laws of matter and force upon which rests the whole fabric of science. All nature is one, but for convenience of classification we have divided our knowledge into a number of sciences which we usually regard as quite distinct from each other. Along certain lines, or, more properly, in certain regions, these sciences necessarily abut on each other, and just there lies the weakness of the specialist. He is like a wayfarer who always finds obstacles in crossing the boundaries between two countries, while to the

* Part of the Address delivered before the American Association for the Advancement of Science at its Brooklyn meeting, Aug. 16, 1894, by the retiring President, Professor Harkness. From *Science*.

traveller who gazes over them from a commanding eminence the case is quite different. If the boundary is an ocean shore there is no mistaking it; if a broad river or a chain of mountains it is still distinct; but if only a line of posts traced over hill and dale, then it becomes lost in the natural features of the landscape, and the essential unity of the whole region is apparent. In that case, the border-land is wholly a human conception of which nature takes no cognizance, and so it is with the scientific border-land to which I propose to invite your attention this evening.

To the popular mind there are no two sciences further apart than astronomy and geology. The one treats of the structure and mineral constitution of our earth, the causes of its physical features and its history, while the other treats of the celestial bodies, their magnitudes, motions, distances, periods of revolution, eclipses, order, and of the causes of their various phenomena. And yet many, perhaps I may even say most, of the apparent motions of the heavenly bodies are merely reflections of the motions of the earth, and in studying them we are really studying it. Furthermore, precession, nutation, and the phenomena of the tides depend largely upon the internal structure of the earth, and there astronomy and geology merge into each other. Nevertheless the methods of the two sciences are widely different, most astronomical problems being discussed quantitatively by means of rigid mathematical formulæ, while in the vast majority of cases the geological ones are discussed only qualitatively, each author contenting himself with a mere statement of what he thinks. With precise data the methods of astronomy lead to very exact results, for mathematics is a mill which grinds exceeding fine; but, after all, what comes out of a mill depends wholly upon what is put into it, and if the data are uncertain, as is the case in most cosmological problems, there is little to choose between the mathematics of the astronomer and the guesses of the geologist.

If we examine the addresses delivered by former presidents of this Association, and of the sister—perhaps it would be nearer the truth to say the parent—Association on the other side of the Atlantic, we shall find that they have generally dealt either with the recent advances in some broad field of science, or else with the development of some special subject. This evening I propose to adopt the latter course, and I shall invite your attention to the present condition of our knowledge respecting the magnitude of the solar system, but in so doing it will be necessary to introduce some considerations derived from laboratory experiments upon the luminiferous ether, others derived from experiments upon ponderable matter, and still others relating both to the surface phenomena and to the internal structure of the earth, and thus we shall deal largely with the border-land where astronomy, physics, and geology merge into each other.

The relative distances of the various bodies which compose the solar system can be determined to a considerable degree of approximation with very crude instruments as soon as the true plan of the system becomes known, and that plan was taught by Pythagoras more than five hundred years before Christ. It must have been known to the Egyptians and Chaldeans still earlier, if Pythagoras really acquired his knowledge of astronomy from them as is affirmed by some of the ancient writers, but on that point there is no certainty. In public Pythagoras seemingly accepted the current belief of his time, which made the earth the centre of the universe; but to his own chosen disciples he communicated the true doctrine that the sun occupies the centre of the solar system, and that the earth is only one of the planets revolving around it. Like all the world's greatest sages, he seems to have taught only orally. A century elapsed before his doctrines were reduced to writing by Philolaus of Crotona, and it was still later before they were taught in public for the first time by Hicetas, or, as he is sometimes called, Nicetas, of Syracuse. Then the familiar cry of impiety was raised, and the Pythagorean system was eventually suppressed by that now called the Ptolemaic, which held the field until it was overthrown by Copernicus, almost two thousand years later. Pliny tells us that Pythagoras believed the distances to the sun and moon to be respectively 252,000 and 12,600 stadia, or, taking the stadium at 625 feet, 29,837 and 1,492 English miles; but there is no record of the method by which these numbers were ascertained.

After the relative distances of the various planets are known, it only remains to determine the scale of the system, for which purpose the distance between any two planets suffices. We know little about the early history of the subject, but it is clear that the primitive astronomers must have found the quantities to be measured too small for detection with their instruments, and even in modern times the problem has proved to be an extremely difficult one. Aristarchus of Samos, who flourished about 270 B.C., seems to have been the first to attack it in a scientific manner. Stated in modern language, his reasoning was that when the moon is exactly half full, the earth and sun as seen from its centre must make a right angle with each other, and by measuring the angle between the sun and moon, as seen from the earth at that instant, all the angles of the triangles joining the earth, sun, and moon would become known, and thus the ratio of the distance of the sun to the distance of the moon would be determined. Although perfectly correct in theory, the difficulty of deciding visually upon the exact instant when the moon is half full is so great that it cannot be accurately done even with the most powerful telescopes. Of course, Aristarchus had no telescope, and he does not explain how he effected the observation, but his conclusion was that at the instant in question the distance between the centres of the sun and moon, as seen from the earth, is less than a right angle by 1/30th part of the same. We should now express this by saying that the angle is 87 degrees, but Aristarchus knew nothing of trigonometry, and in order to solve his triangle he had recourse to an ingenious, but long and cumbersome geometrical process which has come down to us, and affords conclusive proof of the condition of Greek mathematics at that time. His conclusion was that the sun is nineteen times further from the earth than the moon, and if we combine that result with the modern value of the moon's parallax, viz., 3,422·38 seconds, we obtain for the solar parallax 180 seconds, which is more than twenty times too great.

The only other method of determining .the solar parallax known to the ancients was that devised by Hipparchus about 150 B.C. It was based on measuring the rate of decrease of the diameter of the earth's shadow cone by noting the duration of lunar eclipses, and as the result deduced from it happened to be nearly the same as that found by Aristarchus, substantially his value of the parallax remained in vogue for nearly two thousand years, and the discovery of the telescope was required to reveal its erroneous character. Doubtless this persistency was due to the extreme minuteness of the true parallax, which we now know is far too small to have been visible upon the ancient instruments, and thus the supposed measures of it were really nothing but measures of their inaccuracy.

The telescope was first pointed to the heavens by Galileo in 1609, but it needed a micrometer to convert it into an accurate measuring instrument, and that did not come into being until 1639, when it was invented by Wm. Gascoigne. After his death, in 1644, his original instrument passed to Richard Townley, who attached it to a fourteen-foot telescope at his residence in Townley, Lancashire, England, where it was used by Flamsteed in observing the diurnal parallax of Mars during its opposition in 1672. A description of Gascoigne's micrometer was published in the *Philosophical Transactions* in 1667, and a little before that a similar instrument had been invented by Auzout in France, but observatories were fewer then than now, and so far as I know J. D. Cassini was the only person beside Flamsteed who attempted to determine the solar parallax from that opposition of Mars. Foreseeing the importance of the opportunity, he had Richer despatched to Cayenne some months previously, and when the opposition came he effected two determinations of the parallax ; one being by the diurnal method, from his own observations in Paris, and the other by the meridian method from observations in France by himself, Römer, and Picard, combined with those of Richer at Cayenne. This was the transition from the ancient instruments with open 'sights to telescopes armed with micrometers, and the result must have been little short of stunning to the seventeenth century astronomers, for it caused the hoary and gigantic parallax of about 180 seconds to shrink incontinently to ten seconds, and thus expanded their conception of the solar system to something like its true dimensions. More than fifty years previously Kepler had argued from his ideas of the celestial harmonies that the solar parallax could not exceed 60 seconds ; and a little later Horrocks had shown on more scientific grounds that it was probably as small as 14 seconds, but the final death-blow to the ancient values ranging as high as two or three minutes came from these observations of Mars by Flamsteed, Cassini, and Richer.

Of course, the results obtained in 1672 produced a keen desire on the part of astronomers for further evidence respecting the true value of the parallax, and as Mars comes into a favourable position for such investigations only at intervals of about sixteen years, they had recourse to observations of Mercury and Venus. In 1677 Halley observed the diurnal parallax of Mercury, and also a transit of that planet across the sun's disc, at St. Helena, and in 1681 J. D. Cassini and Picard observed Venus when she was on the same parallel with the sun; but although the observations of Venus gave better results than those of Mercury, neither of them was conclusive, and we now know that such methods are inaccurate even with the powerful instruments of the present day. Nevertheless, Halley's attempt by means of the transit of Mercury ultimately bore fruit in the shape of his celebrated paper of 1716, wherein he showed the peculiar advantages of transits of Venus for determining the solar parallax. The idea of utilising such transits for this purpose seems to have been vaguely conceived by James Gregory, or perhaps even by Horrocks; but Halley was the first to work it out completely, and long after his death his paper was mainly instrumental in inducing the governments of Europe to undertake the observations of the transits of Venus in 1761 and 1769, from which our first accurate knowledge of the sun's distance was obtained.

Those who are not familiar with practical astronomy may wonder why the solar parallax can be got from Mars and Venus, but not from Mercury, or the sun itself. The explanation depends on two facts. Firstly, the nearest approach of these bodies to the earth is for Mars 33,870,000 miles, for Venus 23,654,000 miles, for Mercury 47,935,000 miles, and for the sun 91,239,000 miles. Consequently, for us Mars and Venus have very much larger parallaxes than Mercury or the sun, and of course the larger the parallax the easier it is to measure. Secondly, even the largest of these parallaxes must be determined within far less than one-tenth of a second of the truth, and while that degree of accuracy is possible in measuring short arcs, it is quite unattainable in long ones. Hence, one of the most essential conditions for the successful measurement of parallaxes is that we shall be able to compare the place of the near body with that of a more distant one situated in the same region of the sky. In the case of Mars that can always be done by making use of a neighbouring star, but when Venus is near the earth she is also so close to the sun that stars are not available, and consequently her parallax can be satisfactorily measured only when her position can be accurately referred to that of the sun, or, in other words, only during her transits across the sun's disc. But even when the two bodies to be compared are sufficiently near each other, we are still embarrassed by the fact that it is more difficult to measure the distance between the limb of a planet and a star or the limb of the sun, than it is to measure the distance between two stars, and since the discovery of so many asteroids, that circumstance has led to their use for determinations of the solar parallax. Some of these bodies approach within 75,230,000 miles of the earth's orbit, and as they look precisely like stars, the increased accuracy of pointing on them fully makes up for their greater distance, as compared with Mars or Venus.

After the Copernican system of the world and the Newtonian theory of gravitation were accepted, it soon became evident that trigonometrical measurements of the solar parallax might be supplemented by determinations based on the theory of gravitation, and the first attempts in that direction were made by Machin in 1729 and T. Mayer in 1753. The measurement of the velocity of light between points on the earth's surface, first effected by Fizeau in 1849, opened up still other possibilities, and thus for determining the solar parallax we now have at our command no less than three entirely distinct classes of methods which are known respectively as the trigonometrical, the gravitational, and the photo-tachymetrical. We have already given a summary sketch of the trigonometrical methods, as applied by the ancient astronomers to the dichotomy and shadow cone of the moon, and by the moderns to Venus, Mars, and the asteroids, and we shall next glance briefly at the gravitational and photo-tachymetrical methods.

      *   *   *   *   *

The theory of probability and uniform experience alike show that the limit of accuracy attainable with any instrument is soon reached; and yet we all know the fascination which continually lures us on in our efforts to get better results out of the familiar telescopes and circles which have constituted the standard equipment of observatories for nearly a century.   Possibly these instruments may be capable of indicating somewhat smaller quantities than we have hitherto succeeded in measuring with them, but their limit cannot be far off because they already show the disturbing effects of slight inequalities of temperature and other uncontrollable causes.   So far as these effects are accidental, they eliminate themselves from every long series of observations, but there always remains a residuum of constant error, perhaps quite unsuspected, which gives us no end of trouble.   Encke's value of the solar parallax affords a fine illustration of this.   From the transits of Venus in 1761 and 1769 he found 8·58 seconds in 1824, which he subsequently corrected to 8·57 seconds, and for thirty years that value was universally accepted.   The first objection to it came from Hansen in 1854, a second followed from Le Verrier in 1858, both based upon facts connected with the lunar theory, and eventually it became evident that Encke's parallax was about one-quarter of a second too small.

Now, please observe that Encke's value was obtained trigonometrically, and its inaccuracy was never suspected until it was revealed by gravitational methods, which were themselves in error about one-tenth of a second and required subsequent correction in other ways.   Here, then, was a lesson to astronomers who are all more or less specialists, but it merely enforced the perfectly well known principle that the constant errors of any one method are accidental errors with respect to all other methods, and therefore the readiest way of eliminating them is by combining the results from as many different methods as possible.   However, the abler the specialist the more certain he is to be blind to all methods but his own, and astronomers have profited so little by the Encke-Hansen-Le Verrier incidents of thirty-five years ago that to-day they are mostly divided into two great parties, one of whom holds that the parallax can be best determined from a combination of the constant of aberration with the velocity of light, and the other believes only in the results of heliometer measurements upon asteroids.   By all means continue the heliometer measurements, and do everything possible to clear up the mystery which now surrounds the constant of aberration, but why ignore the work of predecessors who were quite as able as ourselves ? If it were desired to determine some one angle of a triangulation net with special exactness, what would be thought of a man who attempted to do so by repeated measurements of the angle in question while he persistently neglected to adjust the net? And yet until recently astronomers have been doing precisely that kind of thing with the solar parallax.

I do not think there is any exaggeration in saying that the trustworthy observations now on record for the determination of the numerous quantities which are functions of the parallax could not be duplicated by the most industrious astronomer working continuously for a thousand years. How, then, can we suppose that the result properly deducible from them can be materially affected by anything that any of us can do in a lifetime, unless we are fortunate enough to invent methods of measurement vastly superior to any hitherto imagined ? Probably the existing observations for the determination of most of these quantities are as exact as any that can ever be made with our present instruments, and if they were freed from constant errors they would certainly give results very near the truth. To that end we have only to form a system of simultaneous equations between all the observed quantities, and then deduce the most probable values of these quantities by the method of least squares. Perhaps some of you may think that the value so obtained for the solar parallax would depend largely upon the relative weights assigned to the various quantities, but such is not the case. With almost any possible system of weights the solar parallax will come out very nearly $8.809'' - 0.0057''$, whence we have for the mean distance between the earth and sun 92,797,000 miles with a probable error of only 59,700 miles ; and for the diameter of the solar system, measured to its outermost member, the planet Neptune, 5,578,400,000 miles.

It is the habit of centipedes to carry their young, clasped by means of their legs, to all parts of the underside of the body, though generally the young are clustered in dense masses. When the young are thus bunched together, the body is coiled upon itself at that part ; and the contrast between a centipede in this position, says Mr. J. J. Quelch—who describes the centipede's method in *Nature*—and a scorpion carrying her young upon her back, just as a small oppossum does, is a marked one.

On the Preparation of Tooth Sections.*

By W. B. Tolputt, L.D.S.

MY subject this evening is the grinding and preparing sections from hard tissue. I shall speak principally, if not wholly, of the tissue with which I have most to do, and with which I am most familiar, namely, a tooth. The enamel of the tooth is the hardest tissue known in animal life; and that treatment which is most applicable to a tooth will apply also to a bone, and, in fact, to all hard tissues.

There are two ways of obtaining tooth sections : one from the tooth as it is, and the other by decalcification. It may be necessary on occasions that one should know of a ready and effective method of making sections of teeth or other hard tissues when desirous of examining the internal structure. And by this I mean, not a simple slice which may tell you anything or nothing, but such a section as will show all and everything, regular and irregular, which a good section should show. It is my desire to give a few simple directions relative to what may be found in most text books on histology, and, at the same time, to supplement them with some practical suggestions which I have found useful, and which may be of utility to any one wishing to make sections of hard tissues, and of teeth in particular. If we take any ordinary section of a tooth purchased at a dealer's, unless it comes from a very expert and painstaking preparer, what do you see? Many times a specimen more or less transparent, with the tubular structure of the dentine obliterated, or if not entirely obliterated, covered by patches of translucency which mar the general appearance as well as detract from the perfect utility of the section. The edges may be fractured and jagged, presenting a very untidy appearance, and, taken altogether, giving but a very meagre presentation of all the beautiful and instructive detail which characterises a well-made specimen. Even in those sections made with all care by ourselves, unless we adopt certain precautions, we may have all this detail present in the earlier stage of a section's existence, and yet be doomed to disappointment and

* A Paper read at the Sheffield Microscopical Society.

annoyance in its examination after a year or two, by the gradual disappearance of its tubular structure.

It was this experience which induced me to adopt various expedients for obviating this annoying result, and although my subject may be considered by some as well-worn, and as well threshed out, I am hopeful enough to think that, by describing the methods I adopt, I may perhaps be the means of assisting some to attain results which will be regarded in after years with satisfaction.   I wish to be very plain and practical, therefore, if my communication appears somewhat of the character of the cookery book of recipes, I hope your pardon will be extended to me.   Cookery books, although perhaps not the highest class of literature, are nevertheless very useful in their results, and are therefore not to be thoroughly despised.

Text books, in treating of this subject, advise first that "thin slices should be cut from the tooth with a saw."   Now, however desirable it may be to cut a tooth into as many sections as possible in order to be enabled to trace the various phases of structural change throughout its extent, I think I need not remind those who may have attempted this method of the numbers of saws broken, to say nothing of those blunted and worn out in cutting through the enamel of one tooth.   A lapidary's wheel has also been recommended for cutting the rough sections.   This would cut but few sections out of many teeth, the number of sections depending upon the thickness of the wheel used; and, furthermore, we may not possess lapidary wheels.   With care, two or three sections may be cut from a tooth by first cutting through the enamel by wetting a new thin gold file with turpentine and soft soap, or turpentine and camphor, and then using a broad frame saw for cutting through the dentine.   There is no difficulty after the enamel is passed.   It may oftentimes be grooved by a thin corundum wheel on the lathe, and the section cut by the saw afterwards.   The plan I adopt may be a very wasteful one, but till we get a ready means of cutting through the enamel, I am afraid I must continue to recommend and to adopt it.

I take a tooth and hold it against the side of a revolving fine corundum wheel till one side is ground flat; then polish that side to the most perfect polish it is capable of receiving on a piece of buff leather with some putty powder on it. Afterwards, take a piece of stout glass, put a little old, and consequently tough, Canada balsam on it, warm it, and spread it a little larger than your section. Let the balsam cool down until it is "tacky"; then press the polished side of the tooth into close contact with the glass and clip. When quite cold the grinding may proceed, as in the first part of the operation, till you get the required thinness, when that side may also be polished. The hard balsam round the section supports and protects the edges, so that they will not be fractured and made jagged and untidy. By not putting the tooth on to the glass plate until the balsam is somewhat cooled, you prevent the polished surface from being covered with fine cracks, which might remind you of a dinner plate which a careless cook has over-heated till the glaze is cracked in all directions; it also prevents the balsam from running into the tubular structure of the dentine.

As the process I adopt for mounting the section is applicable to all sections of hard tissues, I shall reserve my remarks upon it till I have mentioned another plan of grinding down the rough slices, which I can recommend from long experience on account of its readiness, cleanliness, and the perfect parallelism of the . sections produced by it. Having a slice of dental or other hard tissue of moderate thickness, place it between two plates of ground glass, or between a plate of glass and a flat stone, such as an Arkansas or Lake Washita stone, with water and a pinch of levigated pumice powder, and by a rotary motion of the upper glass gradually rub the section down till it is thin enough for examination with even the highest powers of the microscope. But towards the end of the process be careful to watch it, for as the glass and stone, or the two glasses, get closer together, and the section thinner, one turn more of the upper glass will sometimes result in the total disappearance of an hour's work, and you will be eligible to take rank amongst beings of a very high order if an explosion of your private opinion does not occur. Having ground your section sufficiently thin by either of the before-mentioned plans, it remains to be mounted in a suitable medium for examination.

Of all the media recommended, none fulfil the requirements in so satisfactory a manner as Canada balsam if certain precautions are observed, of which I shall speak presently.   Canada balsam is not, strictly speaking, soluble in alcohol, but is converted by it into a white pulverulent condition.   Therefore the plate, having the thin section attached to it, may be placed in alcohol, and, after a few hours' soaking, the thin section is easily detached without fracture, but will be found coated with this altered Canada balsam, every particle of which must be removed with a clean camel-hair brush kept constantly wetted with spirit.   Unless this is done, the section will look messy and muddled when it is mounted permanently.   Having got it quite clean, it may be placed in clean absolute alcohol till you want to mount it.   It might be considered that all this camel-hair-pencil work could have been dispensed with by placing the section into some complete solvent of the balsam, such as chloroform, benzole, or turpentine; but it must be remembered that, by so doing, we should bring about the very thing we have been trying to prevent. We want to mount our section without the highly refractive balsam running into the minute structure and rendering it invisible ; and that is the reason I recommend the treatment by alcohol.

There are two good methods of mounting bone and teeth in Canada balsam, which, while securing the advantage we are desirous of attaining, also preserve, in the highest degree, the visibility of their histological details.   That which I practice is the simpler.   Take your section out of the absolute alcohol and let it partially dry, *carefully protecting it from dust or other contamination.*   When nearly dry, give it a good soaking in filtered distilled water, that the tubular structure, or any minute spaces like lacuna or canaliculi, may become filled with water; afterwards dry its surfaces by wiping them with a clean warm finger, so that all moisture is taken from them, when the section may be mounted in rather firm balsam, with very little fear of the structure being swallowed up in translucency.   The reason blotting paper is not used for preliminary drying is, that the fibres from it adhere to your section and disfigure its appearance.

The second method of mounting is to plunge the section for a moment into alcoholic solution of white shellac, and, quickly withdrawing it, the alcohol evaporates, leaving the porous structure completely occluded and protected from the balsam, however liquid it might be.   I think that both these methods are productive of such satisfactory results that I can commend them to your careful attention if you should at any time wish to preserve specimens of dental histology.

The methods of preparing sections of teeth by decalcification, though not new, have come in vogue of late.   Many specimens are better prepared by softening the hard parts, and cutting, after suitable preparation, in a microtome.   There is no doubt that decalcifying processes are most useful in many instances.   Chromic Acid has for years been used for this latter purpose.   It is claimed for it that it hardens the soft parts while it softens the hard, both changes taking place simultaneously.   It is useful for fully-developed human teeth, also for developing jaws of mammals in which a fair amount of dentine, enamel, and bone, have been already formed.   The following is, though seemingly rather complicated, wasteful, and troublesome, a very successful method :—

1.—A tooth is placed in a solution of Chromic Acid (crystals), 10 grains, to Water, 8 ounces.   It should here remain for two days, at the end of which a fresh solution must be used.

2.—After two days immerse it in a solution of Chromic Acid, 20 grains, to Water, 8 ounces, for four days.

3.—Then in Chromic Acid, 2 ounces ;   Water, 4 ounces ; Hydrochloric Acid (2% solution), 4 ounces.   The latter should be added a few minutes after the Chromic Acid solution is made.

4.—Remove the tooth to fresh solutions, made up according to the last formula, every fourth day until it is sufficiently soft (this takes from eleven to twelve days).   By using the Chromic Acid as mentioned, the advantages derived from the employment of fresh re-agents are assured.

5.—Immerse the tooth for half-an-hour in a large quantity of alkaline solution, and wash under the tap for twenty-four hours.

The tissue is then ready for gum solution and cutting with the microtome.   Embedding in paraffin may of course be substituted for the gum.   Picric acid, in a saturated solution, is often employed for the purpose of decalcifying.   But for a ready solution that is generally to hand, and the method I myself employ, there is

nothing so effectual as a saturated solution of common alum, with
about half a drachm of Hydrochloric acid added to each ounce
of solution.   Steeping the tooth (or teeth) in this for about three
weeks leaves the tooth with a consistency of cork.   If it is now
soaked in glycerine for a few days it may be embedded and cut
into thin sections by any of the usual microtomes.   I prefer this
to either Picric or Chromic acid, because it does not stain the
hands, and I believe does not produce so much granularity as
the other processes do.

Proceeding to the staining stage, Logwood is first on the list,
and can be prepared so as to be certain of working perfectly.   The
way to prepare the extract is to dissolve it up in alcohol, add a few
drops of this to a saturated solution of potash alum, and, when
mixed, add a few crystals of phenol.   This will give a brownish,
purple stain.   Another very useful stain is Borax and Carmine ;
it will penetrate deeply into the tissue.   The preparation consists
of $\frac{1}{2}$ a drachm of Carmine and 2 drachms Borax, dissolved in
distilled water.   Aniline dyes form most useful stains with these ;
all sorts of shades can be brought out.   Another useful stain is
Osmic Acid.   The real use of Osmic Acid is to bring out any fat.

In staining with Chloride of Gold, it really does not matter
whether the section be fresh or not, although many text books say
it is useless to attempt to stain tissues which have been deprived
of life more than an hour.   The method described by Dr. Bödecker
has given the best results.   He immerses the section, whether
cut from a decalcified tooth or ground down from a hard one, in
a solution of Carbonate of Soda for an hour.   It is then placed
in a solution of Chloride of Gold, which must be neutral, and
left in the dark for another hour ;   then put in the Carbonate of
Soda for a few moments, and transferred to a 1 per cent. solution
of Formic Acid, and kept warm over a bath for about an hour and
a half. . Finally, the section should be mounted in Glycerine Jelly
—not in Canada balsam.   Sections which have been decalcified
by Chromic Acid take longer to stain than those that are fresh,
but the whole process only occupies from three to four hours,
instead of at least twenty-four hours as in the old method, and the
result will be found far more satisfactory.   The usual needles, or
any steel instrument, must not be used for manipulating the sec-

tions ; some non-metallic substance, such as a quill tooth-pick, should be used instead.

Having stained the specimen, the next step is to wash it, dehydrate, and clear it.   Clearing means removing all those substances which are of a lower specific gravity than the medium the specimen is to be mounted in.   The clearing agents are : oil of cloves, oil of cedar, oil of bergamot, and oil of sandal wood.   Oil of cloves is by far the most commonly used.

Some Remarks on Clarification, and also on a New Clarifier for Microscopical Purposes.*

By Edwin A. Schultze.

(Read October 19th, 1894.)

ONE of the most important points to be observed in microscopical research is the requisite clarification of the object under examination.   As simple, however, as this appears, the views of the different authorities who have investigated the subject do not appear to coincide.   In studying plant sections, clarification, according to Dippel, is resorted to to produce the necessary transparency of the object, which on account of its contents does not permit of a thorough investigation of the inner structure.   The usual clarifiers mentioned by Dippel are: Liquor potassæ (in some cases to be used with acetic acid and ammonia) ; potassic alcohol, carbolic acid (when necessary to be used after treatment with alcohol or mixed with turpentine); and finally chloral hydrate (8 parts dissolved in 5 of water).   Zimmermann, in his work *Botanische Mikrotechnik*, discriminates between the action of the above named chemical clarifiers and the clarification by means of more or less refractive mounting media, especially essential oils, balsams, and resins.   Of these latter he mentions Canada balsam, dammar and Venetian turpentine.   Tschirch, in his " Pflanzen anatomie," speaks of employing only liquor potassæ and

* Translation from the German of Dr. Wilhelm Lenz, Zeitschr. f. wiss. Mikr. xi. 16 (1894), for the Journal of the *New York Microscopical Society*.

potassium chloride, with acetic acid for clarifying plant sections.

Moeller, to conclude, very correctly places water at the head of his list of clarifying reagents; then follow glycerine, alkaline solutions, Javelle water, and finally chloral hydrate. There appears to be a difference of opinion among the authorities on this last mentioned clarifier, which Schimper frequently and highly recommends. Behrens, in his extensive catalogue of clarifiers, calls attention to chloral hydrate " as being only useful for animal preparations" without naming his authority. Moeller mentions chloral hydrate (5 parts in 2 of water) as having been recommended by A. Meyer, and refers to a publication of this last named authority in the " Archiv der Pharmacie," but in which chloral hydrate does not appear. Moeller says, furthermore, that chloral hydrate is preferable to diluted liquids, inasmuch as it leaves the fecula unchanged.

Such is, however, not the case, for it is easy to demonstrate that chloral hydrate causes every kind of starch to swell and change into a pasty substance after about twenty-four hours' time. The formation of this transparent paste does not in any manner interfere with the admirable clarifying properties of chloral hydrate.

I have used chloral hydrate (8 parts dissolved in 5 of water) to a great extent myself as a clarifier for plant sections, and have also recommended its use to those students interested in microscopy at the Fresenius College in this city, and believe I have thoroughly investigated its advantages and disadvantages while comparing it with other clarifiers. A short communication may, therefore, be acceptable on this as well as on a similar clarifier lately discovered by me, and which, I think, is somewhat superior to chloral hydrate.

As already stated, Dippel has explained in a masterly manner the reason for clarifying plant sections. The means for obtaining this result are, 1, the removal of troublesome substances, and 2, the saturation of the object under examination with a body, the refractive power of which is so nearly equal to that of the cell walls, that the light may penetrate as much as possible; but still sufficiently different to admit of the cell walls appearing sharply outlined under the microscope. The division of clarifiers into chemical and physical by Zimmermann, seems to have originated in the above, yet I cannot agree with him for the reason that most of the clarifying solutions act chemically as well as physically, in consequence of which they must be carefully examined and used accordingly. This is especially the case with chloral hydrate. Its excellent double action, chemical and physical, and consequently the ease with which it may be used, has rendered it a very desirable clarifier.

It is remarkable that as yet so little is known of the physical properties and refractive power of the chloral hydrate solution. According to my researches a solution of 8 parts c. p. crystallised chloral hydrate in 5 parts of water has a specific gravity of 1·3677 at 15° C. The refraction of sodium light is 1·4272, and the divergence 0·0078. I have not been able to discover the refraction of pure cellulose.

If we consider, however, that pure cellulose is clarified almost to invisibility in Canada balsam, while remaining clearly perceptible in glycerine of less and styrax of greater refractive power, we may conclude that the refraction of cellulose is about that of Canada balsam. The mean refraction of the latter is said by Behrens to be 1·535.

There are a few drawbacks which must be mentioned in using chloral hydrate as a clarifier. While the action of the alkaline solutions may be nullified with acetic acid, and any turbidity removed with ammonia, this is not the case with chloral hydrate. Fecula and almost the entire contents of the cells are more or less dissolved by chloral hydrate and rendered transparent. In treating these clarified objects with water or glycerine they become turbid.

These defects are far less perceptible in a solution that I have been using for the past six months, and which is composed of equal parts of sodium salicylate and water. This solution has a specific gravity of 1·2315 at a temperature of 17° C. Refractive power 1·4497 and divergence 0·01318 at 15·4° C. According to my former observations the refractive power of the chloral hydrate solution would equal that of a 67 per cent. solution of glycerine. The salicylate solution, as I shall call my new liquid, shows a refraction equal to that of a solution of 82 per cent. of anhydrous glycerine. The penetrating power of my solution is greater than that of chloral hydrate on account of its low specific gravity. Its refractive power is superior to that of chloral hydrate, without, however, approaching too closely that of cellulose.   It is, therefore, for both reasons, more serviceable as a clarifier than chloral hydrate.

As far as its chemical clarification is concerned the action in swelling the fecula is very prompt.   No starch grains were perceptible in sections of ginger after they had laid in the solution about two hours.   The colouring of the cell walls is also less changed by salicylate than chloral hydrate.

The latter also gradually causes the cell walls to swell, and, in the case of soft structures, they are sometimes rendered partially imperceptible.

In using salicylate I have not discovered this defect; it appears, on the contrary, to have a hardening effect on soft tissues.

If, as an instance, several thick sections of ginger-root, after having lain about twenty-four hours in salicylate, and the superfluous liquid carefully removed with filter paper, be mounted in glycerine and examined, the cell walls will be found to be but little swollen, and their colour, especially that of the cork cells, well preserved.

All the thin strata appear sharply outlined.

In balsam mounts the details become even more clear and beautiful.

In objects which have been treated with chloral hydrate the cell walls will be found more swollen, their colour fainter, their outlines less clearly defined, and the structural details, although in general discernible, somewhat obliterated.

Chloral hydrate is, consequently, only serviceable in examining thick objects, and when these are to be decoloured as much as possible.   Dyes containing tannin are more influenced by chloral hydrate than by salicylate.   The best way to eliminate them is by means of alkaline solutions and afterwards, when necessary, with Javelle water.

Fecula, when swollen in chloral hydrate, forms a transparent jelly, which becomes considerably turbid when a small quantity of water or glycerine is added; a turbidity which increases and forms a thick deposit if this supply is increased.

The jelly produced by fecula swollen in salicylate is much more transparent, does not become cloudy through the addition of concentrated glycerine, and remains, when treated with water, moderately translucent. The swelled fecula reacts with iodine in both cases, but producing two different colours, viz. : the chloral hydrateared, and the salicylate a pure blue. Salicylate with iodine would, for this reason, be valuable for discovering traces of starch.

Salicylate in general has the advantage of containing a neutral, non-corrosive, non-poisonous, and non-hygroscopic salt, its principal chemical virtue being an extraordinary power for swelling starches. The chief feature, however, which renders salicylate so admirably useful in microscopy, is its miscibility with phenols, notably oil of cloves. If one drop of oil of cloves be mixed with 10 of salicylate an opalescent liquid will be obtained, which becomes clear by adding a second drop of oil of cloves. If this process of adding oil of cloves be continued, about the 20th drop will produce a cloudy solution, which may be clarified by a single drop of salicylate. A liquid may thus be obtained with a refractive index of $1\cdot5$, which is about that of cellulose.

WEATHER-CHARTS.—Atmospheric pressure, and consequently the height of the barometer, varies not only from place to place, but also, in any one place, from day to day, and even from hour to hour. All those movements of the air that we know as *winds* are due to more or less local differences of atmospheric pressure ; variations in the force and direction of the wind are among the chief factors in weather ; and so the barometer, by showing whether the pressure is on the increase or on the decrease, whether rapidly or slowly, enables us to foretell changes of weather. The actual height of the mercury being of far less importance than the direction and rate at which it is moving, the words " change," " fair," " set fair," etc., usually placed by the side of the instrument, are of little value. The chief points in weather-charts are, therefore, the readings of the barometer at various places at short intervals of time—usually daily. Lines joining places which have at a given time the same barometric pressure are called *isobars* or *isobaric lines*. These are generally drawn for every tenth of an inch, and the distance between two successive lines gives us the *gradient*. By this we mean the slope of the atmosphere, just as a railway engineer uses the term for the slope of the ground. An engineer will speak of a gradient of 4 in 100, meaning a rise of 4 feet in 100 feet of distance ; in weather-reports the units of distance is a degree of 60 geographical miles, and the vertical scale is in hundredths of an inch of the barometer. Hence a gradient of 4 means a rise of 4/100ths ($= 1/25$th) of an inch in a degree.—*Cassell's New Popular Educator for Dec.*

Co=operation in Plants.*

By Geo. Clayton, F.C.S., of the Northern Col. of Pharmacy.

THE matter I propose to speak about is symbiosis, also to touch on parasitism, though more with the object of distinguishing between it and symbiosis, and also to give illustrations of pure partnership between one plant and another plant, and also between a plant and an animal. A pure parasite, as most of you will know, is a plant or animal which, living on another plant or animal called its " host," withdraws nutriment from it, without in any way contributing to the support of its host. In the plants yellow-rattle, cow-wheat, etc., we have good examples of plants parasitic on other plants. Here I must mention that parasites must not be confounded with epiphytes or saprophytes. Epiphytes being plants like the orchid, which are attached to other plants, but derive no nutriment from them, getting their nourishment from the air by means of their pendulous roots. The saprophytes being plants like the toad-stool, which live on dead or decaying vegetable matters. A true parasite then lives on its living host, having mechanical attachment with it and withdrawing nutriment from it, but giving the host-plant nothing in return. The common yellow rattle, whose yellow flowers are seen in every pasture field, lives on the roots of the clover, and takes its sustenance from it.

But in a great many cases it is most difficult to decide for certain that the host-plant does not get some advantage from the parasite which drains its juices. Should this be the case, however, the latter would no longer be a parasite, and the relationship between the two would be one of mutual assistance—a partnership for the benefit of both. Until lately, for example, the mistletoe was regarded as a true parasite; but some botanists now state that this plant, the mistletoe, contributes food stuff (starch, cellulose, etc., which it has manufactured in its leaves) to the oak, apple, or poplar tree to which it is attached. These botanists point to the enlargement, corpulence, or general robustness of the apple stem in the particular place where the mistletoe is attached, and argue

* From the *British and Colonial Druggist.*

that if the mistletoe were a true parasite the tissue of the apple would be attenuated where the mistletoe was joined. At the present time, we cannot state with certainty the relations between the host tree and the mistletoe, but the tendency of research in similar instances has been to prove reciprocity between plants formerly regarded as parasite and host; so that probably this, too, may be eventually established as an instance of co-operation in plants.

However, in a very great, many cases, it is distinctly proved that there are plants living in union with each other, and which may be said to be in partnership, or in true co-operation, each plant handing over to its fellow material that the other specially requires, and which alone it would have great difficulty in obtaining. This is symbiosis, bringing me to the principal subject of my paper.

Symbiosis in plants then may be defined as "the associated existence of two or more plants for purposes of nutrition. Unlike parasites, two symbiotic plants living in union each supplies its partner with material which the partner requires." A give-and-take, or reciprocity system, being the rule of their combined existence.

I propose to divide the illustrations of symbiosis or co-operation into three divisions :

1.—The symbiosis of green-leaved, flower-bearing plants with fungi.

2.—Symbiosis of algæ with fungi.

3.—Cases of animals and plants in symbiotic relationship.

The first division of my subject is the symbiosis of flowering or phanerogamous plants with fungoid partners. The fungus, as most of you will know, is a cryptogamic or flowerless plant, containing no green colouring matter in its cells. The mushroom is, of course, the best known example; but by far the greater number of fungi, instead of possessing the solidity of the mushroom, are composed of greyish-white, stringy threads, termed myceloid filaments, and these are the kind I wish to refer to in particular. The union of the two partners takes place underground, the roots of the larger plant, the tree, being woven over by the thread-like filaments of its partner—the fungus. The root that descends from the germinating seed into the ground becomes entangled with the myceloid filaments of the fungus already existing in the soil; henceforward the connection continues until death. As the root grows onward the myceloid fungus grows with it, accompanying it like a shadow whatever its course. The ultimate ramifications of roots of trees a hundred years old, and the young roots of year old seedlings, are both covered by the myceloid mantle in the same manner. Gardeners for years have experienced great difficulty in rearing various specimens of rhododendron, winter green, broom, heath. This difficulty is now traced to the fact that the roots of the rhododendron, for example, is unable to find its symbiotic partner in some soils in which it is transplanted, or through which the new roots are piercing their way; the transplanted rhododendron, therefore, perishes on account of its being unable to assimilate the necessary materials from the soil. On the other hand, the roots of the healthy, vigorous rhododendron are found to be invested by the fungoid partner.

The partnership or symbiosis, therefore, in the instance just mentioned is between the rhododendron and the fungus which is attached to its roots. The larger tree providing the fungus with starch and other organic materials elaborated in the green leaves of the tree, whilst the smaller plant supplies the rhododendron with moisture and mineral matters which the fungus absorbed from the ground. The filaments of the fungus grow in sinuous curves, forming a felt-like coat round the root. In colour the filaments are mostly brown or grey, and through the tangle of fungus filaments the root may often be discerned here and there, whilst in some instances the fungus even penetrates some distance into its partner. The fungoid plant is provided externally with hyphæ, or hair-like processes, which pierce the earth, taking up nutriment, and handing it over to its larger partner.

A further proof that a symbiotic partner is indispensable to certain plants, is found in the fact that when the attempt to rear seedlings of the beech and fir is made in a soil destitute of humus (that is, fungus filaments), the seed, after germinating and growing a short time, perishes; but if soil or mould, recently dug from the ground in a wood, and known to contain the living mycelium of the fungus, be placed at the root of the seedling, it at once begins to grow vigorously, owing to it having "connected on" with the smaller subterranean partner, which feeds it from the ground. The number of plants having symbiotic relations of the kind described is very large ; most of the *Ericaceæ*, including heathers, rhododendron, and similar plants ; *Coniferæ*, including firs of all kinds ; *Salicaceæ*, including poplars, willows ; and all *Cupuliferæ*, like the oak, beech, etc.

It is of great interest to note that the chief species of flowering plants which are symbiotic are gregarious in character, and, like the oak, fir, and heather, form large forests or moors, and one is filled with wonder at the magnitude of the immense colonies of subterranean fungi which must exist, interlacing themselves at the roots of such forests of trees.

It will also be plain why there is such a profusion of fungi of all kinds in forests around the roots of certain large trees. We cannot yet state precisely the exact species of fungi which contract union with these plants, or whether there is a selective affinity between flowering tree and fungus. But this much is known—the flowering plant, rather than be at a loss for a fungoid partner, will unite with other fungi than it is usually connected with.

Then again there are instances of certain trees living in symbiotic union with another plant, and at the same time beset by animal and vegetable parasites.

For instance, the common black poplar has its roots in symbiotic union with a fungus for purposes of nutrition. Then other parts of the roots of the poplar will have the parasite toothwort feeding on them ; again, on the upper branches of the same tree may be found the mistletoe, whose presence on the poplar tree is due to the bird, the missel thrush, which, through eating the berries of the mistletoe, disseminates the seeds. The mistletoe, in turn, will have lichens attached to it.

In all, I suppose the poplar tree has nearly fifty plants or animals living in symbiotic or parasitic association with it.

I now come to the second part of my paper, namely, the co-operation or symbiosis of algæ with fungi. Almost the first place, if not the most interesting one, in symbiotic communities, ought to be assigned to the lichens, to which, as will be well known, the Iceland moss and the litmus plant belong. These organisms, the lichens, which were once classed as separate plants, are now acknowledged to be of a composite character.

Each lichen, being comprised of, first, a fungus, made up of a web of myceloid threads, with, second, an alga in its interior, this combination of alga and fungus thus forming the one lichen plant. I must state at this point, to make myself clear, that an alga is a cryptogamic or flowerless plant, without any distinction between stem and leaf, and containing green colouring matter. The ordinary seaweeds are examples of algæ, also the green slime found in stagnant water ; but the algæ which live in partnership with fungi, forming lichens, are very minute plants, numbers of them being unicellular, but all of them containing green colouring matter inside their cells ; they mostly belonging to the natural order *Nostocinæ*.

The myceloid threads of the fungus being most exterior fulfil the function of gathering from the air moisture, and from the ground or substratum mineral matters ; whilst its partner the alga, owing to its having chlorophyl or green colouring matter, manufactures starch and other organic chemicals. Thus here again the partners supply each other with matters necessary for the life of both. Then as to the creation of a lichen. A simple experiment will illustrate the facility by which a lichen may be created. If a sheet of white blotting paper moistened with water be exposed for several hours to the wind in the country, and then its surface carefully examined with a microscope, numerous particles like dust will be found deposited on the paper ; these particles consisting of cell groups of algæ, pollen, grains, spores of mosses and fungi, etc. All these bodies would, under natural circumstances, have been deposited in the crevices of the bark of trees, on rocks, stones, etc., and we can easily see, if in these places the little algal cell groups meet with their partner the fungus, the latter will embrace and enmesh them, and thus create a confederacy known as a lichen. Thus we can easily understand how it is that the bark of trees, rocks exposed to moisture, etc., become clothed with these lichens.

A most interesting proof of this union is afforded by the fact that a celebrated botanist, M. Bornet, has actually synthesised a lichen—that is, created one. Commencing with certain definite algæ, he sowed them, in company with certain definite fungi, in a favourable place. The two separate plants amalgamated and interwove their cells, with the result that a lichen was formed.

It is necessary to add, however, that lichens as we find them, growing on bark of trees, etc., do not in all cases owe their original appearance on the locality where found to a fresh union of an alga with a fungus, but there is another mode of distribution of lichens. The distribution of numerous lichens is brought about by the wafting with the wind of already completed social colonies to places often situated a great distance from the locality where the first union of alga and fungus was contracted, the process being as follows :—In a matured lichen, several cells all bound together, called a soredium, detach themselves from the parent plant, and are blown by the wind to a fresh spot, where, stranding, they establish themselves, commence to grow, and eventually enlarge into a full-sized lichen.

Now, after reviewing the various cases of symbiosis mentioned as occurring between two plants, and also in inquiring into the habits of some animals, we find that we may take a step further and say that many animals and plants have symbiotic relations. I now arrive at the third division of my subject, namely :—Cases of co-operation between certain animals or insects and plants. In such a partnership as this the plant partner is materially assisted, sometimes with food, protection from enemies, etc., by the animal, and in return contributes food, shelter, etc., to the animal.

Although the two partners may not always live in union (though, no doubt, they do sometimes), yet the reciprocity between them is most perfect, establishing symbiosis conclusively. As the first instance of this kind of co-operation, the well-known action of insects in distributing pollen from flower to flower may be cited, the insect getting as its own share in the transaction the honey from the nectaries of the flowers it visits, while the flowers gain by being cross-fertilised. Then again, many sea anemones have seaweeds growing on them, the anemone receiving material from the weed, and in return contributing matter for the support of the weed. But probably the most interesting example of partnership existing between the animal and plant is that of the fig tree and the wasp. It may not be generally known that in the fig growing countries there are two varieties of fig cultivated. One variety which bears true or edible figs, called the *Ficus* variety, and another variety which is productive of barren or worthless figs, called the *Caprificus* variety.

The inflorescence of the fig, most of you will know, is termed a hypanthodium, being pear-shaped externally, but hollow in its interior, where there are numerous flowers supported on a concave receptacle.

The flowers at the base of the inflorescence are usually pistillate, or female flowers, those at the top male or staminate. Now the female flowers of the *Caprificus* or barren variety have short styles, and they mature long before the male flowers, which are at the summit of the inflorescence. The female flowers in the *Ficus* or edible fig, have very long styles, each bent like a hook. Now the insect, really a species of wasp, enters the inflorescence of the *Caprificus*, or barren fig, proceeds to the female flowers, sinks its ovi-depositor down to the short style, and deposits the egg inside the ovary. The larva from the egg increases in size, and fills the whole of the ovary, which thus becomes abortive, producing what is termed a gall-flower, so that in the *Caprificus*, or barren fig, all the ovaries of the female flowers have been converted into gall-flowers, hence its barrenness. The insects inside the galls, when hatched, eat their way out from the gall, creep to the top of the inflorescence, and on emerging have to forcibly brush through the male flowers, which are now ripe and discharging their pollen; thus, the wasp on emerging is covered with pollen. These insects, seeking a place to deposit their ova, arrive at the inflorescence of the fertile fig, into which they descend and proceed to the pistillate flowers. These pistillate flowers have different styles to those of the barren fig, and the wasp is unable, in nearly all cases, to deposit the egg in these ovaries, on account of the great length of the style and its hooked nature; therefore the wasp, in its fruitless endeavours to enter, passes from flower to flower, dabbing the pollen brought from the other fig on the stigmas of these flowers, so causing the cross-fertilisation which is so imperative for the formation of good individuals. The labour of the insect is rewarded probably, after visiting some hundreds of flowers, by finding a flower suitable to lay its eggs in. The plant gains by being cross-fertilized; the insect gains in return shelter for its young. The two therefore co-operate. In order that this action may be facilitated, the fig growers take branches from the *Capri-ficus* variety and hang them on the trees of the true fig, the operation being termed "Caprification."

Then the case of the Yucca lily and Yucca moth is another interesting instance of symbiosis. The Yucca lily is a handsome plant, growing in California, and bearing pendulous flowers in shape like the tulip. These flowers are visited by the female moth, known as the Yucca moth. This insect seems to make a special object of collecting the pollen from the stamens, rolling this pollen into round balls, and carrying it to another flower. Arriving there the moth, by means of a long tube, bores a hole right through the wall of the ovary, and lays its eggs inside the ovary near the young ovules, then creeping up the style the moth places the pollen which it brought from the other flower on the stigma, and even forces this pollen down the style as far as it can. The ovules are, of course, fertilised by this pollen, which, coming from another flower, causes the ovules to ripen into sound succulent seeds. The eggs meanwhile, becoming hatched, form young moths inside the ovary, and these young insects eat up as food many of the succulent seeds before they escape out to the light. Therefore, there is no doubt that, in return for the depositing of this pollen and consequent fertilising of its ovules, the Yucca plant repays the moth by sacrificing some of its seeds as food to the young moths. Yuccas grown where there are no moths produce no seeds, through no fertilisation taking place.

I might enumerate many other examples of decided symbiosis between plants and animals, but will be content with the four or five mentioned, as I think these will suffice to prove that between certain animals and plants we have a most interesting system of reciprocity existing. It will, however, be useful to state that the formation of galls on the oak, willow, etc., caused by certain insects laying their eggs, cannot be regarded as furnishing an instance of symbiosis, the advantage, as far as can be seen at present, being all on the side of the animal, so these gall excrescences must therefore be regarded as parasitic structures.

Another Constituent of the Atmosphere.*

IN announcing the recognition of Argon, not quite two months ago, as a constituent of the atmosphere, we said : " The new element will be a fresh instrument of research." The prediction has already been realised. At the annual meeting of the Chemical Society, held Wednesday, March 27, the Faraday medal was presented to Lord Rayleigh for his discovery of Argon, and no sooner was this done and the presentation suitably acknow-ledged, than Professor Ramsay, the fellow-worker with Lord Rayleigh in that discovery, surprised the chemists present with the announcement that he had found another new element associated with Argon, and he proceeded briefly to explain the nature of his remarkable find, which has in it features as romantic as the detection of Argon itself, and must have highly important consequences to the future of chemical and physical enquiry.

We say romantic advisedly. It is surely a wonder-awakening circumstance when the first evidence of something we possess on earth comes to us from the sun, and then that the chemist should find this substance in a very rare earth, and at length, by the use of the most refined methods of science, arrive at the fact that this material is probably one of those that make up the very air we breathe. Such was, in short, the meaning of Professor Ramsay's communication to the Chemical Society. A ray of light comes from the sun and is passed through a small prism of glass—the spectroscope. In the image of that ray, broken into its several colours, there are certain lines which show the existence in the sun of iron, of sodium, of hydrogen, and various other elements, most of which exist on this planet and behave in the same way towards light. But there are some lines that seem to betoken the presence in the sun of a substance we know nothing of. Nature is constant. The lines are always there, and the physicist does not hesitate to give a name to the hypothetical element—of which there was not a vestige of evidence to be had elsewhere. On the faith of rays of light, decomposed by a suitably mounted glass prism, he

---

* From *Daily Telegraph*, March 28, 1895.

INTERNATIONAL JOURNAL OF MICROSCOPY AND NATURAL SCIENCE.
THIRD SERIES. VOL. V.

asserted that there existed in the sun a substance which he called on that account Helium.   This element Professor Ramsay now believes he has proved to be not only one of our terrestrial elements, but allied with Argon, the new component of the atmosphere.

There were two or three suspicious characteristics about Argon. In the first place, it was most unsocial—it would combine with no other substance ; and, secondly, it did not fit into Mendeléeff's great law, which connects the weights of the atoms and molecules of matter with their qualities.   "So much the worse for the law," said some ; but the older chemists were hardly willing to see the breakdown of one of the grandest generalisations of modern times.   Professor Ramsay's discovery probably gets rid of both these suspicions attaching to the new gas.   In order to find out whether there was not something in the world with which Argon would keep company, he was examining an extremely rare earth found in Norway, and known as cleveite, after its discoverer, the Swedish chemist, Cleve.   When this mineral is treated with weak sulphuric acid, it gives off a gas which hitherto has always been regarded as nitrogen.   The Professor found, by very close examination, that it was not nitrogen at all, but Argon ; and, moreover, there was associated with it another gas, which also upon rigorous scrutiny he found to be, to use his own words, "a gas which had not yet been separated."   He submitted it to Professor Crookes, who is eminent as a spectroscopist as well as a chemist, and the result is to show almost with certainty that the gas thus found is none other than Helium.   Just as the nitrogen of the atmosphere was shown by Lord Rayleigh and Professor Ramsay to hide within it another gas—Argon ; so Argon is now proved to contain another element until now thought to exist only in the sun.   The spectroscope told us that 91,250,000 miles away, on the bright incandescent solar surface, where everything known on earth would be melted into vapour by fervent heat, there was present a gas never met with here ; aud now, by dint of laborious research, the chemist finds it, in an extremely rare earth it is true, but probably a constituent of the atmosphere that supports all terrestrial life.

The discovery was made and confirmed by Mr. Crookes's spectroscopic examination quite recently.   Terrestrial Helium has yet to be fully examined ; its weight, specific heat, chemical affinities, and other properties, to be ascertained. Professor Ramsay believes that when these qualities are known they may explain the anomalous position of Argon in its relation to other bodies. The question arose when that strange gas was detected—Is it an element, is it one body or a combination, or is there some other mixed with it, or is it nitrogen in what is called an allotropic condition? The last suggestion may be put aside. It is now pretty clear that "atmospheric Argon did contain some other gas not heretofore separated," and that this gas is Helium. It becomes so much the more probable that Argon will be found to fit into the system of elementary bodies according to Mendeléeff's law.

What, it may be asked, does it matter whether the list of elements extend to sixty-five, sixty-seven, or any other number? To chemical theory it is all-important, and chemistry is the most fruitful of all sciences in its contributions to human well-being. These elements are the pillars of the material universe. It is a favourite peroration with the popular orator to refer to the fall of empires and the crash of worlds. If the latter cataclysm should ever happen, the one thing that would stand the shock would be the ultimate atom. If, for example, the earth could suddenly be stopped in its orbit and should fall headlong into the sun, as it would, every drop of its water would be resolved into oxygen and hydrogén, and every mineral and metal it holds would be vaporised; but the ultimate atoms of nitrogen, carbon, oxygen, iron, and the like would be unchanged and unchangeable ; ready to enter into fresh combinations that would endure through new eons of existence. These atoms that belong to the infinitely little, of which millions of millions can find room and verge enough to keep up a ceaseless dance within a single cubic inch, and the force of whose impact we call heat, are profoundly interesting entities. Clerk-Maxwell based upon them an irresistible argument for a Creator. For these, he said, can be derived by no system of evolution, and each one of them changeless in weight, and size, and properties, bears the impress of "a manufactured article."

The Flora of the Wabash Valley.*

THE composition of the remarkable forests which, in spite of the terrible inroads that have been made in them during the last twenty-five years, still cover considerable portions of the region in southern Illinois and Indiana watered by the Wabash River and its tributaries, was first made known to the scientific world by a paper published in 1882 in the fifth volume of the *Proceedings of the United States National Museum*, by Dr. Robert Ridgway, the ornithologist of the Smithsonian Institution.

In a second paper on the Wabash Silva, recently published in the seventeenth volume of the *Proceedings of the National Museum*, Professor Ridgway shows that the number of indigenous arborescent species in the Wabash valley south of the mouth of White River is one hundred and seven, or more than a quarter of all the arborescent species in N. America north of Mexico, and even this number can be slightly increased, as one or two species of *Cratægus*, overlooked by Professor Ridgway, grow near Mount Carmel.

Some idea of the surprising richness of the forest-flora in this region can be obtained by an examination of Dr. Ridgway's list of trees growing on restricted areas. On a tract of seventy-five acres he found fifty-four species of trees, and another of twenty-two acres contained forty-three species. On a tract of forty acres one mile south-east of Olney, in Richland County, Illinois, what the author modestly calls an imperfect survey of the woods shows thirty-six species. The nearest approach to such a concentration of tree species in a restricted area is in central Yezo, where Professor Sargent found sixty-two species and varieties of trees growing in the immediate neighbourhood of Sapporo at practically one level above the sea.

The height attained by these Wabash valley trees is as remarkable as the number of species in the forest. Individuals of forty-two species reach a height of one hundred feet, and those of twenty-one species grow to the height of one hundred and thirty feet. Individuals of one hundred and fifty feet high of thirteen of these species have been measured. A specimen of *Quercus*

* From *Garden and Forest*, March 13, 1895.

Texana, called *Quercus coccinea* by Dr. Ridgway, the tallest of the Wabash oaks, and perhaps the tallest oak in North America, measured one hundred and eighty feet, and a tulip-tree one hundred and ninety feet ; a Pecan, the tallest hickory, one hundred and seventy-five feet ; a Cottonwood (*Populus monolifera*), one hundred and seventy feet ; a Bur Oak (*Quercus macrocarpa*), one hundred and sixty-five feet ; while, in addition to the trees already mentioned, a Liquidambar and a Black Oak attained a height of one hundred and sixty feet. The size of the trunks of some of these trees, measured at three feet above the surface of the ground, is hardly less remarkable than their height. A Sycamore (*Platanus occidentalis*) girted thirty-three and a third feet ; a Tulip-tree twenty-five feet ; a White Oak twenty-two feet ; a Black Walnut twenty-two feet ; a Black Oak twenty feet ; and a Texas Oak twenty feet. In comparison with such trees, the inhabitants of eastern forests, where trees one hundred feet tall are extremely rare, appear like pigmies, and persons familiar only with forests of the Atlantic seaboard can form no idea of the magnificence of these trees, the last remaining vestiges of the forests which covered the valley of the Mississippi when the white man first floated down its placid waters.

This region is the home of some of our most beautiful and valuable trees. On the bottom-lands of the rivers the Pecan and the great western Hickory (*Hicoria laciniosa*) grow with all the Swamp White Oaks, the Pin Oak, the Texas Oak, and that remarkable form of the Spanish Oak, which, usually an upland tree, sends up on these bottom-lands a tall, beautiful shaft covered with pale bark, which might readily be mistaken for the trunk of one of the White Oaks.

No other American forest-scene is more beautiful, and certainly no other forest of deciduous trees, for it must be remembered that in all this great collection of trees there is not a single species with evergreen leaves is more interesting. No picture can give an idea of the stateliness and grandeur of these noble trees, or of the luxury of the annual and perennial plants that cover the forest-floor with almost impenetrable thickets.

The Use of Parasitic & Predaceous Insects.*

By Clarence B. Weed.

THERE has recently been much discussion concerning the ing utilisation of parasites and predaceous insects in destroy injurious species. A knowledge of the conditions under which such insects act would render it evident that we cannot hope to exterminate any species of noxious insect by means of its parasites alone; and many too sanguine expectations have been aroused- But, on the whole, parasitic and predaceous insects are of immense service to man. Without them many plant-feeding species would multiply to such an extent that the production of certain crops would require vastly more effort than it does now. To say, as has been said, that parasitic and predaceous insects have no economic value, is to put the case too strongly. Take, for example, two crop pests of the first class—the army worm and the hessian fly. The history of a century shows that these insects fluctuate in numbers; that there are periods of immunity from their attacks, followed by seasons when they are overwhelmingly abundant. It is universally acknowledged that in the case of the hessian fly, this periodicity is due almost entirely to the attacks of parasites, and in the case of the army worm to the attacks of parasites, predaceous enemies, and infectious diseases. Remove these checks and what would be the result? The pests would keep up to the limits of their food supply and would necessitate the abandonment of the culture of the crops on which they feed. Take another case :—Professor J. B. Smith has argued that "under ordinary conditions neither parasites nor predaceous insects advantage the farmer in the least," and to prove it he cites this instance :—" Fifty per cent. of the cutworms found in a field early in the season may prove to be infected with parasites, and none of the specimens so infested will ever change to moths that will reproduce their kind. Half of the entire brood has been practically destroyed, and sometimes even a much larger proportion; but—and the 'but' deserves to be spelled with capitals— these cutworms will not be destroyed until they have reached their full growth, and have done all the damage to the farmer that they could have done had they not been parasitised at all. In other

American Naturalist.

words, the fact that fifty per cent. of the cutworms in his field are infested by parasites does not help the farmer in the least."

But obviously it does help the farmer very greatly *the next season*, for it reduces by half the number of cutworms he will have to contend with. As a matter of fact, cutworms fluctuate in numbers in a way quite similar to the army worm, and the fluctuations are largely due to parasitic enemies. I have seen regions where cutworms were so abundant that grain-fields were literally cut off by them as by a mowing machine, and the following season the worms were so scarce as to do practically no damage.

But Professor Smith is right in saying that as a general rule there is too great a tendency to rely on natural enemies to subdue insect attack. It is nearly always safer to adopt effective measures in keeping pests in check than to trust to the chance of their natural enemies subduing them.

Microscopical Technique.
COMPILED BY W.H.B.

Koch's Solution of Methylene Blue.
Saturated alcoholic solution of Methylene Blue   1 drachm.
Solution of Caustic Potash, 10 per cent.   ...   12 minims.
Distilled Water   ...   . 200 drachms.
The stain is very active in a feebly alkaline state.

Artificial Reproduction of Anhydrite from Evaporation of Salt Solutions.—Brauns * has produced anhydrite in microscopic crystals by bringing upon an object-glass a large drop of a saturated solution of sodium or potassium chloride, or a mixture of the two salts, and placing to one side of this a drop of calcium chloride solution, and on the other side a drop of Epsom salt solution. The three drops are joined to one another by narrow paths and evaporated. During the diffusion of the liquids which takes place, calcium sulphate is formed and appears in crystals of both gypsum and anhydrite along with the crystals of the chlorides. When a little water is added to a group of anhydrite crystals, they are dissolved to re-crystallise as gypsum. By properly regulating

* *Neues Jahrbuch f. Min.*, etc., 1894, II., pp. 256—264, in *American Naturalist*, Feb., 1894.

the amount of water added, *Knaüel* of gypsum may be formed with a corroded core of anhydrite. Although anhydrite has been frequently produced artificially, none of the methods heretofore used have simulated its production in nature from the evaporation of sea-water.

An Excellent Etching Fluid for Glass has recently been published in the *Central Zeitung fur Optiker und Mechaniker*, xii., p. 57. It consists of two solutions which are to be kept in separate vials, and prepared according to the following directions :—

No. 1.—Dissolve sodium fluorite   ... 36 grams.
    in distilled water ...   ... 500 cubic centimeters.
    and add potassium sulphate ... 7 grams.
No. 2.—Dissolve zinc chloride   ... 14 grams.
    in distilled water ...   ... 500 cubic centimeters.
    and add conc. hydrochloric acid 65 grams.

These solutions can be kept in ordinary glass vessels, hence no gutta percha bottles are necessary. When wanted for use equal volumes are mixed in a suitable vessel, best in a hollow paraffin cube, which is easily made by cutting a hole in a piece of paraffin with a knife. A small quantity of Indian ink is recommended as an addition in order to enable the writer to see what he has written. This fluid is said to be superior to the more difficult preparations, and it is said that the finest hair lines have been etched on glass with this medium.—*Meyer's Druggist.*

Bacteriological Examination of Blood and Tissues.—Inghilleri[*] gives a new rapid double staining method for bacteriological examination of the blood and other tissues, including the study of phagocytosis and parasites of malaria, which he claims to excel, not only in quickness but in precision. A cover glass preparation (by the usual methods), or a section prepared from the tissues, is placed in chloroform for thirty minutes, and afterwards stained in the following fluid :—1 per cent. solution of eosin in 70 per cent. alcohol, 40 parts ; saturated aqueous solution of methylene blue, 60 parts, the specimen being gently warmed in this fluid for two or

[*] *Central b. fur. Bakteriol.*, May, 1894, in *Brit. Med. Journ.*, June 21, 1894, Epit., p. 13.

three minutes; after which they are ready for immediate observation (for example, blood, etc.), or after dehydrating, clearing, and mounting as usual.

Borofuchsin as a Stain for Tubercle Bacilli.*

—Prof. Lubimoff describes in the *Meditsinkoe Obozrenie* a new stain for tubercle bacilli, which he calls Borofuchsin. It consists of

Fuchsin	7½ grains.
Boracic Acid ...	7½ grains.
Absolute Alcohol ...	4 drachms.
Distilled Water	5 drachms.

When prepared in this way, it has a slightly acid reaction. It is quite clear and not liable to be spoilt by being kept, and consequently it is always ready for use. The sputum is dried on a cover-glass, and stained by being heated in contact with the borofuchsin for one or two minutes. The stain is then washed out by treatment with dilute sulphuric acid. The specimen is then washed with alcohol, and subsequently immersed for half a minute in a saturated alcoholic solution of methylene blue. After being washed in distilled water and dried, the examination of the specimen is made in oil of cedar or in a solution of Canada balsam. In exactly the same way sections of tuberculous organs may be stained after hardening in spirit, only in such cases the steps of the operation must be somewhat more prolonged. The main difference between this and other staining processes for Koch's bacilli is that when borofuchsin is used the process of washing it out with sulphuric acid is an almost instantaneous one. All other bacilli are, as when other stains are used, rendered colourless and invisible, the tubercle bacilli alone being seen.

For mounting preparations cleared with chloral hydrate which it is desired to retain in their transparent condition, Geoffrey suggests (*Journ. de Botanique*, VII., 55, 1893) a solution of 3 to 4 grms. of pure glycerine in 100 grms. of 10 per cent. chloral hydrate. This can be used like glycerine, with the added convenience that it hardens at the edge of the cover, so that the cover can be cemented without tedious cleaning.

* *Lancet.*

Preparing Large and Thick Sections.—Dr. H. Scheneck recommends (*Bot. Cent.*, liv., April, 1893), a method of preparing unusually large and thick sections for permanent preservation so as to be useful for lecture demonstrations and for examination with the magnifier. The sections are first thoroughly permeated by glycerin by prolonged soaking; the superfluous glycerin is drained off and the section dried with filter paper; it is then placed in an abundance of a thin solution of Canada balsam in xylol, and covered with a large cover-glass. The glycerin does not mix with the balsam, nor is it withdrawn from the object, which remains perfectly clear. The method is applicable to sections of stems of large size, such as tree ferns, palms, etc., whether woody or herbaceous. Of course, suitable size of slides and covers have to be obtained.

Notes.

RELAXING INSECTS FOR CABINET.—J. P. Mutch writes to the *Entomologist's Record* that "Rectified wood naptha, obtainable from any chemist, containing a trace of white shellac, say ten grains to the ounce, applied to the underside of the extreme base of the wings by means of a very fine sable brush, within a few seconds renders the wings quite pliable; the insect is then placed on the setting-board and set to the required position, braces being used if necessary. In from twelve to twenty-four hours the specimen is ready for the cabinet, showing no trace of the manipulation it has undergone. The shellac is recommended to prevent any possible future springing or drooping, but the pure naptha produces an equally satisfactory effect so far as the relaxing goes. The old tedious process of damping may thus be obviated, and the delicate colours left uninjured."

FOSSIL MICROBES.—A recent communication to the *Académie des Sciences*, M.M. Reynault and Bertrand stated that in examining some coprolites of the Permian period, they noted the presence of a considerable quantity of microbes of different kinds—isolated rodlets and diplo-bacilli, strepto-bacilli, vibrios, and filaments. There were also mucedinea, with mycelium and detached spores. The Permium bacterium is said not to resemble any of the forms known to exist at present. M.M. Reynault and Bertrand throw out the suggestion that possibly there may be only one species of bacterium which is polymorphic.—*Brit. Med. Journal.*

IMPRESSIONS IN SULPHUR.—M. Lepirre, a French artist, states that in demonstrating that sulphur melted at about 150° can be cooled in paper, he happened to use a lithographic card, of which the edges were turned up. Upon taking away the card, it was discovered that the lithographed characters were clearly and distinctly impressed upon the cooled surface of the sulphur, remaining thus after hard friction and washing. By repeated experiments in this direction, he has succeeded in obtaining results of a very satisfactory character, removing the paper each time by a mere washing and rubbing process. It is found, in fact, that sulphur will receive impressions from, and reproduce in a faithful manner, characters or designs in ordinary graphite crayon, coloured crayon, writing ink, typographical inks, china ink, lithographic inks—whether coloured or uncoloured varieties—and others. He also states that it will reproduce with remarkable exactitude maps.

According to *Nature*, in a paper read at the Botanical Congress at Genoa last year (1893), Professor Saccards calculates the number of species of plants at present known as 173,706, distributed as follows :—Flowering plants, 105,231 ; ferns, 2,819 ; other vascular cryptogams, 565 ; mosses, 4,609 ; hepaticæ, 3,041 ; lichens, 5,600 ; fungi, 39,603 ; algæ, 12,178. Professor Saccards thinks it probable that the total number of existing species of fungi may amount to 250,000, and that of all other species of plants to 135,000.

ADHERENCE OF METALS TO GLASS, ETC.—C. Margot finds that aluminium, magnesium, cadmium, and zinc possess the property of marking glass and other substances containing much silica, in such a way that neither rubbing nor ordinary washing will remove the marks. Taking advantage of this curious property, he constructed pencils and wheels of aluminium, by means of which designs can be drawn upon glass, and it is understood that mirrors, etc., now being sold in London, which reveal figures and various devices when breathed upon, are prepared in this way. The effect is much the same as when sketches are drawn on glass with French chalk, but more permanent.—*Arch. Soc. Phys. et Nat. de Genève* and *Journ. de Pharm.* (6), I., 263.

Reviews.

PRACTICAL BOTANY FOR BEGINNERS. By F. O. Bowers, D.Sc., F.R.S., etc. Cr. 8vo, pp. xi.—275. (London : Macmillan & Co. 1894.)

This work contains, in an abridged form, the elementary and more essential parts of the text of the larger *Course of Practical Instruction in Botany.* We have, first, a List of Apparatus required for ordinary work in the botanical laboratory ; Full Instructions for making Preparations and the Adjustment of the Microscope for Work ; II.—Practical Exercises on the Structure of the Vegetable Cell, involving Simple Methods of Preparation ; and III.—Common Micro-Chemical Reactions ; followed by Practical Directions for the Study of the types. There are two Appendices, giving a List of Reagents, their Preparation and Uses ; and a List of the Reactions of Bodies commonly found composing the tissues of plants. A very useful book.

A SYNOPSIS of the Genera and Species of *Museæ.* By J. G. Baker, F.R.S., F.L.S., etc. 8vo, pp. 34. (London: H. Frowde. 1893.) Price 1/6 net.

This is a reprint from the *Annals of Botany*, VII., and gives a full description of the Genera, Sub-Genera, and Species of the *Museæ*, together with the varieties and sub-species.

THE AMATEUR'S HANDBOOK OF GARDENING, with a Calendar of Garden Operations for each Month in the Year, with special contributions by Eminent Gardeners. 8vo, pp. viii.—124. (Liverpool : Blake and Mackenzie. 1894.)

Herein is contained a large amount of practical information—useful both to the amateur and the practical gardener. It is said that every book which interprets the secrets of gardening, and every chapter that elucidates the mysteries of the cultivation of plants and flowers is a contribution to the wealth and happiness of mankind, and such the author has endeavoured to make this book. The Calendar of Gardening Operations will be found useful.

DIE NATURLICHEN PFLANZENFAMILIEN. By A. Engler and K. Prantl. Parts 109 to 112. (Leipzig: Wilhelm Engelmann. London : Williams and Norgate. 1894—95.)

The four parts before us of this exhaustive and beautifully illustrated work contain descriptions of the Bignoniaceæ, by K. Schumann ; the Mucorineæ, Entomophthorineæ, Hemiascineæ, Protoascineæ, Protodiscineæ, Helvellineæ, and Pezizineæ, by J. Schröter ; the Araliaceæ, by H. Harms ; the Jungermaniaceæ and Anthocerotaceæ, by V. Schiffner ; and the Musci, by Carl Müller. In these four parts there are 67 illustrations, composed of 433 figures.

A NEW ENGLISH DICTIONARY on Historical Principles ; founded mainly on the materials collected by the Philological Society. Edited by Dr. James A. H. Murray, with the assistance of many scholars and men of science. (Oxford : The Clarendon Press. London : Henry Frowde.)

The last part of this important work which has reached us covers the words from DECEIT to DEJECT, and forms part of Vol. III. The price of this part is 2/6.

THE PLANET EARTH : An Astronomical Introduction to Geography. By Richard A. Gregory, F.R.A.S. Post 8vo, pp. viii.—108. (London : Macmillan and Co. 1894.)

The author of this plainly written and interesting little book observes, in the Preface, that " The scientific method of observation and induction should be used in elementary astronomy as in other physical sciences. Celestial phenomena must be observed before the theories that explain them can be properly understood." The first chapter deals with star groups and the apparent diurnal motion of the celestial sphere. In the seeond chapter it is shown that all the phenomena previously described can be explained by the fact that the earth is a globe in rotation. The determination of the size and mass of the earth is the subject of the third chapter. Then comes an account of the apparent motion of the planets. And, finally, it is shown in Chap. V. that the appearances are easily explainable on the Copernican theory of the order of the universe. There are 36 illustrations. ————

POPULAR SCIENCE. By John Gall. Post 8vo, pp. 196 (London : Thomas Nelson and Son. 1895.)
This is another of those very useful " Royal Handbooks of General Knowledge," in which the author brings together in Alphabetical or Dictionary form a manual of useful information on various branches of science, his aim being to provide a handbook containing short summaries of facts and principles specially adapted for beginners in the study of science. There are a great number of illustrations. ————

CASSELL'S NEW TECHNICAL EDUCATOR. Monthly, price 6d.
No. 29 (the last part to hand) contains chapters on Steel and Iron ; Coachmaking; Coal-mining ; Allotment and Cottage Gardening ; Design in Textile Fabrics ; Building Construction ; Gothic Stonework ; Weaving ; Engineering Workshop Practice ; Civil Engineering ; Gas and Oil Engines ; Printing ; Electrical Engineering ; and Dyeing of Textile Fabrics. There are a number of illustrations. The frontispiece, "Bridges," shows some of the essential details in Bridge Construction. ————

LECTURES ON THE DARWINIAN THEORY, delivered by the late Arthur Milnes Marshall, M.A., M.D., D.Sc., F.R.S., etc. Edited by C. F. Marshall. 8vo, pp. xx.—236. (London : David Nutt. 1894.)
We regret that in our notice of this book in our last issue we quoted the price as 6/- ; it should have been 7/6. ————

A TEXT-BOOK OF SOUND. By Edmund Catchpool, B.Sc. Lond. Crown 8vo, pp. viii.—203. (London : W. B. Clive. 1894.) 3/6.
This is one of the University Tutorial Series. In it the author has tried to keep in view the following aims :—I.—To include all the facts of which a knowledge is expected in elementary examinations ; II.—To present a clear picture of the external physical processes which cause the sensation of sound ; and III.—To keep before the reader the distinction between phrases which describe actual processes or conditions, and those which, while they facilitate the prediction of real processes and real phenomena, do not themselves stand for any physiological condition or event. There are 73 illustrations. ————

INTENSITY COILS : How Made and How Used. By " Dyer." Seventeenth edition. 8vo, pp. 79. (London : Perkin, Son, and Rayment, 99 Hatton Garden. 1891.) Price 1/-
This useful little book gives descriptions of Electric Light ; Electric Bells ; Electric Motors ; The Telephone ; The Microphone ; and The Phonograph. There are 184 illustrations. ————

BEGINNER'S GUIDE TO PHOTOGRAPHY, Showing How to Buy a Camera and How to Use it. Crown 8vo, pp. 119.

THE MAGIC LANTERN : Its Construction and Use. Cr. 8vo, pp. 82. (London : Perkin, Son, and Rayment. 99 Hatton Garden, E.C.)
Two very useful little books, bound in cloth and published at 6d. each ; will be found to give a large amount of information on the subjects of which they treat. They contain a number of illustrations.

INTRODUCTION TO THE THEORY OF ELECTRICITY, with numerous examples. By Linnæus Cumming, M.A. Third edition. Cr. 8vo, pp. xv.—326. (London : Macmillan and Co. 1885.)
In the work before us, geometrical, as distinguished from analytical methods have been employed, and although a large proportion of the propositions involve the ideas of the Doctrine of Limits, the use of the notation of the Calculus has been avoided. In the present (third) edition Chap. 1 has been to a great extent rewritten, and a chapter on Thermo-Electricity has been added.

THE MEDICAL ANNUAL and Practitioner's Index : A Work of Reference for Medical Practitioners. Cr. 8vo, pp. 650. (Bristol : J. Wright and Co. London : Simpkin, Marshall, and Co. 1895.) Price 7/6.
The 13th annual issue of this well-known work is before us, in production of which some thirty-five editors and contributors have been engaged. This volume, we believe, is in all respects equal to any of its predecessors. There are 24 plates and 72 illustrations in the text. We notice that the treatment of Diphtheria by the Anti-Toxic Serum has received special attention. The Dictionary of New Remedies and Review of Therapeutic Progress for 1894 occupy the first place, and are followed by a chapter on Electro-Therapeutics by A. D. Rockwell, A.M., M.D. ; Anti-Microbic Treatment, by Prof. Alfred H. Carter, M.D., F.R.C.P. A Dictionary of New Treatment in Surgery forms the Second Section. Section 3 treats of Sanitary Science, etc. etc.

TRAVAUX D'ELECTROTHERAPIE GYNECOLOGIQUE. Par Le Dr. G. Apostoli. Vol. I. 8vo, pp. 720. (Paris : Société d'éditions Scientifiques. 1894.)
Dr. Apostoli has set himself the task of collecting together the most important papers that have appeared on his method of treating fibroid tumours, etc., by electricity. In this first volume all the papers are by practitioners of various nationalities, but translated into French, with footnotes by the Doctor. If the results are as brilliant as described in the large number of cases whose clinical history is minutely given, it would seem to be unjustifiable, if not criminal, to subject a patient to the risk of hysterectomy without a previous trial of this method.

PHYSIOLOGY FOR BEGINNERS. By M. Foster, M.A., M.D., F.R.S., and Lewis E. Shore, M.A., M.D. Cr. 8vo, pp. xii.—241. (London : Macmillan and Co. 1894.)
This finely illustrated little work is intended for those who, without any previous knowledge of the subject, desire to begin the serious study of Physiology. It is written in an elementary style, so as to form an introduction to more advanced treatises ; a few chemical and physical facts are given in the earlier chapters. There are upwards of a hundred good illustrations.

THE ANIMAL WORLD : A Monthly Advocate of Humanity. Vol. XXV. (London : S. W. Partridge and Co.)
The Animal World is issued by the Royal Society for the Prevention of Cruelty to Animals. It is full of good pictures and thoroughly interesting reading matter.

BAND OF MERCY (issued by the Royal Society for the Prevention of Cruelty to Animals). Vol. XVI. Cr. 4to, pp. 96. (London: S. W. Partridge and Co. 1894.)

All children who are fond of dumb animals must be pleased with this prettily illustrated volume. Besides a number of pictures, music, etc., it is full of interesting tales and anecdotes relating to animals, birds, and insects.

THE BOOK OF THE FAIR. Parts VIII. and IX. (Chicago, Ill., U.S.A.) Price $1.

This interesting publication, by Hubert Howe Bancroft, is the only work of the kind published regarding the great Exposition. It is to be a full and complete history and description of the World's Fair at Chicago, organisation, buildings, and exhibits, covering the whole ground, and is as full in detail as can be within the limits assigned—namely, 1,000 imperial folio pages of pictures and print, to be issued in 25 parts, of 40 pages each. The publication began soon after the opening of the Exposition. *The Book of the Fair* is published by the Bancroft Company, Auditorium Building, Chicago.

These numbers commence with a description of the WOMAN'S DEPARTMENT, which is followed by an account of the MACHINERY. The section allotted to AGRICULTURE is commenced. The illustrations are very numerous and most excellent.

THE AMERICAN ANNUAL OF PHOTOGRAPHY and Photographic Times Almanack for 1895. 8vo, pp. 438. (New York: The Scovill and Adams Co.) Price: Stout paper covers, 50 c.; cloth, $1.

The volume before us is the Ninth Annual Volume of the series, of which we feel bound to say each has been better than its predecessor. This volume, besides a number of beautiful plates and some eighty or more illustrations in the text, contains a large amount of interesting matter, written by men who have earned for themselves honourable distinction in the history of photography.

EUROPEAN BUTTERFLIES AND MOTHS. By W. F. Kirby, F.L.S., F.E.S., etc. (London: Cassell and Co.) Price 6d. monthly.

This beautifully illustrated periodical has reached Part 10. Each part contains a coloured plate, showing several species of Lepidoptera in their various stages.

A HAND-BOOK TO THE BRITISH MAMMALIA. By R. Lydekker, B.A., F.R.S., V.P.G.S., etc. Edited by R. Bowdler Sharp, LL.D., F.L.S., etc. Crown 8vo, pp. xiv.—339. (London: W. H. Allen & Co. 1895.) 6/-

This beautifully illustrated volume is one of "Allen's Naturalist's Library" series, and gives descriptions, not only of those mammals now found in Britain, but it contains brief notices of the species exterminated within historic period, and a further section devoted to the fossil forms. The work is throughout handsomely got up, both as to binding and letterpress, and contains in addition 32 excellent coloured plates.

A HAND-BOOK TO THE PRIMATES. By Henry O. Forbes, L.L.D., F.Z.S., etc. Vols. I. and II. Cr. 8vo, pp. xv.—286 and xv.—296. (London: W. H. Allen and Co. 1894.)

The first of these volumes contains an account of the *Lemuroidea* and the *Anthropoidea* as far as the group of the Macaques of the family *Cercopithecidæ*. The second volume continues with the latter genus, and contains the rest of the Monkeys and the Apes, as well as a summary of the geographical distribution of the species of the order Primates. There are in the two volumes 29 coloured plates and 8 coloured maps.

A HAND-BOOK TO THE MARSUPIALIA AND MONOTREMATA. By
Richard Lydekker, B.A., F.G.S., etc. Cr. 8vo, pp. xvii.—302. (London :
W. H. Allen and Co. 1894.)
This volume of Allen's Naturalist's Library gives a scientific and yet popular account of the Australian Mammals. It also gives a description of some of
the more generally interesting extinct representatives of the Order. There are
in this volume 38 coloured plates and several wood engravings.

A HAND-BOOK TO THE BIRDS OF GREAT BRITAIN. By R.
Bowdler Sharpe, LL.D. Vol. I. Cr. 8vo, pp. xxii.—342. (London : W.
H. Allen and Co. 1894.)
The aim of the author has been to furnish the student with a useful guide,
which shall give him some idea of the characters, colour, geographical distribution, nests, and eggs of the birds of his native country, together with notes on
the habits of the different species. In this volume there are 31 coloured plates,
besides wood engravings.

A HAND-BOOK TO THE ORDER LEPIDOPTERA. By W. F.
Kirby, F.L.S., F.Ent.S., etc. Part I., Butterflies. Vol. I. Cr. 8vo, pp.
lxxiv.—261. (London : W. H. Allen and Co. 1894.)
This volume is devoted to the great family, *Nymphalidæ*, with its many
sub-divisions, and includes about half the known butterflies, the British and
Foreign species being described in their proper order. The introduction to this
volume will be found very helpful to the lepidopterist, as it describes the anatomy of the Butterfly and Moth, as well as gives hints on collecting and preserving them.
We are exceedingly pleased with all the volumes we have seen of Allen's
"Naturalist's Library." Every volume is written in a popular and thoroughly
interesting manner, and is beautifully illustrated ; the binding also is handsome
and good. The price of each volume, we believe, is 6/- We unhesitatingly
recommend all our readers to secure a set of this work.

SCIENCE PROGRESS. No. 13, March, 1895. (London: The
Scientific Press.) Price 2/6, or 25/- per annum, post free.
The first part of Vol. III. contains the following articles :—Antitoxin, by
E. Klein, M.D., F.R.S. ; Foreign Work amongst the Older Rocks, by J. E.
Mare, F.R.S. ; Insular Floras, Part IV., by W. Bottingley Hemsley, F.R.S. ;
Peptone, by W. D. Halliburton, M.D., F.R.S. ; Budding in Tunicata, by W.
Garstang, M.A. ; The Reserve Material of Plants (continued), by J. Reynolds
Green, M.A., D.Sc. ; with Appendices :—I., Notices of Books ; II., Chemical
Literature for January, 1895.

THE STORY OF THE STARS Simply Told for General Readers.
By George F. Chambers, F.R.A.S. Foolscap 8vo, pp. 192. (London :
George Newnes. 1895.) Price 1/-
In writing this little book, Mr. Chambers thought of those rapidly growing
thousands of men and women of all ranks who are manifesting an interest in
the facts and truths of Nature and Physical Science. He treats of the Brilliancy and Distances of the Stars, their grouping into Constellations, the number of the Stars, etc. etc. It is a very readable book, and supplies a large
amount of information.

THE MODEL STEAM-ENGINE : How to Buy, How to Use, and
How to Construct. By "A Steady Stoker." Cr. 8vo, pp. 96. (London :
Houlston and Sons. 1895.) Price 1/-
All boys are interested in Steam-Engines. In this little book they will find
all the various parts fully described, with instructions for constructing one on a
small scale for themselves.

The Influence of Light on Life.

By W. Thomson.

THE object of this paper is to gather together from the world of nature a few examples of the influence which light exerts upon life, so that by the consideration of its effects in individual instances we may the more easily be led to a more perfect appreciation of its influence upon all living things. That light plays an important part in the present economy is, of course, admitted at once by everyone; but it requires, at least by some of us, a little consideration before we can thoroughly grasp the full truth of such a statement as that made by Prof. Frankland in the *Nineteenth Century* for May, 1894 :—" Almost every exhibition of force with which we are acquainted on the earth is wholly due, either directly or indirectly, to the influence of the sun through those rays which reach us after traversing 93,000,000 miles of space. Of the whole energy sent forth by the sun, we know that only an excessively minute fraction—not more than one twenty-two hundred millionth part (1/2,200,000,000)—is intercepted by the earth at all. But in spite of the enormous distance traversed and the small fraction of this radiant energy which we are only able to 'trap,' modern science has shown us that we owe almost everything that we have to its agency, and the remarkable manner in which it controls the phenomena of our earth is daily receiving fresh illustration and support."

The commonest phenomena are very often those to which we give the least attention. When an earthquake suddenly occurs, men are forced to give heed to it ; but as the sun continues from day to day exerting its mighty influence, and silently keeping in continual motion the wheels of life, there are few who regard it with more than a common observation. Night follows day, week succeeds week, the sun rises, the sun sets, and few are there who consider for one moment what would take place if the sun set and rose no more (that is, if continual night were to be the order of things) ; or what would take place if the sun rose and set no more (that is, if continual day supplanted the alternating order of day and night).

As we look round about upon the gardens of the world, and observe the green leaves of the trees, the herbage of the fields, the beautifully coloured flowers, and the myriad varieties of form into which the vegetable kingdom is divided, we see them as they have grown and been developed under the influence of day and night. The same also applies to the animal world. In a general way, we may say that all animals and all plants spend half their lives in light and half in darkness, and that their present condition has been arrived at under the influence of an equal amount of light and darkness. Now, as each plant requires a certain quantity of light for its successful growth, it is pretty evident that if that quantity be unduly increased or unduly decreased, the consequences thereof may be of a serious nature to the plant. By considering what takes place, either in exceptional cases in the world of nature or by actual experiment, when the normal quantity of light is greatly interfered with, we shall, perhaps better than in any other way, understand the influence of light, and learn how important it is that in our daily life we should not forget the beneficial results which follow from a right relation between the rays of the sun and the living things upon this planet, or ignore the consequences which may follow if that relation be not attended to. The force or reason of these statements may perhaps become more manifest as we proceed.

The carbon which a plant requires in order to build up its cells and tissues is obtained from the carbonic acid gas which is contained in the air. Carbonic acid gas, as we all know, consists of carbon and oxygen; but as it is only the carbon which the plant requires, it is therefore necessary that the compound should be broken up, so that the requisite carbon may be obtained. Now, living plants possess this wonderful power, and are able to extract the carbon and set free the oxygen, but they are only able to exert this power *under the influence of light*. During the day, therefore, this process goes on without interruption; but during the night the plant is no longer able to continue it, as the influencing power, the light, is of course absent. No sooner, however, does the morning light once more bathe the plant than once again it is enabled to appropriate the carbon.

If we sow some seeds in a pot and place it in a dark cupboard we shall find that in due season these seeds will sprout, but we shall also find that the small seed-leaves will remain of a yellowish colour. If left in the dark, the young plants will, perhaps, lengthen out a little, but, after the food which was stored up in the seed has been exhausted, no increase of the plant's substance will take place. They will become swollen with water and soon droop down and die. If at the same time as we make that experiment we also place a pot containing seeds in the light, we shall find that the young plants will bear *green* leaves, and that, instead of becoming water-logged and limp, they will every day fix more carbon in their tissues, so that the plant substance will daily increase. In these two experiments we have the same seeds, the same quantity of water provided, and the same air, containing in each case the same quantity of carbonic acid; but the great difference in the circum- stances is, of course, that in one case we have no light at all and the plant dies, while in the other case the plant enjoys the influence of the mysterious light, and therefore lives and grows. If we removed the plant which was trying to grow in the dark into the light, we should find that its colourless leaves would soon become green, and that in every way the plant would turn from its state of sickness to one of verdant health. The seeds germinated in the dark, of course, because the necessary food was present in the seed, and light was therefore not required for the fixation of carbon. Indeed, light is objectionable at that stage, as it tends to fix the carbonic acid which the young sprout wishes to throw off.

The same proof of the necessity of light for plants is obtained by exposing one glass of spring water to the sunshine, and hiding another one in a dark chamber. In the former glass, green films of algæ life will by-and-by appear, owing to the action of light on the germs which were present in the water, whereas, as might be expected, no such growth will appear in the other glass. From these simple experiments we see that it is absolutely necessary, in order to have vegetation, that we must have light, and we find our experiments fully corroborated in Nature, for in coal-mines and caves where there is no light at all there is no plant life at all. We may just call to mind, in passing, that in the cultivation of .celery the altering effect of darkness is taken advantage of; the bottom part of the plant's stem is protected from the light, and is thereby rendered more succulent and desirable as an article of food.

Different species of plants require different amounts of light. Some are very greedy and some are content with a very little indeed. In the deepest parts of the sea, where perpetual darkness reigns, we are told that there are no plants at all ; but in shallower waters where there is a certain degree of light, although the quantity may be very small, seaweeds are to be found in abundance. In Messrs. Kerner and Oliver's book on *The Natural History of Plants*, a certain moss *(Hookeria splendens)* is referred to, which lives chiefly in hollow tree-trunks. The light in such situations is necessarily often very feeble, yet this moss is noticeable for its glossy green. On examination the leaves are found to be very thin and delicate, but the cells of the leaves which are turned to receive the scanty light are convex and act somewhat like lenses, and condense the light on to the granules of chlorophyll which lie behind these convex cells. By such a beautiful contrivance as this are these humble plants able to concentrate sufficient light to enable them to secure for themselves the carbon which they require.

We all know that in shady nooks and sheltered corners quite different species of plants may be found from those which may be found in exposed places and in the full blaze of sunlight. No one, for instance, would ever think of looking for a rose in the darkest parts of the woods, nor would we think of looking for the healthiest ferns or mosses in such places as would be most congenial to the rose. We may find fungi in dark—indeed, altogether dark—places; but then we know that fungi are different from other plants, as they do not produce chlorophyll, they being producers of carbonic acid instead of decomposers of it.

We have now seen that the presence of a *moderate supply* of light has an influence in sustaining the life of certain plants, and even in bringing about a luxuriance such as we often see in grottoes of ferns ; we have seen that even a *very little supply* of light indeed has an influence in sustaining such plants as possess the power of gathering up the scanty rays; and we have also seen that when the supply *fails altogether*, then life ceases.

We shall now consider what would take place if a plant were kept continually in light, or if it received a great deal more light than the normal quantity.   As we cannot very well, ourselves, perform an experiment to prove what happens, we must borrow from Professor Stokes.   Three pots of mustard-seed were sown at the same time.   One was kept under the usual conditions of daylight by day and darkness by night ; another was kept in darkness during the day and was exposed to the electric light during the night.   At the end of the experiment the plants in these two pots were found to be much alike.   The third pot, however, was · exposed to daylight during the day and to the electric light during the night.   At the end of the experiments the plants in this pot were found to be stouter looking, to have larger leaves, and to be of a darker green than the plants in the other two pots, but they were not so tall.   We gather from this experiment that the plants in the third pot were hastened forward by having gathered into a short space of time an amount of light which, in the ordinary course of things, would have been spread over a much longer period.

Professor Bailey, an American, has recently been making numerous experiments as to the influence of the electric light upon plants, and he has found that when plants were kept under continuous light—that is, sunlight by day and electric light by night— growth was expedited.   In the case of some cabbages grown under these conditions, maturity was reached about two weeks before others which were cultivated under the usual conditions.   It was also observed that if plants were placed *too* near the electric light it proved very injurious to them.   The electric light contains certain rays which are harmful to vegetation, but it has been proved that this injurious property may be avoided to a great extent by passing the light through transparent glass.   These experiments show that the electric light affects different plants in different ways, but that the general result is a hastening of the growth.   Such an unnatural state of things, however, must surely prove injurious to the plant if continued for very long without a sufficient period of rest or darkness being allowed.

Dr. Carpenter tells us that the same annual plant, in arriving at its full development, requires everywhere the same amount of light and heat, so that at the equator it will come to maturity so much the sooner.   The more the light and heat the quicker they obtain the sufficient quantity for their development.   It is worthy of note that, though, as is the case when fruit-trees are transported from this country to another clime of greater light and heat, a great acceleration takes place in the growth, the period of life is much shortened.   Prof. Stokes points out how rapidly vegetation grows in the Arctic regions when summer has fairly set in, and how that rapidity is the result of the long summer days, the sun, indeed, remaining above the horizon for the greater part of the twenty-four hours.   Vegetation in those northern latitudes, however, gets a long rest during the winter, and thus is compensated for the extra activity induced by the continuous sunlight of the summer days.

It would, perhaps, now be well to call attention to the behaviour of chlorophyll granules under varying quantities of light, and when we consider how directly dependent upon light the chlorophyll is, and how entirely the chlorophyll depends upon that influence in order to perform its work and supply the plant with nutriment, we are not altogether surprised to find that the granules endeavour to accommodate themselves to any changes which may occur.   Those who have closely examined their behaviour inform us that the granules in some cases alter their shapes, while in other cases they arrange themselves in the cells in positions which are most suitable to receive the proper quantity of light.   If the light is just sufficient for their needs, the granules will (as in the Ivy-leaved Duckweed) spread themselves over the cell-walls (Fig. 1).

If the light is altogether absent, then the granules crowd up close under the epidermis, as though striving hard to get nearer to the source of light which has been temporarily shut off from them, and take up a position as exposes the largest surface possible (see Fig. 2).   If, however, the light is *too strong*, then the granules station themselves along the upright walls––a position which, it is evident, will be most conducive to shelter, as thereby the least possible surface of chlorophyll will be exposed (Fig. 3).   These movements of the protoplasm or chlorophyll granules are most interesting, and the matter becomes even more interesting when we remember that the leaves, twigs, and even branches of plants also take up positions in accordance with the degree of illumination.

The tendency of the growth of a plant is always in the direction of the most favourable light. Leaves and twigs all grow in such a manner that they will not, in an undue degree, interfere with the proper supply of light to the other leaves and twigs on the same shrub or tree. We all know that when we look beneath the surface of a clump of ivy or thick bushy shrub, we find that the green leaves are mostly all outside, spreading themselves out in the sunshine. Anyone who has grown a geranium in the house knows very well that the leaves, and indeed the whole plant, bend towards the window in search of light, and microscopists know that if a tube containing *Volvox globator* be left standing in the room for a short while, all the organisms will congregate on that side of the tube which is turned towards the light. In the same way does vegetation everywhere grow according to the light, and take up a position in the shade or in the sun according as its nature demands.

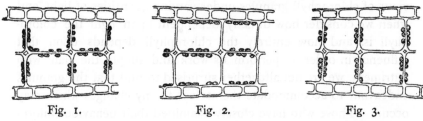

Fig. 1.   Fig. 2.   Fig. 3.

Where an excess of carbonic acid is present in the air, a plant can bear a greater amount of sunshine than is possible under ordinary circumstances. Many plants, however, are unable to stand an excessive light and so perish, but some species are provided with protection. For instance,* on the Mediterranean shores, the leaves and stems of plants which grow on exposed situations are provided with coverings of hair, whereby the green is almost hidden from sight, which serve as a protection from the fierce light of the sun. In some plants a violet colouring matter is found near the surface of the leaves, so that, coming between the chlorophyll and the light, the excess of illumination is counteracted. Many other devices with the same object have been observed, but these will suffice for our purpose. That these hairs and the colouring matter are for protection from excess of light is evident, for when the same species are grown in diffused light, the hairs and

*See Kerner and Oliver.

colouring matter are not developed.   From this we can see that
light has a good deal to do with the distribution of plants, as,
unless they are provided with protective appliances, they will never
be able to grow in any situation where the light exceeds a certain
fixed standard.

The weed which grows in ponds is often of a less green colour
on the surface of the pond than it is a little beneath the surface,
and this is due to the bleaching action of the sun, for when the
light is very strong the chlorophyll loses its bright green colour.
If a piece of black paper be fixed on a green leaf exposed to
strong light, it will protect the part so covered, and when the paper
is again removed the places which were uncovered will be seen to
be quite pale in comparison to the part which was sheltered.   It is
to this bleaching action of light that the disappearance of the
green in autumn is said to be due.   I steeped several ivy leaves in
methylated spirits and obtained a beautiful bright green solution,
due, of course, to the chlorophyll having been dissolved out of the
cells.   Part of this solution was then exposed to the light, and in
a day or two I found that the green colour had quite disappeared,
and that the liquid had assumed a brownish hue.   This, then, is
the bleaching action just referred to, and which is very noticeable
in cabbage and other leaves which have become separated from
the stock.

Some plants turn the edges of their leaves to the light when it
is too strong, and when the quantity is agreeable they turn the flat
surfaces to the light, an arrangement which, like other contrivances
to avoid undue light—such as the folding up of leaves, etc.—keeps
down at the same time an unnecessary waste of moisture, for
it is through the light that the stomata are caused to open
and allow the process of exhalation to proceed.   As too little
light tends to impede exhalation, too much light tends to cause
danger in exactly the opposite direction.   It is light which causes
unripe or acid fruit to be changed into ripe or sweet fruit ; and it
is light, as we so well know, that controls the opening and closing
of the daisy, or "day's eye," and so many other flowers of the
field.

Experiments have been made with a view of determining which
rays of the spectrum produce the greatest influence on plants, and

it has been ascertained that the red, yellow, and orange rays are the most powerful in causing growth, whereas the blue and violet rays are almost entirely without effect in inducing the chlorophyll to perform its function. On the 26th January I sowed some seeds (canary, hemp, and rape) on pieces of flannel kept moist with water. These seeds were then exposed to the light under different coloured glasses, with the following result :—Under the yellow glass the growth was strong and rapid ; under the red it was nearly the same ; under the green the growth was, during part of the time, very much delayed, but eventually it gained strength ; and under the blue the growth was the slowest, and the green of the plants did not appear to be of so bright a hue as in the other cases. Although the green glass was the most transparent, yet the growth was not nearly so rapid as under the much less transparent red glass. Prof. Stokes ascribes to the greenish-yellow the most active power, but my experiments, just quoted, appear to support Messrs. Kerner and Oliver's conclusions that the red, yellow, and orange rays are those which the chlorophyll requires to fulfil its functions. It should, perhaps, be remarked that experiments with coloured glasses are not so conclusive as experiments performed with colour rays obtained directly from a prism.

There seems, however, to be no end to the instances which might be quoted of the influence of light upon *plant* life ; indeed, the greater puzzle seems to be to find something which is not due to its influence. Our present object, however, besides trying to show how plant-life is altogether dependent on light, is to endeavour more particularly to show that an increase or a decrease in the normal amount of light, or an alteration in its quality, has a great influence on plants ; that there is a certain fixed quantity which is most acceptable to each species, and that when that fixed quantity is exceeded or diminished to any great extent, evil consequences must follow. I think that these conclusions may with safety be deduced from what has just been said, and it therefore seems reasonable to conclude that, if investigations are properly conducted, we ought to be able to ascertain with a great degree of nicety, the exact supply that is best for each species, and so be able to measure out, in cultivating a plant, just as much as will exert the highest and best influence.

When we refer to the greatest influence for good which light exerts on *plant* life, we are almost naturally led to ask, Is it possible that light also exerts an influence on *animal* life ?   And if so, surely there must also be in that kingdom a certain measure of the subtle influence which also proves most beneficial.

Such a consideration, then, forms the second part of our paper. We have just seen that light is essential to the very life of a plant, but it can scarcely be said in exactly the same way that light is *absolutely* essential to the existence of *all* animals.   It is true that animals are altogether dependent upon plants, either directly or indirectly, for the supply of food, and, as plants are altogether dependent on light, therefore animals could not exist without light, but, putting that indirect dependence to one side, we do not find that the total withdrawal of light is always followed by the death of the creature in the same way as we found that the total withdrawal was the death of the plant.

In animals we have no chlorophyll to deal with, except *perhaps* in a few instances, such as *Hydra viridis, Euglena*, and *Stentor;* and, therefore, if we find light agreeable or necessary to the life of an animal, it must be for some other reason than that for which a plant requires it.   In Professor Semper's book on *Animal Life* (International Scientific Series), he goes so far as to suggest that animals may be at least as dependent as plants on the *direct* influence of light, although the nature of the relation may be altogether different.   I daresay that most of us have read the story regarding Van Helmont, a celebrated alchemist doctor in the time of Louis XIV., and the wonderful power which he attributed to the rays of the sun.   "Scoop out," he wrote, "a hole in a brick ; put into it sweet basil, crushed; lay a second brick upon the first, so that the hole may be perfectly covered.   Expose the two bricks to the sun. and at the end of a few days the smell of the sweet basil, acting as a ferment, will change the herb into real scorpions."   Present-day science scarcely ascribes such an influence as that to the sun's light ; but it certainly does present us with many wonderful results, and results which I daresay would have surprised Van Helmont quite as much as his production of scorpions surprises us.

Dr. Carpenter, in his *Physiology*, states that if cockroaches are reared in an entire absence of light, they will grow all right, but they will be colourless. Whether that is really the fact or not, I can't quite say, as I have been unable to find any other experiment confirming it. In a somewhat old edition of the *Encyclopædia Britannica* an experiment by a Dr. Edwards is quoted. Dr. Edwards said that if the spawn of frogs be kept in total darkness, the eggs will not hatch, and further that if tadpoles be kept in the dark they will not develop into the mature frog. They will increase in weight, but their tadpole state will be preserved. This experiment seems almost too simple to allow of any mistake being made, but that we cannot pin our faith to it implicitly is manifest when we find Prof. Semper quoting an almost similar experiment, in which the larvæ of frogs were reared in completely dark cellars without discovering any difference in their development beyond a retardation due to the diminished warmth. If it were only the proper season of the year, we could easily prove which of these two scientific gentlemen is correct; but, in the meantime, we must just set the two experiments like an equal negative and positive quantity against each other. In support of the non-development of frog-spawn in the dark, it is only fair to say that Dr. Carpenter states that the eggs of silkworms are not hatched nearly so well in the dark as in the light.

If, however, these experiments are somewhat inconclusive, we can get more certain information by examining the creatures which are found inhabiting mines or caves. In some of these caves—as, for instance, the famous caves of Kentucky—the darkness is complete, a single ray of light never entering their silent depths. Travellers who have visited such places tell us strange stories of the sights which are there to be seen. At present we do not concern ourselves with the caves themselves. But let us suppose that we have quitted the daylight and dived into the dark passage which leads to the innermost recesses of some great cave. After passing, in our journey, many mysterious passages which strike away into unknown darkness, and crossing great chasms yawning into unfathomable depths, we are at last led by our guide into a central hall, and there, by the aid of our torches, is seen a silent lake occupying the centre. In the dark waters of that lake the traveller may see a number of strange-looking fish gliding swiftly about. We are told that the skin of these fish is of a whitish fleshy colour and transparent enough to allow of the heart and liver being seen within. They have four legs and a long eel-like body, and they breathe both by gills and lungs. On closer examination two black spots may be seen upon the head. These black spots are situated at some distance beneath the skin, and prove on investigation to be all that remains of the creature's eyes. For generations these animals have lived in perpetual darkness, and as they have therefore never had opportunity to exercise the power of their eyes the faculty of vision has been lost, and by the law of degeneration the very organ itself has been changed, bidding fair to be, in the course of time, entirely absorbed or extinguished as unnecessary. We thus see at once that certain life is possible under the entire absence of light, but we also see that vital changes have been made.

Other animals, besides those just referred to, are also to be found in dark caves, and all of them either half-blind or altogether blind. A spider discovered in these regions has developed very long antennæ, thus making up for its loss of sight by unusually prominent powers of touch. A blind rat, also, which inhabits those caves has grown very long and sensitive whiskers. These are undoubtedly very strong evidences of the result caused by the absence of light, and they are strengthened by the finding of blind fish at very great depths in the sea where no light ever finds its way. Like most observations of this kind, there are several exceptions, but they do not affect us at present, as we are dealing more particularly with the general question, which is clear enough.

In the caves just referred to, it is noted further that the prevailing colour of the animals and insects is white—a condition which, it seems fair to conclude, has been brought about by their continual dwelling in such a dark abode. We know, however, that the colours of animals are not dependent entirely upon light for their production, as, in the case of a brightly painted butterfly, it is complete in its beauty as soon as it leaves the chrysalis stage, although the pupa may have been buried in the earth, and, therefore, been developed quite in the dark. But if this butterfly spent all its days in the dark, and succeeding generations did the same, we should no doubt find that its colours would disappear in the course of time, till, if it could exist at all, nothing but a whitish sort of tinge would remain. From this we may gather that a temporary withdrawal of light is not always followed by a loss of the colour-forming power, but that a continued absence of it, as the perpetual darkness of caves, destroys it and bleaches the creatures. It may just be mentioned in passing that it is said that the pigment colours in flowers (not the chlorophyll, of course) can be produced in darkness.

Now, although we find it stated that darkness does not, immediately, hinder the production of colours in animals, yet there can be little doubt that light has an influence, greater or less, in affecting the brilliance or intensity of the colours. That such is the case appears pretty manifest when we contrast the splendid colours of many of the tropical insects and animals with the more sombre hues of the creatures in this part of the world, and remember how much more powerful is the influence of the solar rays in the tropics than it is with us. We seem almost naturally to link the brilliance of tropical light with the brilliance of the plumage of the tropical birds which we are acquainted with on the shelves of museums. If some of these birds, with their rich and gorgeous plumage, be brought to and reared in this country under artificial heat, Dr. Carpenter tells us that they are much longer in attaining their full degree of colour, and that they never exhibit the same amount of brightness as when brought up under the influence of the tropical sun.

But we do not require to take birds as an example of the variation in colour according to the light, for we only require to compare the skin of any man whom we may find working in the fields beneath the summer's sun with our own skin, and we at once see a very great difference. We know also that a very short exposure to the sun is sufficient to spangle the faces of some people with freckles, and that we are not at all surprised when a traveller returns from foreign climes with his complexion changed into a very decided brown hue.

To refer again to Dr. Carpenter, he says that there can be no doubt that the prolonged influence of light, during one generation after another, tends to make such a hue permanent. Whether this accounts altogether for the black skins of negroes or not, we will not stay to consider, as other matters would have to be dealt with which are not connected with our subject; but we may just remark that there does not appear to be any very prominent reason why we should not ascribe the blackness of skin to be due to a long-continued habitation in sunny regions. At any rate, we know that we can obtain brown hues from the sun's influence, and that anyone who is for long shut out from the light, as a prisoner in a dark cell or a workman in a mine, becomes pasty-looking and very pale-faced.

Now, leaving out of consideration altogether Darwin's theory of Natural and Sexual Selection as the cause of colours in animals, we can fairly consider it established that light has to a certain extent a *direct* influence on the pigment. Professor Semper calls attention to a most interesting influence which light exerts on some animals in an indirect way. The animals referred to are those which have the power of changing their colour to suit the surroundings, and the Professor asserts that this is brought about by the influence of the surrounding light acting upon the retina of the eye. For instance, if a fish possessing this power rests over red sand, then the impression of the red through the eye upon the brain will cause the brain to effect certain contractions in the pigment cells on the surface of the creature's body, whereby all the different coloured pigment-cells will be contracted except the *red* ones, and the fish will consequently bear then a similarity to the red sand. That, at any rate, is the explanation which has been put forward, and by that theory it is supposed that the red light is unable to excite the red pigment-cells (chromatophores) to contract. And so with other colours. If the light is blue, then all the pigment-cells except the blue ones will contract, and the skin will thus be blue.

It is somewhat interesting to find Darwin, in his *Descent of Man*, explaining that the colours of the shells of some Mollusca are no doubt influenced, to a certain extent, by the amount of light, and pointing out that the lower surface of the shells is generally less highly coloured than the upper and exposed surface. When we thus pass in review the influence which light, or the absence of light, has in altering colours, in changing, it may be, the manner of a creature's life, in causing the degeneration of the organ of vision, and even in preventing development or causing

death, we are not far wrong if we presume that therefore the delicate organisation of man and the mystic power of his mind will
also be influenced according to the quality or quantity of the light.

It is common experience that a person's spirits are depressed
or elevated according to the weather.   If the day is very dark and
cloudy, then with many people everything is dismal and the feeling
is that of oppression.   It "disposes much all hearts to sadness,"
as Cowper wrote.   Gloomy clouds fill the mind, and there is a
general sensation of misery; indeed, I think I have seen it stated
somewhere that during a long season of dark days the number of
suicide cases was increased.   No sooner, however, does the darkness give way and the sun once more shine forth, than its cheering
influence is at once manifest.   Men's faces wear a more peaceful
and happy expression, and a feeling of satisfaction pervades their
being.   So pleasant, indeed, is the light to man that to "bask in
the sunshine" is regarded by most as a high species of enjoyment.
If we had to live for any extended period in darkness, we should
no doubt become very wretched, if, indeed, we could exist at all.
A good light by day and darkness by night is best suited to our
requirements, and when this natural order is interfered with then
we suffer accordingly.

When the night is unduly prolonged, as happens during the
winter season in the Arctic regions, we find that it is altogether
objectionable.   Dr. Kane, in his diary of his Arctic Expedition,
has given some very interesting records of his experience of the
long Arctic winter.   He wrote:—"Noonday and midnight are
alike, and except a vague glimmer in the sky that seems to define
the hill outlines to the south, we have nothing to tell us that this
Arctic world has a sun.   The influence of this long, intense darkness was most depressing.   Even our dogs, although the greater
part of them were natives of the Arctic circle, were unable to
withstand it.   Most of them died from an anomalous form of
disease, to which I am satisfied the absence of light contributed
as much as the extreme cold."

In contrast to the baneful influence of the absence of light, we
must refer to the opinion which has been expressed, that the well-
formed bodies of many tribes where little clothing is worn is due
to the large amount of sunshine which is thus allowed to operate

freely upon them, and that the noted absence of deformity is also
in some measure due to the same cause. Although many causes
may, no doubt, assist in such results, yet light is supposed to play
an important part. It has been proved by experiments that when
infants are kept in the dark their temperature is decreased, and
this must affect their growth. If we wish to find ill-formed bodies
we shall more surely come across them in narrow, dark streets and
confined situations than amongst the individuals of a nomadic
tribe, who spend their existence in the open air. If light, therefore, has an influence on a healthy body, we should reason correctly
if we concluded that it has also an influence on an unhealthy body.

Dr. Carpenter quotes the report of a medical gentleman, who
observed that in a large barracks at St. Petersburg the cases of
disease were three times as numerous on the dark side of the
building as they were on the light side ; and, further, he tells us
that it has been found that, in a certain London hospital, residence
in the wards which looked south was much more conducive to the
welfare of the patients than residence in the wards which looked
north.

I read some time ago that Insanity had been treated by taking
advantage of the influence of light, and after enquiry in a reliable
quarter was obliged with the following information :—" Dr. Pritchard
Davies, who has made experiments on the photo-chromatic treatment of Insanity, has expressed his opinion as follows :—' Failures
are very many and far outnumber the cures. The list of those
upon whom it had no effect whatever is a very long one, and in
many the improvement was but slight. Still, I am convinced that
it has materially benefited some, and those not slight cases, but
cases which had resisted other treatment and given great trouble.'
Dr. Davies believes that it is most beneficial in hysteria, moral
insanity, acute mania, and even in cases where, though the disease
is of long standing, there are lucid intervals. The experiments
were made in the blue room." It appears that the system consists,
so far as I can make out, in keeping the patients in a room in
which all the light is of a certain colour. Though the result is not
very satisfactory, yet it is sufficient for our present purpose, as it
shows us that this light treatment has *some* effect, and it hints to
us that it may be possible, if only the right key could be found as

to quality and quantity, to literally use light beneficially in the treatment of the Insane. That coloured light should have any effect at all may perhaps appear a little strange until enquired into.

Without entering into the question of the wave-lengths and degrees of refrangibility of the different colour-rays of the spectrum, it will be sufficient to call attention to one or two examples of the influence of coloured light. Turkeys are irritated by red and bulls also, a fact which gladiators take advantage of, we are told, when they wish to excite to rage the bulls with which they are to combat. If, then, red excites some animals, why should not some other colour soothe them? In the treatment of the Insane, just mentioned, red is recognised as an exciting colour, blue and violet as saddening ones, and green as a soothing one.

Prof. Semper informs us that under green light frogs disengage more carbonic acid than under a red light; and that under a green light tadpoles will not develop into frogs; while white rabbits are said to be most certainly and easily reared in a white reflected light. Sir John Lubbock found that Water-Fleas (*Daphnia*) preferred yellow or green light to white light, and in his book on *Ants and Bees* he quotes a great many experiments which he had made with the view of ascertaining which colours were most preferred by ants. We cannot here quote the different results he obtained under varying conditions, but we can relate one of the experiments. He covered the box which contained the ants with strips of coloured glass—green, yellow, red, and violet—and then counted the number of ants which took up their position under the various colours. After half-an-hour he changed the position of the glass slips, putting the red where the green was before, and so on, and the ants then re-arranged themselves. In twelve experiments it was found that the total number which stayed under the red glass was 890, under the green 544, under the yellow 495, and under the violet only 5. Although the violet glass was as dark in shade as the red, and darker than the green, yet it is remarkable that the ants had a decided objection to the violet light. As a rule, Sir John found that the ants collected in the shadiest place in the box, but they did not prefer the shade of violet glass, although it was much more shady to our eyes than the green or yellow glass.

These references are perhaps sufficient to show that different coloured lights influence animals in different ways.

It is scarcely necessary to mention, before we pass to another matter, the influence of darkness in inducing sleep among so many animals. Even when darkness comes on at unusual times—such as during an eclipse of the sun—it has been noticed that the birds ceased singing and, like many other creatures, went to sleep.

I now wish to place before you a brief *resumé* of some recent investigations on the influence of light on certain micro-organisms as set forth by Professor Percy Frankland. Although the effects which have already been mentioned are interesting, yet the following results of sunlight upon microbes are almost more interesting and wonderful. Messrs. Downes and Blunt put some test-tubes, containing a liquid which was swarming with bacteria, in such a position that the tubes were exposed to the *direct* rays of the sun for several hours daily. The result of this exposure was that when the light was most favourable the development of the bacteria was entirely *prevented*, and when the light was not so strong the growth of the microbes was only *retarded*. In tubes which had been screened from the light, the bacteria were found to have *increased* in numbers. The *direct* rays of the sun were proved to be most effectual in destroying the organisms, but diffused daylight was found to have a considerable damaging effect upon them.

As a result of experiments which were made with a view of finding out which rays were most active in bringing about such a novel effect, it was ascertained that bacteria which were exposed to the blue and violet rays had their growth entirely prevented, whereas those which were exposed to the red rays of the spectrum had their development delayed. It is worthy of particular notice that the rays which exert the greatest destroying power upon the bacteria are the same as those which we have just stated were most obnoxious to the ants—indeed, the same rays which most powerfully affect the photographer's plate. This destroying action of light is said to be a process of oxidation, as the action is increased when the supply of oxygen is increased, and diminished when the oxygen is diminished. · For instance, Anthrax bacilli which were exposed to the direct rays of the sun in the presence of *air*, were killed in two and a-half hours; whereas Anthrax bacilli which were exposed to the sun in vacuum were still alive after fifty hours' exposure.

A very striking experiment is quoted by Frankland in illustration of this power of the sun. A glass tray is filled with a jelly, in which are mixed the germs of, say, Typhoid bacillus. On the back of the tray is pasted a cover of black paper, out of which has been cut the word "Typhoid." The glass tray is then exposed, upside down, to the sun ; that is, so that the jelly will receive the sun's rays only through the spaces where the letters have been cut out. After being exposed to the sun for two or three hours, the glass tray is put into a dark cupboard and kept at such a temperature as will permit of the germs in the jelly developing and multiplying. Now, it is manifest that, if the sun has killed the germs in the places where it shone upon them through the cut-out letters, no development will take place throughout the whole word "typhoid," and it is also clear that the germs will grow in all the other parts of the jelly which were screened from the sun by the black paper. This, according to Frankland, is what takes place, the word "typhoid" being quite free from bacterial growths, while the other parts are thickly populated.

In addition to this killing, or bactericidal, power. as it is called, it has been found that, if the exposure to the light is not sufficient to destroy the microbes, it may greatly alter their character. For instance, in the case of a microbe which produced red pigment, the sun's influence changed its nature so as to prevent it from any longer producing the red pigment ; and, further, Dr. Palermo, of Naples, placed some cholera bacilli in the sunshine, and found that when they had been sunned for three or four hours they were perfectly harmless, and that the animals which were inoculated with them suffered no harm. Indeed, it was discovered that guinea pigs which had been inoculated with the sunned and therefore innocuous bacilli were rendered proof against the virulent or unsunned bacilli, and thus acted as a vaccination. When the bacilli were not "sunned," they killed guinea pigs in eighteen hours, as is usual.

Frankland says, in pointing out the fact that in the Thames water the number of micro-organisms was often twenty times as numerous in winter as in summer, that the presence or absence of sunshine no doubt plays some part in producing that result, and he relates how two German students proved, by examining the water in a certain river from six o'clock in the evening till six o'clock in the morning, that the numbers were increased enormously during the dark hours of night, and that when the light returned in the morning the numbers were reduced again. An experiment is quoted which showed that in a certain small quantity of water, where there were at the surface 2,100 bacteria, three hours' sunshine *reduced* the number to only nine, whereas in darkness a similar number *increased* to 3,103. The destroying influence of light falling *perpendicularly* does not extend to much depth in the water. The organisms at the bottom of the quantity just referred to were scarcely lessened in numbers. They were, however, prevented from increasing.

It may just be fair to state that I tried the experiment of "sunning" an infusion of hay to see if the *Bacillus subtilis* would be destroyed or hindered in the development; but I regret to say that they appeared to develop more rapidly than some which were kept in the dark. My trial was a failure, but it cannot be regarded as proving anything, for I feel sure it was not conducted properly. The amount of hay in the tube would no doubt prove a sufficient shelter for the organisms from the sun, and would therefore defeat the very object in view. I might have tried the experiment again, but was compelled to give it up in disgust, as the sun would not shine when it was required to do so. When the experiment was so far in train the sun would suddenly disappear for, perhaps, a couple of days, and then reappear when it was too late. It is satisfactory to note, however, that other gentlemen have had equal troubles on this score, as they occasionally speak in a sort of resigned tone of the advantages of experimenting under a more favourable sky than we are blessed with. However, this does not alter the wonderful results which have been found to follow insolation (that is, exposure to the sun's rays), and it certainly ought to move us to consider the possibilities which may be within our reach when our knowledge of the subject is further advanced, and to regard the sun as a benefactor whose bounty can scarcely be fully appreciated.

No doubt the purity of the air, and to a certain extent also of the water, is due to this marvellous power of light, for from what we have just seen it seems reasonable to conclude that the sun is continually destroying the germs of disease which exist around us. No doubt the "sun bath," which I have been told is sometimes resorted to in diseases of the lungs, is found advantageous for such reasons as have just been given. It would be interesting to compare the state of health of a community during a season of dull and rainy weather with the state during a time of sunshine, as, other things being equal, we might well expect to find our conclusions as to the sun's beneficial influence confirmed by such a comparison.

An effort has been made in this paper to confine our attention to the action of light as it *directly* influences the *vegetable* and *animal* world, and to avoid an extension of the subject to the influence of light upon other phenomena in the world of Nature. Although its influence upon other forces and matters is of great importance in an indirect manner in affecting plant and animal life, still our subject is sufficiently complete in itself without referring to them here; at any rate, it is sufficiently large for one evening's consideration. We have gleaned the various particulars here given from many writers, as well as from common observation, and have taken some pains to confirm, from as reliable sources as possible, the different experiments and results which have been quoted.

We have seen that almost the whole vegetable kingdom requires, as absolutely necessary to existence, a sufficient supply of light; that animals are largely dependent, in a direct manner, upon the same subtle influence, if, indeed, not also as dependent upon it as plants; that man's physical and mental parts rebel against the loss or absence of light, and are influenced in various ways by its alteration; and that the germs of disease which might multiply beyond all knowledge are kept in check and subjugated by the same friendly power.

When we focus all these facts to a point, we feel inclined to repeat the dying words of Goethe and ask for "more light," and we are almost forced to recognise, in wide streets, large windows, and plenty of sunshine and fresh air, a method or plan by which Nature will help us most effectually to preserve a sound mind in a sound body.

As we glance backwards, we see clearly that everything seems
to point definitely in the same direction, and to assert most positively that it is light which strews the earth with life in its myriad
forms ; that it is light which sustains and regulates the various
processes and phenomena of the vegetable kingdom, and also,
indirectly if not directly, of the animal kingdom ; that it is the
alteration of the quality or quantity of light which ushers in a
crowd of evil conditions ; and finally that it is the entire absence
of light which forbids the existence of life, and turns a garden of
paradise into a wilderness of death.

Predacious & Parasitic Enemies of Aphides
(including a Study of Hyper=Parasites).
By H. C. A. Vine.   Plates XII. & XIII.

DIFFERING widely in almost every detail of form and
structure from the Aphidivorous Insects which have been
considered in the previous sections, some of the Neurop-
TERA resemble them in their carnivorous habits, the chief nutriment of the larvæ of three families at least consisting of the juices
of Aphides, which, pierced by their tremendous mandibles, are
almost immediately reduced to an empty skin.   Insect for insect,
these larvæ are probably greater destroyers of Aphides than any
of the Syrphidæ or the Coccinellidæ, although the comparative
rarity with which they occur no doubt renders their aggregate
effect much less than that of either of the latter families.

Out of the many thousand species of insects indigenous in
Britain, not more than seven hundred are claimed for this very
beautiful order, even by those naturalists who include within it
some of the following groups :—*Mallophaga* (Bird lice), *Thripidæ*
(Thrips, Black fly), *Thysanura* (Bristle-tails), *Collembola* (Spring-
tails), and *Trichoptera* (Caddis flies) ; all of which, by different
hands, have been placed in other orders.   Apart from these, less
than three hundred and fifty species of the *Neuroptera* can
be said with certainty to be natives of Britain, and many of them
are very rarely met with.

Proposals have been made from time to time for the removal of the most typical families of this order, as originally constituted, to the *Orthoptera*; and since the time of Erichson they have been so described by many naturalists, probably without taking due account of their anatomy and development, under the designation of "Pseudo-neuroptera." I shall hope, in a later part of this section, to examine the anatomy, and possibly the embryology, of one or two genera, with a view to the question of classification. Independently of this proposed change, sections of the *Neuroptera* have been bandied about by different writers to and from the *Orthoptera* and the *Diptera*, while Professor Westwood declined even to admit bird-lice and spring-tails among the *Insecta*. At present the *Neuroptera*, as arranged by Kirby and other leading Entomologists, includes fifteen families or sub-families. Some writers recognise only a lesser number, and Miss Omerod, who classes it in eleven families, truly says in her "Injurious Insects" that there is among them scarcely a leading characteristic which does not meet with an exception.

The Order is of Linnean origin, and Fabricius, the pupil of the great Swede, in attempting to produce a classification which should rival that of his master, constituted in the section *Odonata*, containing the dragon flies, a division which has been generally accepted by those who retain these species among the *Neuroptera*. The other principal division, the *Neuroptera Planipennia*, includes the family of *Hemerobiinæ*, among which the aphidivorous species of this order are to be found, the majority of the larvæ of this section being carnivorous. The various divisions of this interesting and beautiful order, including all those families which are commonly placed with it, comprise, throughout the world, about four thousand species, but only a very limited number find a home within the seas of Britain. A notable feature in the order, which has occasioned much difference of opinion as to the true place of many species of the section *Odonata*, is the imperfect nature of their metamorphosis. In both families of Aphis-eating insects which have been examined in these pages, the series of transformations undergone has been that known as "perfect metamorphosis," in which an active larva has been succeeded by an inert pupa, which, in turn, produces a perfect winged form.

But among the *Neuroptera*, in many species, the larvæ of which live an aquatic life, the pupal form is characterised by an equal amount of activity, and the imago emerges from the pupa still enveloped in a delicate casing, which is thrown off after a short interval. Other deviations from the normal metamorphosis also occur, and will be of interest when, later on, we examine the classification and internal anatomy, though, at present, we are only concerned with the habits and characters of those species which are Aphis-eaters, and among these the transformations are fairly normal.

The elegance and brilliancy, the ferocity and strength of wing which distinguish some members of this order have always rendered them attractive objects to those who take an interest, even in a casual way, in the observation of nature.

Mr. R. Maclachlan, writing in the pages of the *Ento. Transactions*, in 1868, in those valuable monographs which are the standard authority on the *Neuroptera*, says of them :—"The larvæ of these delicate insects play a great part in the economy of nature, and must be considered as benefactors of the human race in no small degree. With those of *Coccinella* and *Syrphus* they help to counteract the extraordinary fecundity of the Aphides, and although their numbers are seldom so great as those of the *Coccinella*, yet, from their activity and from the short time they take to extract the juices of their prey, they must destroy innumerable multitudes of these pests of the horticulturist." In another place he refers to the well-established habit of some of these insects to feign death when in peril. " Most of these insects fall down on one side and feign death when disturbed, the legs being then doubled up, the head drawn under the thorax, and the antennæ concealed."

In the early mornings of hot summer days and late in the afternoons, and even sometimes in the heat of the day, the great Dragon Fly (*Libellula*) or the elegant and slender Demoiselle (*Calypteryx* or *Agrion*) may often be found in the neighbourhood of sheltered streams, and occasionally in open pastures or about roadside hedges, in rapid flight after some insect, by the capture of which they will satisfy their carnivorous instincts. Among the weeds of the neighbouring streams, the larvæ follow an aquatic existence, destroying and making an increasing war upon the other denizens of the water whom they are sufficiently powerful to overcome.

But while in the more open spaces the quick movements and untiring flight of these striking flies attract the eye, a more beautiful variety of the same order, of heavier flight and less metallic, though more elegant colouring, will often be seen beneath an overhanging oak or an aged elder, fluttering clumsily from bough to bough, and sometimes leaving upon the leaves or stems where it has pitched a white spot resembling a speck of mildew. The insect belongs to the sub-family of *Chrysopidæ*, which, with the *Hemerobiidæ* and *Coniopterygidæ*, constitute the family *Hemerobiinæ*, the larvæ of which are Aphis eaters.

The larger of these flies, and especially the *Chrysopidæ*, are known popularly as "lace-wing flies," from the gauzy and peculiarly web-like texture of their ample wings, and sometimes as "golden eyes," from the metallic, brassy appearance of their very prominent eyes. The exquisite grass tint general in this sub-family which pervades even the nervures of the delicate wings, prevents their being readily detected, except when in actual flight, but the presence of the female may often be judged by the white spots already mentioned upon the leaves, or by a line of small white excrescences upon some adjacent twig, such spots, in truth, being but groups of the curious pedunculated eggs which have attracted the attention of many generations of naturalists. The female fly of *Chrysopa perla*, the species which being most common in Britain is most conveniently studied, when about to oviposit, touches with the extremity of her abdomen the surface of a suitable leaf, leaving thereon a minute speck of glutinous 'substance secreted by special glands, subservient to the organs of reproduction. She next lifts her body, and from the leaf to the ovipositor there stands a tiny white pillar of whalebone-like flexibility, to the extremity of which an egg, escaping from the ovipositor, seems to attach itself and remains erect, the patient insect re-commencing its task and continuing until about ten or a dozen eggs are similarly placed. Such a group is shown on Plate XII. at Fig. 1, where may be observed the curious precaution taken by the fly lest the foundation of her supports should be

insecure. In some instances, it would seem, one point of attachment does not suffice, and the egg-carrying shaft is supported upon
two oblique buttresses of the same material. Sometimes, perhaps,
the surface of the leaf is unsuitable at the point which the fly has
selected, and then two or four buttresses yoked together afford the
necessary base for the attachment of the stem, and occasionally
such a foundation bears two or more stems, with their attached
eggs.

Dr. Fitch, in his report on *Noxious and Beneficial Insects*, thus
describes the process of oviposition:—"Nature has furnished
these insects with a fluid analogous to that which spiders are provided with for spinning their webs, which possesses the remarkable
property of hardening immediately on being exposed to the air.
When ready to drop an egg, the female touches the surface of the
leaf with the end of her body, and then elevating the latter, draws
out a slender, thread-like cobweb, half an inch long or less, and
places a little oval egg at its summit. Thus, a small round
spot resembling mildew is formed upon the surface of the leaf,
from the middle of which arises a very slender, glossy, white thread,
which is sometimes split at its base, thus giving it a more secure
attachment than it would have if single."

The interesting and beautiful nature of this arrangement,
which produces the effect when seen under a low magnifying
power of a bunch of seed pearls, each mounted upon a thread,
has led to the expenditure of much ingenuity in the endeavour to
account for so peculiar a mode of oviposition. It is the generally
accepted view that it has arisen from the necessity of protecting
the eggs against the voracity of their own kind; but I am inclined
to think, from my own observation, that various other insects are
equally likely to be the agents of destruction, especially as the
larva of *Chrysopa perla* is sufficiently strong to attack and devour
Aphides larger than itself three or four hours after its emergence
from the egg, and therefore could have little need to become an
egg-eater, unless in the scarcity or absence of its usual food.

I have observed in several instances the larvæ of *Coccinella
bipunctata* eat the eggs of its own kind, and the small *Myina* larva
is known to destroy the eggs of many insects. Dr. Fitch has
shown in his Report on *Noxious and Beneficial Insects*, already

quoted, that when first hatched the larvæ of *Chrysopidæ* live often on the eggs of *other* insects, an observation which is confirmed by a writer in *Science Gossip* for the year 1890—Mr. T. W. Wonfor, of Brighton—who says that in the July of that year he saw a larva of the Lacewing fly feeding on the eggs and young larvæ of the Dot Moth (*Manestra persicariæ*). It is, therefore, pretty certain that while the young larvæ of several insects which may frequent the haunts of *Chrysopa* are egg-eaters, nothing has been observed to show that the larvæ of the latter destroy the eggs of their own species. It seems not improbable that the frequency with which *Chrysopa* is met with in comparison with the other *Hemerobiinæ* may be due to the preservation of the eggs by the peculiar mode of oviposition described.

If we adopt the classification of Dr. Hagen, we find that out of thirty-two species of *Hemerobiidæ* which he describes, no less than fifteen are placed under *Chrysopa*, whilst *Hemerobius* claims seven only, and the remaining ten species are divided among five genera.

Reaumur (*Memoires sur les Insectes*, Tom. III., Pl. 32, 33) gives four illustrations, apparently of various larvæ of the *Hemerobiinæ*, but the want of detail renders it impossible to say to what genus they belong except one, which is no doubt *C. perla*. He has, with the exercise of a fanciful imagination, often found in the naturalists who laid the basis of our present knowledge, called these curious larvæ 'Aphis lions,' on account of some resemblance in carnivorous habit to the *Myrmeleones*, or 'Ant-lions,' and, indeed, they are at least as formidable to the delicate aphis as the latter are to their especial prey. They differ, however, in this: that whereas the Ant-lion, having digged a pit, waits for the victim to come to him, the 'Aphis lion,' on the contrary, exhibits usually a remarkable activity in seeking its prey, and in case of scarcity of food not hesitating occasionally to attack a larva of its own kind. The struggles arising when two larvæ happen to seize the same aphis simultaneously are of a violent and often very ludicrous character.

The larvæ of *Chrysopa* and *Hemerobius* (which are practically the only aphis-eating genera commonly met with) are long and narrow in body, depressed, segments well marked and successively decreasing in size towards the tail, the last segments, however, sometimes widening out slightly, so that the larva presents a slight resemblance in shape to a fish. The head is proportioned to the body, but small rather than large, and armed with a pair of formidable mandibles, as seen in Pl. XIII., Fig. 1. No good description has been given of the mouth-organs, Reaumur's statement that the mandibles are tubular being erroneous. The labial palpi are long, porrected, and three jointed. The antennæ are long, and present the tapering appearance of a whip-lash. On examination with a fairly high power, the scaly structure of the epidermis is readily seen, as shown in Pl. XIII., Fig. 1, and the extremity, which appears to consist of two delicate joints, terminates in a fine bristle.

The eye will quickly attract attention by its peculiar arrangement of several ocelli, grouped in slight protrusions on either side of the head, and giving not the slightest indication of the extraordinary development which the corresponding organs assume in the perfect insect. Situated distinctly behind the lenses may be seen the mass of pigment cells, which, being dark, is obvious even through the egg-shell before the larva is hatched. The peculiarities of the eye can be observed by reference to Fig. 8, Pl. XIII.

The mandibles, which are deeply grooved on their inner surfaces to form channels within which the hollow maxillæ play freely, bear at each extremity a curious pad, which is depicted at Fig. 7 of the same plate, and which is surmounted by a stiff fine bristle, the presence of which explains in some degree why an aphis, once touched by the mandibles, rarely escapes.

About ten days after oviposition, in favourable weather, the eggs, which at first have a very delicate greenish hue, become slightly opaque, and then darken, and in a few hours the marking of the segments and the pigment of the eye become visible through the thin shell, as shown at Fig. 2, Pl. XII. A slight observation shows that, unlike some other families, the head remains in close proximity to the microphyle, the posterior end of the embryo being turned up so that the entire shape is somewhat that of a flattened C. An examination of the contents of the egg at this stage reveals the fact that the embryo is developed around, as it were, the yolk, which is embraced by the anterior and posterior extremities. At this time the limbs are well developed, and the reddish colour of the young larva is distinctly visible through the shell.

The markings become more pronounced, and a few hours later —usually in the early morning—the éggs split at their apices in a direction parallel to their axis, and the larvæ gradually twist themselves out. They frequently remain seated for a time on the top of the empty eggs, and then creep away to hide themselves for a short time under any neighbouring leaves, after which they commence to attack any aphides that may be at hand. Occasionally, as has been said, at this stage they will pierce and suck out the contents of the eggs of any other insects which they happen to meet with; but it would seem that this generally happens in the absence of a plentiful supply of Aphides, which appear to be their most congenial food.

During the autumn of 1894, I bred from the eggs of *Chrysopa* a number of larvæ, which in due time pupated, and in this present spring should produce the perfect flies. My notes of the progress of one of the hatches may be of interest, and I transcribe them.

August 19th, 1894. —*Hemerobiinæ.*—Under an oak tree about half-a-mile from Midsomer Norton Church, Somersetshire, I took to-day a fine specimen of *Chrysopa perla*, which proved to be a female which had or was about to oviposit. On examination of the branch of the oak from which the fly emerged, I discovered a group of nine stalked eggs recently laid, and the leaf with the eggs was removed, and placed in a suitable breeding glass.

August 28th.—These eggs, which were at first of a beautiful, translucent green colour, when viewed with a magnifier, became now white and opaque, and on careful examination showed a dark red-brown spot near the upper end, and a series of somewhat rectangular spots, in pairs, evidently indicating the segments of the embryo, down one side.

August 29th.—Four or five larvæ hatched between six and seven a.m.; length (extreme), as nearly as I could ascertain, 1/10th of an inch. The eggs were each ruptured in the neighbourhood of the microphyle, and more or less slit down the sides. The colour of the young was greyish black. At 1 p.m. a larva, separated in a glass box, attacked and ate a winged aphis (*Callipterus*

Quercus), a vast number of which were upon the oak whence the eggs were taken. The body of the aphis was more than double the size of the larva. It had great difficulty at first to secure a hold, plunging its mandibles first into one portion and then into another of the victim, as if not secure of a grasp, and when fairly anchored by both mandibles it was several times lifted off the leaf bodily by the struggles and superior weight of the aphis. The larva, however, maintained its grasp, all the time sucking the juices of the victim, and although unable to exhaust the latter so completely as would be the case when fully grown, it left it in a condition of complete collapse, the skin of the abdomen being shrivelled and empty. During the afternoon other larvæ were hatched, making altogether nine from as many eggs. The eggs are shown on Plate XII. at Figs. 1 and 2, and again at Fig. 3, where the manner in which they are split open and the escape of the larva is to be seen. At Fig. 4 is shown the larva above mentioned shortly after sucking the aphis, and details of the same insect are given at Figs. 5, 6, 7, 8, and 9.

August 30th.—Length of larvæ at 9 a.m., 1/6th inch, several aphides having disappeared. During the afternoon one attacked a very large pea aphis, five times as heavy as itself. The aphis, seized behind the cornicles, struggled hard to get free, fixing its legs and pulling. The larva also fixed its legs, apparently utilising the suctorial power which enables it to ascend glass vertically, and also attaching itself by the suction of the anal orifice at the posterior extremity firmly to the surface on which it rested. Thus anchored, it resisted all efforts of the huge aphis to drag itself away, and retained its hold sufficiently to consume the greater part of its victim, though several times, when the hold of the tail was relaxed, lifted off its feet. The larvæ are now all fairly active, especially in the morning and evening, and after dark a match suddenly struck reveals them running about the glass shade with great rapidity. In middle day they frequently creep up the glass, and remain underneath the thick glass at the top.

August 31st.—Length, barely 1/5th inch. The reddish colouring of the segments is now very apparent.

September 2nd.—Length, nearly 1/4th inch—11/48. The larvæ appear to feed chiefly at night. Pea, Oak, and Hazel Aphides were all eaten indiscriminately.

September 4th.—Length, 16/60th of an inch.

September 5th.—Length, 6/20th of an inch   The bulk of the insect increases much faster than the legs, which are now relatively much smaller, and, after feeding the larvæ, are now sometimes so distended and heavy as to be unable to crawl any distance of the vertical sides of the glass cover, and they fall from the height of an inch or two, rolling over on their backs.

September 9th.—Length of larvæ, 1/3rd of an inch.   They seem very sluggish and unable to move.

September 10th.—Larvæ have all cast their skins, emerging from the discarded covering by a longitudinal slit in the dorsal surface of the thoracic segments (I shall figure this cast skin in a later section).   Length now, 5/12th of an inch.   Colours much brighter.   The black colouring of the extremities of the limbs is due to the cast skin in which it is still encased is evident, the fresh skin of the larva being clear.

September 15th.—Larvæ, a bare 5/8th inch long.   Now very voracious, and of a flatter, broader shape than before.   They now eat thirty or forty Aphides apiece during the night, and appear much distended.   They make no attempt to ascend the glass, and appear never to feed in the daytime, unless deprived of food at night.   When fully fed they are very thick, but in a few hours become flat, like a flat-fish in shape.

September 22nd.—The larvæ to-day became pupæ, having attained a length of over 5/8ths of an inch.   The legs are now very small in proportion to the body, which has become very broad.

The pupæ are enveloped in a dense ball of fine silk, nearly spherical in form and attached to the under surfaces of the leaves. The cocoon is less in diameter than the length of the larva, and the pupa is completely hidden within, a contrast to some of the *Hemerobiidæ*, which are so slightly covered as to leave the pupa visible.   The respective pupæ are shown on Plate XIII., at Figs. 9, 9*a*, and 10, 10*a*, where the difference in appearance is well displayed.

Some species of *Hemerobiidæ* have a curious habit of covering themselves partially with the skins of the exhausted Aphides, which, by a peculiar jerking action of the head, they throw backwards, to find a lodgment among the strong bristles of the back and sides.

Such a larva is shown at Fig. 8 on Plate XIII., and, in its natural situation on a stem or leaf, it often exhibits so little of the appearance of an insect as to be taken, until a movement betrays it, for a patch of dry lichen. Probably this habit may assist in protecting it against the sharp eyes of birds.

In some species of this sub-family the larvæ are small and more delicate in structure, rarely attaining to 5/12ths of an inch in length, and from their subdued grey colouring are much less readily seen and captured than those previously described. They are of exceeding quickness of movement, and complete their larval existence in a shorter term than the young of the *Chrysopidæ*. The pupa is but slightly covered with silk in those genera which I have been able to observe, and the fly is as remarkable for the dulness and soberness of its tints as that of the previously described group is for its brilliancy. The wings of the *Hemerobiidæ* are generally thickly fringed with sparse, distinct hairs, which arise also at the junction of the nervures of the wings. The cross nervures of the latter are much fewer than in the *Chrysopidæ*, and constitute an important differential character.

The larva of *Hemerobiidæ* is shown on Plate XIII. at Fig. 3, and the pupa at 9 and 9*a*. Fig. 4 shows the perfect insect which emerged from the pupa, 9*a*. The disproportion between the size of the pupa and the imago is not so marked in this sub-family as in the *Chrysopidæ*, although even here it seems impossible that the delicate wings should be contained in so small a space as that occupied by the pupa.

The most plentiful species of *Hemerobiidæ*, so far as I am aware, appears to be the *H. humulus*, which is found in great numbers in hop gardens, where it performs an essential service to man by the rapidity with which it destroys the fly—*aphis humuli*— one of the great enemies of the hop-grower, and which is with difficulty kept in check by artificial means. Some species, such as that figured, frequent in the larval state fine grass, while others may not uncommonly be taken beneath the leaves of the oak, on the elder, and on the dog rose.

Besides *Hemerobius*, two other genera of the sub-family are believed to be aphidivorous: *Drepanopteryx*, a genus which consists of one species only; *D. Phalænoides* (Linn.), which is very rare, a few specimens only having been taken in the north of England and the southern counties of Scotland. It is very similar to *Hemerobius*, and is said by Dr. Hagen and other writers to be aphidivorous, but there appears to have been but little accurate observation made of its larval habits and development.

The remaining genus, to the larvæ of which the aphis-eating habit is attributed, is *Micromus*. The young resemble those of *Hemerobius*, and are said by Mr. MacLachlan to be " probably aphidivorous," but until opportunity occurs for a more perfect examination of their life-history, it is scarcely safe to assume so much for a certainty. As yet, I have failed to obtain either egg or a living specimen of the larva.

In the typical genus of this group of aphis eaters, *Hemerobius*, the species are, among other features, characterised, as regards the males, by remarkable modifications of the extremity of the abdomen. These are of great interest for purposes of classification, and in a later number will be figured, when the arrangement of the families and genera is reviewed. Unfortunately, nothing like corresponding differences in structure are visible in the female lace-wings; hence, their value for ordinary working purposes is comparatively little.

At the lower part of Pl. XIII. will be seen at Fig. 6 a peculiar looking insect, the appearance of which would scarcely lead to the supposition that it was a Neuropterous larva. The drawing is taken from Curtis's *British Entomology*, and represents the larva of the curious little 'lace-wing,' *Coniopteryx tineiformis*, a fairly typical species of the *Coniopterygidæ*. The fly is frequently to be found in summer on fir trees, in plantations and woods, in company with the various aphides that feed on them; but it is difficult to detect, or at any rate to secure, the larva. Its aphidivorous habits are, however, well established, and for its size there is no reason to doubt that its destructive powers equal those of the larger genera. The larva spin an oval cocoon of densely woven silk, which entirely conceals the pupa formed within it; and although in the earlier generations of the year the escape of the imago is not long delayed, the pupæ of the autumn larvæ retain their condition until the following spring, when the continuance of sunshine quickly causes the appearance of the imago.

The fly is small and does not readily attract attention ; indeed, it may easily be taken as it is found among a group of Aphides, for some large species of that family, a resemblance which is increased by the dusting of the body with a white mealy powder. Once noticed, however, this character—combined with their peculiar neuropterous wing venation and shape and the thread-like antennæ—make their identification easy. The front wings are more irregular in shape than in the families already mentioned (which approach the typical form), and are almost destitute of the transverse veins which form so striking a feature in *Chrysopa* and *Hemerobius*. The longitudinal veins, however, represent very nearly the chief lines of venation in the former, as will be seen when the drawing, now given, of the wings of *Coniopteryx* is compared with an enlarged drawing of the wing of *Chrysopa*, which will be given in the next part. By a comparison of these it seems almost possible to see how the single longitudinal veins of the humbler, although no doubt the more ancient family have given rise to the elegant network which fixes our admiration in the delicate ' Lacewing,' in which, as we shall see hereafter, the firm veins of the former have given place to the fragile vascular structure which supports the latter, and which is provided with a system of circulation of a most marvellous character.

The posterior wings of *Coniopteryx* differ far more from the front wings, both in size and shape, than is the case with other aphis-eating species, but the venation, so far as it persists, follows the same plan as in the front wings. The shape of the hinder wings, with their often narrowed extremities, approaches somewhat that of the same structures in some species of the Hymenoptera. The peculiar character of the venation possesses special interest for the student of the development of species, as its comparatively rudimentary structure suggests the possibility of a survival of the details of an early stage in the growth of the neuropterous group.

These insects were observed by Westwood to differ from the typical species *Hemerobius*, in the slight reticulation of the wings, the slight mealy covering without any appearance of cilia upon the wings, the large size of the terminal joint of labial palpi, obsolete ligula, absence of tibial spurs, and smaller size of posterior wings.

In common with other of the *Hemerobiinæ* they sit with wings

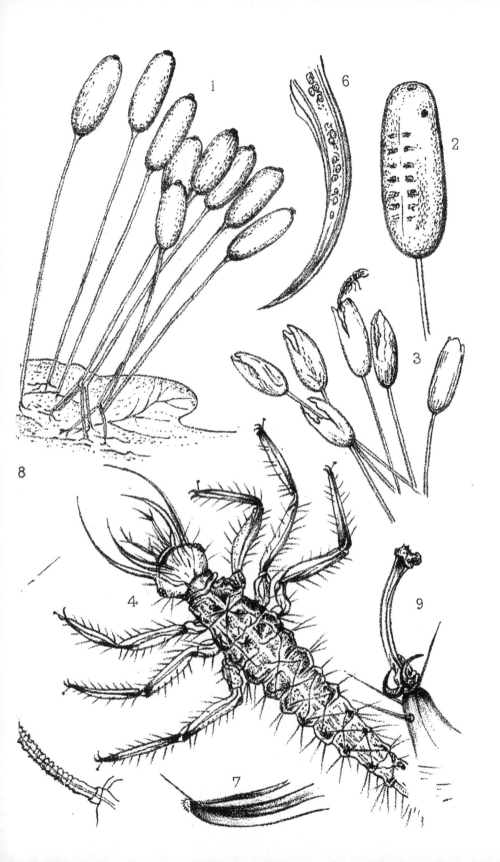

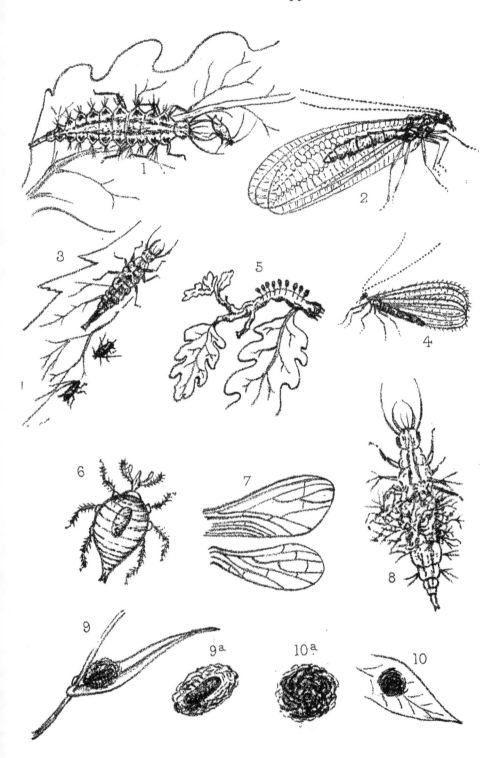

F Phill

deflexed, and when threatened feign death by bending the antennæ under the body.

The sub-family consists of but the single genus *Coniopteryx*, which comprises three species only, as given by Maclachlan.

DESCRIPTION OF PLATES XII. AND XIII.

PLATE XII.

Illustrations of Aphidivorous Larvæ of the *Neuroptera Planipennia*, fam. *Hemerobiinæ.*

Fig. 1.—A group of nine stalked eggs on oak leaf. Taken shortly after laying by the lace-wing fly (*Chrysopa perla*) (3-in. objective).

,, 2.—One of the eggs the day prior to hatching (twelve days after oviposition), showing young larva within (1-in. objective).

,, 3.—Same eggs as Fig. 1, showing larvæ escaping and the empty eggs, split *longitudinally.*

., 4.—Larva at four or five hours after hatching. At this age it is fairly transparent, very active, and will readily attack aphides

,, 5.—One of the long palpi of head (? *maxillary palpi*), showing a structure resembling that of the antennæ of Aphis, and terminated by a long fine bristle.

,, 6.—Mandible, showing hollow channel and the maxilla within, which works up and down in the groove. The latter is seen to contain food.

,, 7.—Extremity of mandible, showing the minute cushion with exceedingly fine spines, by which it is terminated ; also, the long thin bristle at apex, and within, the extremity of the sucking maxilla.

,, 8.—The eye, seen sideways. The eye in the larva consists of several ocelli, grouped together.

,, 9.—The claw and appendages, showing at the extremity of the tarsus the double hooks with strong bristles, and extending between them the slender tendinous process, carrying at its extremity the suctorial (?) pad, which has in its cup-shaped hollow several minute teat-like organs, which are probably the outlets of glands, yielding some secretion which lubricates the pad.

PLATE XIII.

Fig. 1.—Larva of *Chrysopa perla*, nearly full grown, with an aphis in its mandibles.

,, 2.—Imago of same (female).

Fig. 3.—Larva of *Hemerobius*, full grown. The proportion to the larva in Fig. 1 is nearly correct.   The larvæ of many species are very minute.

,,   4.—Imago bred from larva. The difference in wing venation and the hairs on the wings may be observed.

,,   5.—Oak stem, bearing a number of stalked eggs of *Chrysopa*, placed in a consecutive line on the stem, as is sometimes found.

,,   6.—Larva of *Coniopteryx tineiformis*, after Curtis.

,,   7.—Wing venation of the imago of *C. tineiformis*, showing the smaller hind wing and absence of transverse veins.

,,   8.—Larva of *Hemerobius*, laden with the empty skins of slain Aphides ; likened by Reaumur to Hercules elothed in the skin of the Nemean lion.

,,   9, 9*a*.—The cocoon and pupa of *Hemerobius*, showing the slight nature of the silk threads sparsely arranged round the pupa to form the cocoon.

,, 10, 10*a*.—The cocoon of *Chrysopa perla*. The pupa is quite hidden by the dense silk-like texture of the cocoon, which is thick and nearly spherical.

How the Musk-Rat Breathes under Ice.—Animals that breathe by means of lungs can prolong their stay under water only through special anatomical arrangements, or by having recourse to some extraneous means. Mr. W. Spoon, of the Elisha Mitchell Society, who has hunted the musk-rat in winter, asserts that the animal, when obliged to traverse, under ice, a pond so wide that it cannot keep up its breathing, stops from time to time and exhales the air from its lungs. This air, being confined by the ice, becomes oxygenated in contact with the water, and the animal, taking a fresh inspiration, dives in order to begin its swimming again a little further along. It appears that other observers have found that if this air is dispersed through the ice being struck the animal is killed through asphyxia.

The Origin of the Oldest Fossils and the Discovery of the Bottom of the Ocean.*

By W. K. Brooks.

IN the *Origin of Species*, Darwin says that the sudden appearance of species belonging to several of the main divisions of the animal kingdom in the lowest known fossiliferous rocks is at present inexplicable, and may be truly urged as a valid objection to his views.

If his theory be true, he says that "it is indisputable that before the lowest Cambrian stratum was deposited long periods elapsed, as long as, or probably far longer than the whole interval from the Cambrian age to the present day; and that during these vast periods the world swarmed with living creatures. Here," he says, "we encounter a formidable objection; for it seems doubtful whether the earth, in a fit state for the habitation of living creatures, has lasted long enough." "To the question why we do not find such fossiliferous deposits belonging to these assumed earliest periods prior to the Cambrian system I can give no satisfactory answer."

On its geological side this difficulty is even greater than it was in Darwin's day, for we now know that the fauna of the lower Cambrian was rich and varied; that most of the modern types of animal life were represented in the oldest fauna which has been discovered, and that all its types have modern representatives. The palæontological side of the subject has been ably summed up by Walcott in an interesting memoir on the oldest fauna which is known to us from fossils, and his collection of 141 American species from the lower Cambrian is distributed over most of the marine groups of the animal kingdom, and, except for the absence of the remains of vertebrated animals, the whole province of animal life is almost as completely covered by these 141 species as it could be by a collection from the bottom of the modern ocean. Four of the American species are sponges, two are hydrozoa, nine are actinozoa, twenty-nine are brachiopods, three are lamellibranchs, thirteen are gasteropods, fifteen are pteropods,

* A paper read before the University Scientific Association, and printed in the *Journal of Geology*, Chicago. 1894.

eight are crustacea, fifty-one are trilobites, and trails and burrows show the existence of at least six species of bottom forms, probably worms or crustacea. The most notable characteristic of this fauna is the completeness with which these few species outline the whole fauna of the modern sea floor. Far from showing us the simple unspecialised ancestors of modern animals, they are most intensely modern themselves in the zoölogical sense, and they belong to the same order of nature as that which prevails at the present day.

The fossiliferous beds of the lower Cambrian rest upon beds which are miles in vertical thickness, and are identical in all their physical features with those which contain this fauna. They prove beyond question that the waters in which they were laid down were as fit for supporting life at the beginning as at the end of the enormous lapse of time which they represent, and that all the conditions have since been equally favourable for the preservation and the discovery of fossils. Modern discovery has brought the difficulty which Darwin points out into clearer view, but geologists are no more prepared than he was to give a satisfactory solution, although I shall now try to show that the study of living animals in their relations to the world around them does help us, and that comparative anatomy and comparative embryology and the study of the habits and affinities of organisms tell us of times more ancient than the oldest fossils, and give a more perfect record of the early history of life than palæontology.

While the history of life, as told by fossils, has been slow and gradual, it has not been uniform, for we have evidence of the occurrence of several periods when modification was comparatively rapid.

We are living in a period of intellectual progress, and, among terrestrial animals, cunning now counts for more than size or strength, and fossils show that while the average size of mammals has diminished since the middle tertiary, the size of their brains has increased more than one hundred per cent. ; that the brain of a modern mammal is more than twice as large, compared with its body, as the brain of its ancestors in the middle tertiary. Measured in years, the middle tertiary is very remote, but it is very modern compared with the whole history of the fossiliferous rocks, although more of brain development has been effected in this short time than in all preceding time from the beginning.

The later palæozoic and early secondary fossils mark another period of rapid change, when the fitness of the land for animal life, and the presence of land plants, brought about the evolution of terrestrial animals.

I shall give reasons for seeing, in the lower Cambrian, another period of rapid change, when a new factor, the discovery of the bottom of the ocean, began to act in the modification of species, and I shall try to show that, while animal life was abundant long before, the evolution of animals likely to be preserved as fossils took place with comparative rapidity, and that the zoölogical features of the lower Cambrian are of such a character as to indicate that it is a decided and unmistakable approximation to the primitive fauna of the bottom, beyond which life was represented only by minute and simple surface animals not likely to be preserved as fossils.

Nothing brings home more vividly to the zoölogist a picture of the diversity of the lower Cambrian fauna, and of its intimate relation to the fauna on the bottom of the modern ocean, than the thought that he would have found on the old Cambrian shore the same opportunity to study the embryology and anatomy of pteropods and gasteropods and lamellibranchs, of crustacea and medusæ, echinoderms and brachiopods that he now has at a marine·laboratory ; that his studies would have followed the same lines then that they do now, and that most of the record of the past which they make known to him would have been ancient history then.   Most of the great types of animal life show by their embryology that they run back to simple and minute ancestors which lived at the surface of the ocean, and that the common meeting point must be projected back to a still more remote time, before these ancestors had become differentiated from each other.

After we have traced each great line of modern animals as far backwards as we can through the study of fossils, we still find these lines distinctly laid down.   The lower Cambrian crustacea, for example, are as distinct from the lower Cambrian echinoderms or pteropods or lamellibranchs or brachiopods as they are from these of the present day, but zoölogy gives us evidence that the early steps in the establishment of these great lines were taken under conditions which were very different from those which have prevailed, without any essential change, from the time of the oldest fossils to the present day, and that most of the great lines of descent were represented in the remote past by ancestors which, living a different sort of life, differed essentially, in structure as well as in habits, from the representatives of the same types which are known to us as fossils.

In the echinoderms we have a well-defined type represented by abundant fossils, very rich in living forms, very diversified in its modification, and therefore well fitted for use as an illustration. This great stem contains many classes and orders, all constructed on the same plan, which is sharply isolated, and quite unlike the plan of structure in any other group of animals. All through the series of fossiliferous rocks echinoderms are found, and their plan of structure is always the same. Palæontology gives us most valuable evidence regarding the course of evolution within the limits of a class as in the crinoids or echinoids ; but we appeal to it in vain for light upon the organisation of the primitive echinoderm, or for connecting links between the classes. To our questions on these subjects, and on the relation of echinoderms to other animals, palæontology is silent, and throws them back upon us as unsolved riddles.

The zoölogist unhesitatingly projects his imagination, held in check only by the laws of scientific thought, into the dark period before the times of the oldest fossils, and he feels absolutely certain of the past existence of a stem, from which the classes of echinoderms have inherited the fundamental plan of their structure. He affirms with equal confidence that the structural changes which have separated this ancient type from the classes which we know from fossils, are very much more profound and extensive than all the changes which each class has undergone from the earliest palæozoic times to the present day.

He is also disposed to assume, but, as I shall show, with much less reason, that the amount of change which structure has undergone is an index to the length of time which the change has required, and that the period which is covered by the fossiliferous rocks is only an inconsiderable part of that which has been consumed in the evolution of the echinoderms.

The zoölogist does not check the flight of his scientific imagination here, however, for he trusts implicitly to the embryological evidence which teaches him that; still farther back in the past, all echinoderms were represented by a minute floating animal which was not an echinoderm at all in any sense, except the ancestral one, although it was distinguished by features which natural selection has converted, under the influence of modern conditions, into the structure of echinoderms.   He finds in the embryology of modern echinoderms phenomena which can bear no interpretation but this, and he unhesitatingly assumes that they are an inheritance which has been handed down from generation to generation through all the ages from the prehistoric times of zoölogy.

Other groups tell the same story with equal clearness.   Lingula is still living in the sand bars and mud flats of the Chesapeake Bay, under conditions which have not effected any essential change in its structure since the time of the lower Cambrian.   Who can look at a living lingula without being overwhelmed by the effort to grasp its immeasurable antiquity; by the thought that while it has passed through all the chances and changes of geological history, the structure which fitted it for life on the earliest palæozoic bottom is still adapted for a life on the sands of the modern sea floor?

The everlasting hills are the type of venerable antiquity; but lingula has seen the continents grow up, and has maintained its integrity unmoved by the convulsions which have given the crust of the earth its present form.   As measured by the time-standards of the zoölogist, lingula itself is modern, for its life-history still holds locked up in its embryology the record, repeated in the development of each individual, of a structure and a habit of life, which were lost in the unknown past at the time of the lower Cambrian, and it tells us vaguely but unmistakably of life at the surface of the primitive ocean, at a time when it was represented by minute and simple floating ancestors.

Broadly stated, the history of each great line has been like that of the echinoderms and brachiopods.   The oldest pteropod or lamellibranch or echinoderm or crustacean or vertebrate which we know from fossils exhibits its own type of structure with perfect distinctness, and later influences have done no more than to expand and diversify the type, while anatomy fails to guide us back

to the point where these various lines met each other in a common
source, although it forces us to believe that the common source
once had an individual existence.   Embryology teaches that each
line once had its own representative at the surface of the ocean,
and that the early stages in its evolution have passed away and
left no record in the rocks.

If we try to call before the mind a picture of the land surface
of the earth we see a vast expanse of verdure, stretching from
high up in the mountains over hills, valleys, and plains, and through
forests and meadows down to the sea, with only an occasional lake
or broad river to break its uniformity.

Our picture of the ocean is an empty waste, stretching on and
on, with no break in the monotony except now and then a flying
fish or a wandering sea-bird or a floating tuft of sargassum, and
we never think of the ocean as the home of vegetable life.   It
contains plant-like animals in abundance, but these are true animals and not plants, although they are so like them in form and
colour.   At Nassau, in the Bahama Islands, the visitor is taken in
a small boat, with windows of plate-glass set in the bottom, to visit
the "sea gardens" at the inner end of a channel through which
the pure water from the open sea flows between two coral islands
into the lagoon.   Here the true reef-corals grow in quiet water,
where they may be visited and examined.

When illuminated by the vertical sun of the tropics and by
the light which is reflected back from the white bottom, the pure
transparent water is as clear as air, and the smallest object forty or
fifty feet down is distinctly visible through the glass bottom of the
boat.   As this glides over the great mushroom-shaped coral domes
which arch up from the depths, the dark grottoes between them
and the caves under their overhanging tops are lighted up by the
sun, far down among the anthozoa or flower animals and the
zoöphytes or animal plants, which are seen through the waving
thicket of brown and purple sea fans and sea feathers as they toss
before the swell from the open ocean.

There are miles of these "sea gardens" in the lagoons of the
Bahamas, and it has been my good fortune to spend many months
studying their wonders, but no description can convey any conception of their beauty and luxuriance.   The general effect is very

garden-like, and the beautiful fishes of black and golden yellow and iridescent cobalt blue hover like birds among the thickets of yellow and lilac gorgonias.

The parrot fishes seem to be cropping the plants like rabbits, but more careful examination shows that they are biting off the tips of the gorgonias and branching madrepores or hunting for the small crustacea which hide in the thicket, and that all the apparent plants are really animals.

The delicate star-like flowers are the vermilion heads of boring annelids or the scarlet tentacles of actinias, and the thicket is made up of pale lavender bushes of branching madrepores, and green and brown and yellow and olive masses of brain coral, of alcyonarians of all shades of yellow and purple, lilac and red, and of black and brown and red sponges.   Even the lichens which incrust the rocks are hydroid corals, and the whole sea garden is a dense jungle of animals, where plant-life is represented only by a few calcareous algæ so strange in shape and texture that they are much less plant-like than the true animals.

_ The scarcity of plant-life becomes still more notable when we study the ocean as a whole. On land herbivorous animals are always much more abundant and prolific than the carnivora, as they must be to keep up the supply of food, but the animal life of · the ocean shows a most remarkable difference, for marine animals are almost exclusively carnivorous.

The birds of the ocean—the terns, gulls, petrels, divers, cormorants, tropic birds, and albatrosses—are very numerous indeed, and the only parallel to the pigeon roosts and rookeries of the land is found in the dense clouds of sea birds around their breeding grounds, but all these sea birds are carnivorous, and even the birds of the seashore subsist almost exclusively upon animals, such as mollusca, crustacea, and annelids. The seals pursue and destroy fishes ; the sea-elephants and walruses live upon molluscs ; the whales, dolphins, and porpoises and the marine reptiles all feed upon animals, and most of them are fierce beasts of prey.   ′

There are a few fishes which pasture in the fringe of seaweed which grows on the shore of the ocean, and there are some which browse among the floating tufts of algæ upon its surface, but most of them frequent these places in search of the small animals which hide among the plants.

In the Chesapeake Bay the sheepshead browses among the algæ upon the submerged rocks and piles like a marine sheep, but its food is exclusively animal, and I have lain upon the edge of a wharf watching it crunch the barnacles and young oysters until the juices of their bodies streamed out of the angles of its mouth, and gathered a host of small fishes to snatch the fragments as they drifted away with the tide.

Many important fishes, like the cod, pasture on the bottom, but their pasturage consists of molluscs and annelids and crustacea instead of plants, and the vast majority of sea fishes are fierce hunters, pursuing and destroying smaller fishes, and often exhibiting an insatiable love of slaughter, like our own blue fish and the tropical albacore and barracuda.   Others, such as the herring, feed upon smaller fishes and the pelagic pteropods and copepods; and others, like the shad, upon the minute organisms of the ocean; but all, with few exceptions, are carnivorous.   In the other great groups of marine animals we find some scavengers, some which feed upon micro-organisms, and others which hunt and destroy each other; but there is no group of marine animals which corresponds to the herbivora and rodents and the plant-eating birds and insects of the land.

There is so much room in the vast spaces of the ocean, and so much of it is hidden, that it is only when surface animals are gathered together that the abundance of marine life becomes visible and impressive; but some faint conception of the boundless wealth of the ocean may be gained by observing the quickness with which marine animals become crowded together at the surface in favourable weather.   On a cruise of more than two weeks along the edge of the gulf-stream, I was surrounded continually night and day by a vast army of dark brown jelly-fish (*Linerges mercutia*), whose dark colour made them very conspicuous in the clear water. We could see them at a distance from the vessel, and at noon when the sun was overhead we could look down to a great depth through the centre-board well, and everywhere, to a depth of fifty or sixty feet, we could see them drifting by in a steady procession, like motes in a sunbeam.   We cruised through them for more than five hundred miles, and we tacked back and forth over a breadth of almost a hundred miles, and found them everywhere in such abundance that there were some in every bucketful of water which we dipped up, nor is this abundance of life restricted to tropical waters, for Haeckel tells us that he met with such enormous masses of *Limacina* to the north west of Scotland, that each bucket of water contained thousands.

The tendency to gather in crowds is not restricted to the smaller animals, and many species of raptorial fishes are found in densely packed banks.

The fishes in a school of mackerel are as numerous as the birds in a flight of wild pigeons, and we are told of one school which was a windrow of fish half a mile wide and at least twenty miles long.   But while pigeons are plant eaters, the mackerel are rapacious hunters, pursuing and devouring the herrings as well as other animals.   Herring swarm like locusts, and a herring bank is almost a solid wall.   In 1879 three hundred thousand river herring were landed in a single haul of the seine in Albemarle Sound; but the herring are also carnivorous, each one consuming myriads of copepods every day.

In spite of this destruction and the ravages of armies of medusæ and siphonophores and pteropods, the fertility of the copepods is so great that they are abundant in all parts of the ocean, and they are met with in numbers which exceed our power of comprehension.   On one occasion the *Challenger* steamed for two days through a dense cloud formed of a single species, and they are found in all latitudes from the Arctic regions to the equator in masses which discolour the water for miles.   We know, too, that they are not restricted to the surface, and that the banks of copepods are sometimes more than a mile thick.   When we reflect that thousands would find ample room and food in a pint of water, one can form some faint conception of their universal abundance.

The organisms which are visible in the water of the ocean and on the sea bottom are almost universally engaged in devouring each other, and many of them, like the blue-fish, are never satisfied with slaughter, but kill for mere sport.

Insatiable rapacity must end in extermination unless there is some unfailing supply, and as we find no visible supply in the water of the ocean we must seek it with a microscope, which shows us a wonderful fauna made up of innumerable larvæ and embryos and small animals, but these things cannot be the food-supply of the ocean, for no carnivorous animal could subsist very long by devouring its own children.   The total amount of these animals is inconsiderable, however, when compared with the abundance of a few forms of protozoa and protophytes, and both observation and deduction teach that the most important element in marine life consists of some half-dozen types of protozoa and unicellular plants ; of globigerina and radiolarians, and of trichodesmium, pyrocystis, protococcus and the coccospheres, rhabdospheres, and diatoms.

Modern microscopic research has shown that these simple plants, and the globigerinæ and radiolarians which feed upon them, are so abundant and prolific, that they meet all demands and supply the food for all the animals of the ocean.   This is the fundamental conception of marine biology.   The basis of all the life in the modern ocean is found in the micro-organisms of the surface.

This is not all.   The simplicity and abundance of the microscopic forms and their importance in the economy of nature show that the organic world has gradually taken shape around them as its centre or starting-point, and has been controlled by them. They are not only the fundamental food-supply but the primeval supply, which has determined the whole course of the evolution of marine life.   The pelagic plant-life of the ocean has retained its primitive simplicity on account of the very favourable character of its environment, and the higher rank of the littoral vegetation and that of the land is the result of hardship.

On land the mineral elements of plant-food are slowly supplied, as the rains dissolve them; limited space brings crowding and competition for this scanty supply ; growth is arrested for a great part of each year by drought or cold ; the diversity of the earth's surface demands diversity of structure and habit; and the great size and complicated structure of terrestrial plants are adaptations to these conditions of hardship.

At the surface of the ocean the abundance and uniform distribution of mineral food in solution ; the area which is available for plants ; the volume of sunlight and the uniformity of the temperature are all favourable to the growth of plants, and as each plant is bathed on all sides by a nutritive fluid, it is advantageous for the new plant-cells which are formed by cell-multiplication to separate from each other as soon as possible, in order to expose the whole of their surface to the water.   Cell-aggregation, the first step towards higher organisation, is therefore disadvantageous to the pelagic plants, and as the environment at the surface of the ocean is so monotonous, there is little opportunity for an aggregation of cells to gain any compensating advantage by seizing upon a more favourable habitat.   The pelagic plants have retained their primitive simplicity, and the most distinctive peculiarity of the microscopic food-supply of the ocean is the very small number of forms which make up the enormous mass of individuals.

All the animals of the ocean are dependent upon this supply of microscopic food, and many of them are adapted for preying upon it directly, but a review of the animal kingdom will show that no highly organised animal has ever been evolved at the surface of the ocean, although all depend upon the food-supply of the surface.

The animals which now find their home in the open waters of the ocean are, almost without exception, descendants of forms which lived upon or near the bottom, or along the sea-shore, or upon the land, and all the exceptions are simple animals of minute size.   A review of the whole animal kingdom would take more space than we can spare, but it would show that the evidence from embryology, from comparative anatomy and from palæontology, all bears in the same direction, and proves that every large and highly organised animal in the open ocean is descended from ancestors whose home was not open water but solid ground, either on the bottom or on the shore.

Embryology also gives us good ground for believing that all these animals are still more remotely descended from minute and simple pelagic ancestors, and that the history of all the highly organised inhabitants of the water has followed a roundabout path from the surface to the bottom, and then back into the water. When this fact is seen in all its bearings, and its full significance is grasped, it is certainly one of the most notable and instructive features of evolution.

The food-supply of marine animals consists of a few species of microscopic organisms which are inexhaustible, and the only source of food for all the inhabitants of the ocean. The supply is primeval as well as inexhaustible, and all the life of the ocean has gradually taken shape in direct dependence upon it. In view of these facts, we cannot but be profoundly impressed by the thought that all the highly organised marine animals are products of the bottom or the shore or the land, and that while the largest animals on earth are pelagic, the few which are primitively pelagic are small and simple.

The reason is obvious. The conditions of life at the surface are so easy that there is little fierce competition, and the inorganic environment is so simple that there is little chance for diversity of habits.

The growth of terrestrial plants is limited by the scarcity of food, but there is no such limit to the growth of pelagic plants or the animals which feed on them, and while the balance of life is no doubt adjusted by competition for food this is never very fierce, even at the present day, when the ocean swarms with highly orga- nised wanderers from the bottom and the shore. Even now the destruction or escape of a microscopic pelagic organism depends upon the accidental proximity or remoteness of an enemy rather than upon defence or protection, and survival is determined by space relations rather than a struggle for existence.

The abundance of food is shown by the ease with which wan- derers from the land, like sea birds, find places for themselves in the ocean, and the rapidity with which they spread over its whole extent.

As a marine animal the insect, Halobates, must be very mod- ern as compared with most pelagic forms, yet it has spread over all tropical and sub-tropical seas, and it may always be found skimming over the surface of mid-ocean as much at home as a Gerris in a pond. I never found it absent in the Gulf Stream when conditions were favourable for collecting.

The easy character of pelagic life is shown by the fact that the larvæ of innumerable animals from the bottom and the shore have retained their pelagic habit, and I shall soon give reasons for believing that the larva of a shore animal is safer at sea than near the shore.

There was little opportunity in the primitive pelagic fauna and flora for an organism to gain superiority by seizing upon an advantageous site or by acquiring peculiar habits, for one place was like another, and peculiar habits could count for little in comparison with accidental space relations. After the fauna of the surface had been enriched by all the marine animals which have become secondarily adapted to pelagic life, competition with those improved forms brought about improvements in those which were strictly pelagic in origin, like the siphonophores, and those wanderers from the bottom introduced another factor into the evolution of pelagic life, for their bodies have been utilised for protection or concealment and in other ways, and we now have fishes which hide in the poison curtain of Physalia, crustacea which live in the pharynx of Salpa or in the mouth of the menhaden, barnacles and sucking fish fastened to whales and turtles, besides a host of external and internal parasites. The primitive ocean furnished no such opportunity, and the conditions of pelagic life must at first have been very simple, and while competition was not entirely absent the possibilities of evolution must have been extremely limited, and the progress of divergent modification very slow, so long as all life was restricted to the waters of the ocean.

There can be no doubt that floating life was abundant for a long period when the bottom was uninhabited. The slow geological changes by which the earth gradually assumed its present character present a boundless field for speculation, but there can be no doubt that the surface of the primeval ocean became fit for living things long before the deeper waters or the sea floor, and during this period the proper conditions for the production of large and complicated organisms did not exist, and even after the total amount of life had become very great it must have consisted of organisms of small size and simple structure.

Marine life is older than terrestrial life, and as all marine life has shaped itself in relation to the pelagic food supply, this itself is the only form of life which is independent, and it must therefore be the oldest. There must have been a long period in primeval times when there was a pelagic fauna and flora rich beyond limit in individuals, but made up of only a few simple types. During this time the pelagic ancestors of all the great groups of animals were slowly evolved, as well as other forms which have left no

descendants.   So long as life was restricted to the surface no great
or rapid advancement, through the influences which now modify
species, was possible, and we know of no other influences which
might have replaced them.   We are, therefore, forced to believe
that the differentiation and improvement of the primitive flora and
fauna was slow, and that, for a vast period of time, life consisted
of an innumerable multitude of minute and simple pelagic organisms.   During the time which it took to form the thick beds of
older sedimentary rocks, the physical conditions of the ocean
gradually took their present form, and, during a part at least of
this period, the total amount of life in the ocean may have been
very nearly as great as it is now, without leaving any permanent
record of its existence, for no rapid advance took place until the
advantages of life on the bottom were discovered.

We must not think of the populating of the bottom as a physical problem, but as discovery and colonisation, very much like
the colonisation of islands.   Physical conditions for a long time
made it impossible, but its initiation was the result of biological
influences, and there is no reason why its starting-point should
necessarily be the point where the physical obstacles first disappeared.   It is useless to speculate upon the nature of the physical
obstacles ;   there is reason to think one of them, probably an important one, was the deficiency of oxygen in deep water.

Whatever their character may have been they were all, no doubt,
of such a nature that they first disappeared in the shallow water
around the coast, but it is not probable that bottom life was first
established in shallow water, or before the physical conditions had
become favourable at considerable depths.

The sediment near the shore is destructive to most surface
animals, and recent explorations have shown that a stratum of
water of very great thickness is necessary for the complete development of the floating microscopic fauna and flora, and it is a
mistake to picture them as confined to a thin surface stratum.
Pelagic plants probably flourish as far down as light penetrates,
and pelagic animals are abundant at very great depths.   As the
earliest bottom animals must have depended directly upon the
floating organisms for food, it is not probable that they first
established themselves in shallow water, where the food supply

is both scanty and mixed with sediment; nor is it probable that their establishment was delayed until the great depths had become favourable to life.

The belts around elevated areas, far enough from shore to be free from sediment and deep enough to permit the pelagic fauna to reach its full development above them, are the most favourable spots, and palæontological evidence shows that they were seized upon very early in the history of life on the bottom.

It is probable that colony after colony was established on the bottom, and afterwards swept away by geological change like a cloud before the wind, and that the bottom fauna which we know was not the first.   Colonies which started in shallow water were exposed to accidents from which those in great depths were free; and in view of our knowledge of the permanency of the sea-floor and of the broad outlines of the continents, it is now impossible that the first fauna which became established in the deep zone around the continents may have persisted and given rise to modern animals.   However this may be, we must regard this deep zone as the birthplace of the fauna which has survived; as the ancestral home of all the improved metazoa.

The effect of life upon the bottom is more interesting than the place where it began, and we are now to consider its influence upon animals, all whose ancestors and competitors and enemies had previously been pelagic.   The cold, dark, silent, quiet depths of the sea are monotonous compared with the land, but they introduced many new factors into the course of organic evolution.

It is doubtful whether the animals which first settled on the bottom secured any more food than floating ones, but they undoubtedly obtained it with less effort, and were able to devote their superfluous energy to growth and to multiplication, and thus to become larger and to increase in numbers faster than pelagic animals.   Their sedentary life must have been favourable to both sexual and asexual multiplication, and the tendency to increase by budding must have been quickly rendered more active, and one of the first results of life on the bottom must have been to promote the tendency to form connected cormi, and to retain the connection between the parent and the bud until the latter was able to obtain its own food and to care for itself.   The animals which first

acquired the habit of resting on the bottom soon began to multiply faster than their swimming allies; and their asexually produced
progeny, remaining for a longer time attached to and nourished by
the parent stock, were much more favourably placed for rapid
growth. As the animals of the bottom live on a surface, or at
least a thin stratum, while swimming animals are distributed
through solid space, the rapid multiplication of bottom animals
must soon have led to crowding and to competition, and it quickly
became harder and harder for new forms from the open water to
force themselves in among the old ones, and colonisation soon
came to an end.

The great antiquity of all the types of structure which are
represented among modern animals is therefore what we should
expect, for after the foundation of the fauna of the bottom was
laid it became, and has ever since remained, difficult for new forms
to establish themselves.

Most of our knowledge of the sea bottom is from three
sources: from dredgings and other explorations; from rocks which
were formed beyond the immediate influence of continents; and
from the patches of the bottom fauna which have gradually been
brought near its surface by the growth of coral reefs; and from all
these sources we have testimony to the density of the crowd of
animals on favourable spots. Deep-sea explorations give only the
most scanty basis for a picture of the sea bottom, but they show
that animal life may thrive with the dense luxuriance of tropical
vegetation, and Sir Wyville Thomson says he once brought up at
one time on a tangle which was fastened to a dredge over twenty
thousand specimens of a single species of sea urchin. The number
of remains of palæozoic crinoids and brachiopods and trilobites
which are crowded into a single slab of fine-grained limestone is
most astounding, and it testifies most vividly and forcibly to the
wealth of life on the old sea-floor.

No description can convey any adequate conception of the
boundless luxuriance of a coral island, but nothing else gives such
a vivid picture of the capacity of the sea-floor for supporting life.
Marine plants are not abundant on coral islands, and the animals
depend either directly or indirectly upon the pelagic food-supply,
so that their life is the same in this respect as that of animals in
the deep sea far from land.

The abundant life is not restricted to the growing edge of the reef, and the inner lagoons are often like crowded aquaria. At Nassau my party of eight persons found so much to study on a little reef in a lagoon close to our laboratory that we discovered novelties every day for four months, and our explorations seldom carried us beyond this little tract of bottom. Every inch of the bottom was carpeted with living animals, while others were darting about among the corals and gorgonias in all directions. But this was not all, for the solid rock was honeycombed everywhere by tubes and burrows, and when broken to pieces with a hammer each mass of coral gave us specimens of nearly every great group in the animal kingdom. Fishes, crustacea, annelids, mollusca, echinoderms, hydroids, and sponges could be picked out of the fragments, and the abundance of life inside the solid rock was most wonderful. The absence of pelagic life in the landlocked water of coral islands is as impressive and noteworthy as the luxuriance of life upon and near the bottom.

On my first visit to the Bahama Islands I was sadly disappointed by the absence of pelagic animals, where all the conditions seemed to be peculiarly favourable The deep ocean is so near that, as one cruises near the inner sounds past the openings between the islets which form the outer barrier, the deep blue water of mid-ocean is seen to meet the white sand of the beach, and soundings show that the outer edge is a precipice as high as the side of Chimborazo and much steeper.

Nowhere else in the world is the pure water of the deep sea found nearer land or more free from sediment, and on the days when the weather was favourable for outside collecting we found siphonophores and pteropods, pelagic molluscs and crustacea and tunicates and all sorts of pelagic larvæ in great abundance in the open water just outside the inlets.

Inside the barrier the water was always calm, and day after day it was as smooth as the surface of an inland lake. When I first entered one of these beautiful sounds, where the calm, transparent water stretches as far as the eye can reach, while new beauties of islets and winding channels open before one as those which are passed fade away on the horizon, I felt sure that I had at last found a place where the pelagic fauna of mid-ocean could be gathered at our door and studied on shore. The water proved to be not only as pure as air, but almost as empty. At high water we sometimes captured a few pelagic animals near the inlets, but we dragged our surface nets through the sounds day after day only to find them as clean as if they had been hung in the wind to dry. The water in which we washed them usually remained as pure and empty as if it had been filtered, and we often returned from our towing expeditions without even a copepod or a zoea or a pluteus.

The absence of the floating larvæ is most remarkable, for the sounds swarm with bottom animals which give birth every day to millions of swimming larvæ. The mangrove swamps and the rocky shores are fairly alive with crabs carrying eggs at all stages of development, and the boat passes over great black patches of sea-urchins crowded together by thousands. The number of animals engaged in laying their eggs or hatching their young is infinite, yet we rarely captured any larvæ in the tow net, and most of these we did find were well advanced and nearly through their larval life.

It is often said that the water of coral sounds is too full of lime to be inhabited by the animals of the open ocean ; but this is a mistake, for the water is perfectly fit for supporting the most delicate and sensitive animals, and those which we caught outside lived in the house in water from the sounds better than in any other place where I ever tried to keep them, and instead of being injurious the pure water of coral sounds is peculiarly favourable for use in aquaria for surface animals.

The scarcity of floating organisms can have only one explanation. They are eaten up, and competition for food is so fierce that nearly every organism which is swept in by the tide and nearly every larva which is born in the sounds is snatched by the tentacles around some hungry mouth.

Nothing could illustrate the fierceness of the struggle for food among the animals on a crowded sea-bottom more vividly than the emptiness of the water in coral sounds where the bottom is practically one enormous mouth. The only larvæ which have much chance to establish themselves for life are those which are so fortunate as to be swept out into the open ocean, where they can complete their larval life under the milder competition of the pelagic fauna, and while it is usually stated that the larvæ of bottom animals have retained the pelagic habit for the purpose of distributing the species, it is more probable that it has been retained on account of its comparative safety.

These facts show that competition must have come quickly after the establishment of the first fauna on the bottom, and that it soon became very rigorous and led to severe selection and rapid modification ; and we must also remember that life on the bottom brought with it many new opportunities for divergent specialisation and improvement. The increase in size which came with the economy of energy increased the possibilities of variation, and led to the natural selection of peculiarities which improved the efficacy of the various parts of the body in their functions of relation to each other, and this has been an important factor in the evolution of complicated organisms.

The new mode of life also permitted the acquisition of protective shells, hard-supporting skeletons, and other imperishable parts, and it is therefore probable that the history of evolution in later times gives no index as to the period which was required to evolve from small, simple pelagic ancestors the oldest animals which were likely to be preserved as fossils.

Life on the bottom also introduced another important evolutionary influence: competition between blood-relations. In those animals which we know most intimately, divergent modification, with the extinction of connecting forms, results from the fact that the fiercest competitors of each animal are its closest allies, which, having the same habits, living upon the same food, and avoiding enemies in the same way, are constantly striving to hold exclusive possession of all that is essential to their welfare.

When a stock gives rise to two divergent branches, each escapes competition with the other so far as they differ in structure or habits, while the parent stock, competing with both at a disadvantage, is exterminated.

Among the animals which we know best, evolution leads to a branching tree-like genealogy, with the topmost twigs represented by living animals, while the rest of the tree is buried in the dead past. The connecting form between two species must therefore be sought in the records of the past or reconstructed by comparison.

Even at the present day, things are somewhat different in the open ocean, and they must have been very different in the primitive ocean, for a pelagic animal has no fixed home, one locality is like another, and the competitors and enemies of each individual are determined in great part by accidents.

We accordingly find, even now, that the evolution of pelagic animals is often linear instead of divergent, and ancient forms, such as the sharks, often live on side by side with the later and more evolved forms. The radiolarians and medusæ and siphonophores furnish many well-known illustrations of this feature of pelagic life.

No naturalist is surprised to find in the South Pacific or in the Indian Ocean a salpa or a pelagic crustacean or a surface fish or a whale which was previously known only from the North Atlantic, and the list of species of marine animals which are found in all seas is a very long one. The fact that pelagic animals are so independent of those laws of geographical distribution which limit land animals, is additional evidence of the easy character of the conditions of pelagic life.

One of the first results of life on the bottom was to increase asexual multiplication and to lengthen the time during which buds remain united to and nourished by their parents, and to crowd individuals of the same species together, and to cause competition between relations. We have in this and other obvious peculiarities of life on the bottom a sufficient explanation of the fact that since the first establishment of the bottom fauna, evolution has resulted in the elaboration and divergent specialisation of the types of structure, which were already established, rather than the production of new types.

Another result of the struggle for existence on the bottom was the escape of varieties from competition with their allies by flight from the crowded spots, and a return to the open water above ; just as in later times the whales and sea-birds have gone back from the land to the ocean. These emigrants, like the civilised men who invade the homes of peaceful islanders, brought with them the improvements which had come from fierce competition, and they have carried everything before them and produced a great change in the pelagic fauna.

The rapid intellectual development which has taken place among the mammals since the middle tertiary, and the rapid structural changes which took place in animals and plants when the land fauna and flora were established, are well known ; but the fact that the discovery of the bottom initiated a much earlier, and probably more important, era of rapid development in the forms of animal life has never been pointed out.

If this view is correct the primitive fauna of the bottom must have had the following characteristics :

1.—It was entirely animal, without plants, and it, at first, depended directly upon the pelagic food supply.

2.—It was established around elevated areas, in water deep enough to be beyond the influence of the shore.

3.—The great groups of animals were rapidly established from pelagic ancestors.

4.—The animals of the bottom rapidly increased in size and hard parts were quickly acquired.

5.—The bottom fauna soon produced progressive development among pelagic animals.

6.—After the establishment of the fauna of the bottom, elaboration and differentiation among the representatives of each primitive type soon set in and led to the extinction of connecting forms.

Many of the oldest fossils, like the pteropods, are the modified descendants of ancestors with hard parts, and there is no reason to suppose that the first animals which were capable of preservation as fossils have been discovered, but it is interesting to note that the oldest known fauna is an unmistakable approximation to the primitive fauna of the bottom.

The lower Cambrian fossils are distributed through strata more than two miles thick, some, at least, of them showing by their fine grain and by the perfect preservation of tracks and burrows which were made in soft mud, and of soft animals like jelly-fish, that they were deposited in water of considerable depth. The sediment was laid down slowly and gently in water so deep as to be free from disturbance, and under conditions so favourable that it contains the remains of delicate animals not often found as fossils.

While the fauna of the lower Cambrian undoubtedly lived in water of very considerable depth, it was not oceanic, but continental, for we are told by Walcott that "one of the most important conclusions is that the fauna of the lower Cambrian lived on the eastern and western shores of a continent that in its general configuration outlines the American continent of to-day." "Strictly speaking, the fauna did not live upon the outer shore facing the ocean, but on the shores of interior seas, straits, or lagoons that occupied the intervals between the several ridges that ran from the central platform east and west of the main continental land surface of the time."

This fauna was rich and varied, but it was not self-supporting, for no fossil plants are found, and the primary food supply was pelagic. Animals adapted for a rapacious life, such as the pteropods, were abundant, and prove the existence of a rich supply of pelagic animals. All the forms known from the fossils are either carnivorous—like the medusæ, corals, crustacea, and trilobites—or they are adapted—like the sponges, brachiopods, and lamellibranchs—for straining minute organisms out of the water, or for gathering those which rain down from above, and the conditions under which they lived were very similar to those on the bottom at the present day.

Walcott's studies show that the earliest known fauna had the following characteristics :—It consisted, so far as the record shows, of animals alone, and these were dependent upon the pelagic food supply for support. While small in comparison with many modern animals, they were gigantic compared with primitive pelagic animals. The species were few, but they represent a very wide range of types. All these types have modern representatives, and most of the modern types are represented in the lower Cambrian. Their home was not the bottom of the deep ocean, but the shores of a continent under water of considerable depth.

The Cambrian fauna is usually regarded as a half-way station in a series of animal forms which stretches backwards into the past for an immeasurable period, and it is even stated that the history of life before the Cambrian is longer by many fold than its history since.

So far as this opinion rests on the diversity of types in Cambrian times, it has no good basis ; for if the views here advocated are correct, the evolution of the ancestral stems took place at the surface, and all the conditions necessary for the rapid production of types were present when the bottom fauna first became established.

As we pass backwards towards the lower Cambrian, we find closer and closer agreement with the zoölogical conception of the character of primitive life on the bottom. While we cannot regard the oldest fauna which has been discovered as the first which existed on the bottom, we may feel confident that the first fauna of the bottom resembled that of the lower Cambrian in its physical conditions and in its most distinctive peculiarities; the abundance of types, and the slight amount of differentiation among the representatives of these types; and we must regard it as a decided and unmistakable approximation to the beginning of the modern fauna of the earth, as distinguished from the more ancient and simple fauna of the open ocean.

Floscularia Ibooðii.

BY JOHN HOOD, F.R.M.S., Dundee.   Pl. XIV.

THIS large and strange Rotatorian was found by me for the first time in December, 1882, attached to *Sphagnum*, in a ditch on Tents Muir, Fifeshire. The large size of the rotifer, and its possessing a corona unlike any other species of the genus *Floscularia*, together with its great cowl-shaped, dorsal lobe, would make it sufficiently remarkable; but in addition to the two extraordinary finger-shaped processes, perched one on each side of the summit of the dorsal lobe (Pl. XIV., Figs. 1 and 2, *d. p.*), there are two very significant organs possessed by no other species of the genus. It has been, indeed, suggested to me that this species should be removed from the genus *Floscularia* and placed in a new genus by itself. But with the exception of the finger-like processes, the other organs are identical with those of a true Floscule, so I therefore think it should be retained in that genus.

Dr. Hudson described this Rotatorian from living specimens, which I sent him from Dundee on the 25th December, 1882. His description, with two figures, were published in the *Journal of the Royal Microscopical Society*, II. Series, Vol. III., pp. 161—171,

Pl. III. In his description he says :—"I cannot yet hazard a suggestion as to the function that these finger-like processes perform."

As I have since had frequent opportunities of observing this creature in various stages of its development, I desire in the present paper to supplement Dr. Hudson's description. The finger-like processes are hollow tubes open at the points. The upper end of each of these processes is slightly constricted ; the tubes communicate with two sub-spherical spaces lying between the two surfaces of the dorsal lobe. Fine muscular threads pass down and across the tubes ; the threads run down from their base on the dorsal side to the neck below the vestibule. The animal can contract each of these tubes independently of the other, and can put them in different positions ; there is not a trace of setæ at their apex to suggest their function as antennæ or feelers. The lateral antennæ are situated in the same region to those of the other species of Floscularia. Their function is doubtless that of excretory organs.

I have frequently observed that both the adult and young individuals discharge a granular mucous matter through these finger-like processes, which gathers round their free extremities, as shown in Pl. XIV., Fig. 2, *g*, *m*, and is not got rid of until the animal retires into its tube, where it is rubbed off to increase the volume of the creature's fluffy domicile ; when the animal again emerges from its tube, the free ends of the finger-like processes are quite freed from the accumulation.

The cuticle of the trunk and coronal cup of *F. Hoodii* is composed of two layers, an outer and an inner membrane, tough and elastic, and between the two layers is a fluid more or less granular. The granular fluid may be seen in some degree in every species, but more conspicuously in *F. ambigua*, *F. algicola*, and *F. annulata*. In these the fluid is rendered semi-opaque by the large number of granules floating therein ; but in the case of *F. Hoodii* the granules in the fluid are not so numerous, and consequently the creature is more transparent than the other species. It is doubtless by means of this fluid that the lobes of the furled corona are pushed forward and expanded, especially as it has inter-communicating cavities and channels containing fluid, which is driven upwards and downwards by the contractions of the muscles and by the various motions of the body ; the transverse muscles of the trunk force the fluid into definite channels, rendering the lobes tight and stiff, like the ribs of an umbrella.

Doubtless the function of the finger-shaped processes on the back of the dorsal lobe is to excrete a portion of the inter-membranal fluid, which is then utilised in the construction of the fluffy, gelatinous tube; and although no other species is known to possess similar appendages, it cannot be said that the excretion of this fluid is confined exclusively to *F. Hoodii*, for there will be observed on the back of the dorsal lobes of *F. trilobata* and *F. ambigua* a single knob, and on the back and near the top of the dorsal lobe of *F. cucullata* will be seen two slight prominences, which are probably used for a similar purpose. The granules in the mucous matter discharged have a very active swarming motion while they cling round the free ends of the finger-like processes. This swarming motion is not observed in the mucous fluid within the two membranes of the skin in either of the lobes or the trunk. It is only when the mucous matter is set free that the swarming, or Brownian, motion of the granules is observable.

Sometimes when the animal is fully protruded from its fluffy tube, only one of the processes is seen to be fully distended; the other will be observed lying flat on the dorsal lobe; then, in a short time, it also will be slowly raised and distended.

The corona is large and globular, and when fully expanded shows a large, mouth-like funnel of three lobes, two small ones being on the ventral, and one, much the largest, on the dorsal side. The rims of the three lobes are clothed with a double fringe of setæ. The setæ of the outer row are long and are directed outwards. Those of the inner row are much shorter and point inwards. The gap of the mouth-funnel alters frequently. The food of these creatures consists of living animalcules of various dimensions, one or more of which may be seen swimming at the same time within its large circular cavity. To prevent the escape of its victims, when the animalcules are of large size and swift swimmers, the floscule closes the gap by means of the many muscular threads on the corona, so as to reduce the aperture in various degrees, at times even reducing it to a mere slit.

In consequence of the transparency of the corona, the semi-circular wreath of the vibratile cilia at the bottom of the vestibule is easily seen. With the exception of *F. trilobata*, very few, if any, other species afford the same facility.

In a side or lateral view of the floscule, on the dorsal surface, will be seen a pair of transparent ridges, which run up from the back of the trunk to the back of the dorsal lobe. These are suggestive of buttresses between the trunk and coronal head, shown in Pl. XIV., Fig. 1, *r*. Between these ridges there is a deep hollow, bounded above by the dorsum and below by the rounded surface of the body. At the lowest portion of this hollow, which forms the neck, close to the buttress ridges, are two small red eyes, which are present both in the young and in the adult individuals, and are only seen, both at the same time, in a dorsal aspect. One eye only is seen in a lateral view on either side, but cannot be seen at all in the ventral view (see Fig. 1, *e*).

The true rotatory organ of *F. Hoodii*, which consists, as in other species, of a ciliated rim at the base of the mouth-funnel, towards the ventral side, is continued in two curved lines down the vestibule to the lips. This organ is seen with greater facility than in any other species, owing to the exceptional transparency of the creature's large head.

The contractile vesicle is ample, and frequently contains yellowish globules, which appear dark brown with transmitted light. It deposits one to four large eggs in the fluffy tube close to the foot. This species is the most interesting of the genus *Floscularia* and the most hardy. It can be kept in a trough under the microscope for a number of weeks, and the observer will not tire of looking at it. It must be supplied with a change of water daily, which should be taken from an aquarium where there is present an ample supply of food in the form of living Infusorians. The observer never fails to be entertained by watching the creature's manner in feeding. Sometimes three of four small animalculæ will enter the open mouth-funnel at one time. These the floscule will swiftly swallow one after the other in rapid succession. Then, again, a large-sized Infusorian will dash into the cavity, where it will be seen to swim rapidly round and round in the coronal cup, like a minnow newly introduced into a fish-globe. Then, after a

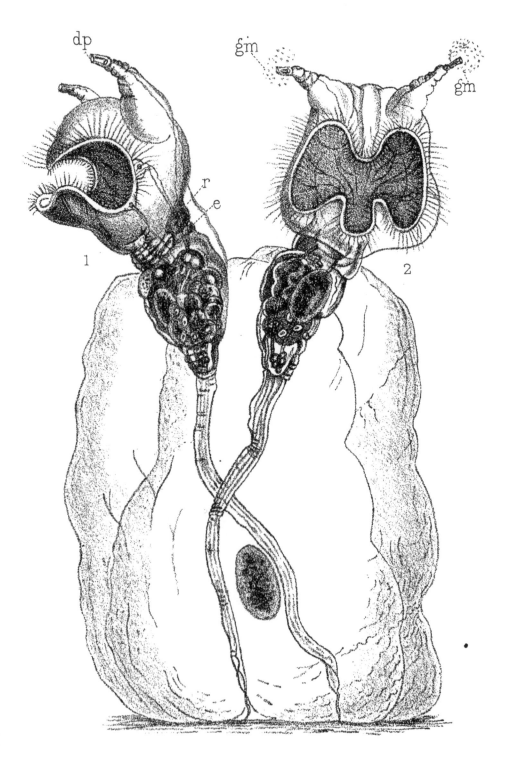

time, the floscule will be observed to be slowly closing the aperture to prevent the escape of the vigorous animalcule, which has made the fatal mistake of entering the alluring flower-like cup.

F. Hoodii inhabits a large, clear, transparent tube, and finds a suitable environment either in a marsh pool or lake. The first examples found were attached to *Sphagnum* collected from a marsh ditch on Tents Muir, Fifeshire ; then, again, on *Myriophyllum*, dredged from the centre of Loch Lundie, Forfarshire ; and on *Ranunculus* from Blacklock, Blairgowrie, Perthshire, and recently on *Ranunculus*, dredged from a lake in co. Mayo, Ireland.

It is to be regretted that this handsome rotiferon is so rare, yet it is fairly well distributed, for I have found it in three counties in Scotland and also in the West of Ireland. Examples have also been found in Würtemberg, Germany, by Herr L. Bilfinger, who found a number on *Myriophyllum* on the 28th of September, 1890, an account of which is published in his List of *Rotatorienfauna Würtembergs*, 1892, and is the only other published record of this animal that I am aware of since the *Monograph on Rotifera*, by Dr. Hudson and P. H. Gosse, published in 1886.

LITERATURE RELATING TO *Floscularia Hoodii :*

C. T. Hudson, LL.D., F.R.S., F.R.M.S.—" On Five New Floscules ; with a Note on Prof. Leidy's genera, *Acyclus* and *Dictyophora*," *Journal of the Royal Microscopical Society*, 1883, pp. 161—171, Pls. III. and IV.

Dr. Hudson and P. H. Gosse, " The Rotifera or Wheel Animalcules," 1886, p. 55.

Von L. Bilfinger, "*Zur Rotatorienfauna Würtembergs*," 1892, p. 110, *Jahreshefte des Vereins für Vaterl. Natur. in Würtemberg.*

EXPLANATION OF PLATE XIV.

Fig. 1.—Lateral view. *d.p.*, Dorsal processes. *e.*, The position of the eyes in the neck. *r.*, Transparent dorsal ridge.   •

,, 2.—Ventral view. *g.m.*, Granular mucous matter excreted from the free ends of the dorsal processes.

Figures by **Dr. Hudson.**

Oliver Wendell Holmes & the Microscope.

IN a recent number of the *New York Medical Journal* (LXI., 1895, pp. 236—9), Dr. Palmer C. Cole relates some of his microscopical reminiscences, one of which gives a view of the Autocrat of the Breakfast Table as a Microscopist. Dr. Cole says: " It was my good fortune to be a student of the Harvard Medical School when Oliver Wendell Holmes was professor of anatomy and physiology. Professor Holmes was an enthusiast in the use of the microscope, and possessed some of the finest of the Spenser lenses, then the best in the world. He had constructed a stand, principally of wood, stable enough to allow the use of high-power objectives, and so simple that any student with the slightest mechanical ability could make one for himself. My first stand I made on this model at a cost of about fifty cents for material.

The coarse adjustment was obtained by means of a pin in the tube carrying objective and eyepiece, which was pushed against a wedge-shaped piece of brass. The fine adjustment was obtained by revolving the pin against the wedge. Over the tube was slipped a cardboard disc, some six to eight inches in diameter, covered with black velvet, allowing both eyes to be kept open during observation. The tube was also lined with black velvet. The tube-support revolved on a pivot, allowing inclination, though we were taught to use it with direct light. In a recent letter to me, Oliver Wendell Holmes refers to this stand as 'a rough wooden contrivance that answered its purpose.' Later, Holmes's lecture-room microscope, for use in his own class, became very popular, and is still extensively used.

Notwithstanding his engagements with others in founding the *Atlantic Monthly*, his medical lectures five days a week, an occasional public lecture, and monthly instalments in the *Atlantic* of that brilliant series of essays, *The Autocrat of the Breakfast Table*, he found time at his own house in Montgomery Place, to give weekly instruction to half-a-dozen chosen students in the use of the microscope. In 1856 and 1857, I was fortunate enough to be one of the chosen half-dozen. We were taught the use of the instrument and the preparation of slides. The slides used from week to week were prepared by Holmes himself, to show the different tissues of the body. The only consideration ever given for these charming and instructive lessons were the grateful thanks of his pupils.

I believe that Oliver Wendell Holmes was the first in America to teach classes of medical students the use of the microscope. If so, to him, and through him to the Harvard Medical School, should be accorded the honour.

Since writing the above, Oliver Wendell Holmes has passed away. None knew him but to love him. His anatomical and physiological lectures were charming, and always filled the amphitheatre; even dry bones in his hands became endowed with life and individuality.

It was my rare good fortune to be occasionally invited to his house, where I spent some delightful hours in his study and workroom. Once I was present when a large package of books arrived from New York for his inspection, which I helped him unpack. When at the bottom of the case we came upon a large folio copy, original edition, of *Vesalius*, his eyes fairly sparkled with joy. Hastily he turned over the leaves, and remarked that, as New Yorkers have been fools enough to allow such a prize to leave their city, it would never return from Boston.

All the world knew him as professor, author, and poet, but few knew his fondness for music, and that he possessed a rare mechanical genius of which he was very proud. He once told me he thought more of the 'gimcracks' he made with his hands than of anything he had ever written; and at that time he was delighting two hemispheres with *The Autocrat of the Breakfast Table.*

In his last autograph letter to me he says: 'My most successful contrivance was a *stereoscope* of a very simple pattern, which had a great run, and has remained popular, I think, to the present time.'"

British Hydrachnidæ.

PART II.　By CHARLES D. SOAR.　Plate XV.

THE next genus we have to notice is the well-known *Arrenu-rus*. I say well known, because this genus was described *in extenso* in a series of articles which appeared in *Science Gossip* some years ago, written by Mr. C. F. George, of Kirton Lindsey. The genus *Arrenurus* contains mites of the most beautiful shapes and colours, the colours being especially brilliant, some of them appearing quite metallic, and look most gorgeous when viewed under reflected light. In shape the males are as different from the females as it is possible to imagine. This genus, when thoroughly worked out, will, I have no doubt, be found to be the most prolific in species of all the British Hydrachnidæ. A great many species are very common. No collector or lover of pond-life but has at some time or other taken some of these interesting and prettily coloured mites home in his bottle, perhaps admired them, and put them on one side and thought no more about them; very few naturalists have thought it worth while to write about them.

If the reader who is interested in this genus will turn to *Science Gossip* for the years 1881 to 1884 (Vols. XVII.—XX.), he will find several papers devoted to this family by Mr. C. F. George, who described and figured about sixteen of the species found in the British Isles. My intention, however, in this paper is to speak of the genus and give its characteristics, so that anyone will know an *Arrenurus* when they find one. The number of known British species I hope to give later on.

Genus ARRENURUS (Dugès).

(A. Dugès, Recherches sur l'ordre des Acariens, etc., *Ann. des Sciences Nat.*, Tom. I., p. 17, 1834.)

Body chitinous, with a maculated appearance in both sexes. On the dorsal side is a depressed line, which is oval in shape in the female and of a horse-shoe shape in the male. The legs are strong and well supplied with swimming bristles. All the tarsi have claws. Mandibles in two distinct portions. Eyes widely separated and near margin of body. On each side of the operculum are numerous copulative pores set in a special plate. Palpi short, the fourth joint longer and stronger, the fifth joint forming

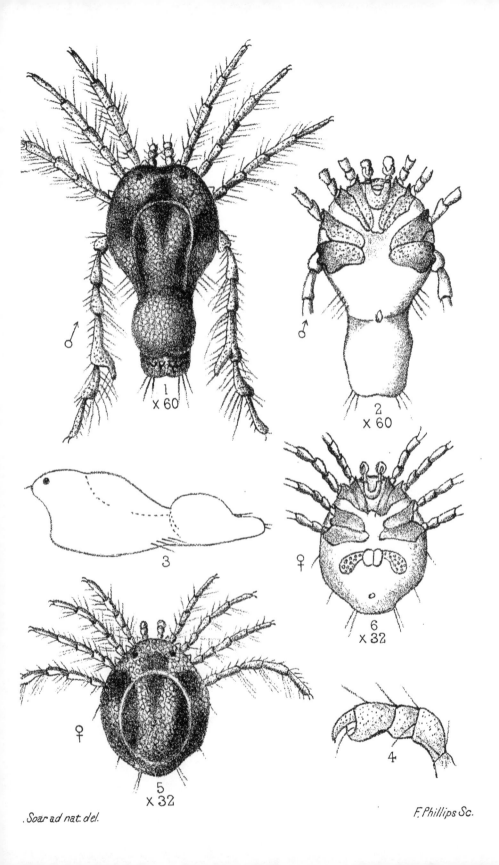

.Soar ad nat. del.

1
X 60

♂

2
X 60

♂

3

6
X 32

♀

5
X 32

♀

4

F. Phillips Sc.

a thumb and finger.   As an illustration of this genus, we give the following description of *A. globator :*—

Arrenurus globator (Müller).

1776.—*Hydrachna globator.*   Müller, Zool. Dan. Prodr., p. 188, No. 2242.
1781.—*Hydrachna globator.*   Ibid., Hydrachnæ, p. 37, Tab. 2, Figs. 1—5.
1793.—*Trombidium variater.*   J. C. Fabricius, Ent. Syst., Tom. II., p. 403, No. 22.
1805.—*Atax variater.*   Ibid., Syst. Antliatorum, p. 369.
1835.—*Arrenurus globator.*   C. L. Koch, Deutschlands Crust., etc., p. 13, Figs. 22, 23.
1854.—*Arrenurus globator.*   Bruzelius Beskr. ö. Hydrachn. som. Forek. i. Skäne, p. 31, Tab. III., Fig. 3.
1879.—Neuman Sveriges Hydrachnides, p. 88, Tab. X., Fig. 2.
1882.—C. F. George, *Science Gossip*, XVIII., p. 272, Figs 194—202.

Male.—This beautiful mite cannot very well be mistaken for any other form.   It has a globular process on the dorsal side of the tail, from which it takes its name.   Colour, a light delicate green, with brown markings.   Eyes, deep crimson.   On the fourth joint of the fourth leg is a peculiar development, often, but not always, seen in the males of the species of *Arrenurus.*   This is a very small mite, its length being only 3/100th of an inch.

Female.—Almost oval in shape, with indentations on each side.   Same colour as the male—pale delicate green, with brown markings.   Length, 1/25th of an inch, being a little larger than the male.

This mite is by no means uncommon; I have found it several times.   The two specimens from which the drawings (Pl. XV.) were made were found at Snaresbrook on one of the Quekett excursions Sept. 29th, 1894.   Having a chitinous epidermis, these mites can easily be mounted in Canada balsam, and if the legs are regularly arranged they make very pretty mounts, the delicate colouring being brought out by the balsam even more brilliantly than in life.   Koch describes about forty species, but some have since been removed to other genera.

EXPLANATION OF PLATE XV.

Fig. 1.—Dorsal view of Male. Fig. 4.—Palpi of Male.
  ,, 2.—Ventral view of do. ,, 5.—Dorsal view of Female.
  ,, 3.—Side view of do. ,, 6.—Ventral view of do.

Bacteria of the Sputa and Cryptogamic Flora of the Mouth.

BY FILANDRO VICENTINI, M.D., Chieti, Italy.

SECOND MEMOIR.

Translated by Professor E. Saieghi.

Recent Bacteriological Researches on the Sputa; the Morphology and Biology of the Microbes of the Mouth.

FURTHER REMARKS ON THE BACTERIA AND BACILLI FOUND IN THE SPUTA.

FRUCTIFICATION BY SPORES (EARS) REPRODUCED IN THE SPUTA.

I WILL now deal with the reproduction of the fructification by spores (spicæ) in the sputa. Hitherto we have been unable to obtain cultures of *Leptothrix*, and microbes of the mouth in general, upon the usual artificial soils; and on this point the authors, from Cornil and Babes to the more recent, are agreed.* It will not be inopportune to point out a case of fructification by ears, reproduced in a specimen of sputum, which induced me to make the present communication.

The presence of filaments, isolated or intertwined, of *Leptothrix*, in the sputa, is well known in clinical microscopy. Leyden and Jaffé recognised even *Leptothrix pulmonaris;* and Hunter Mackenzie exhibits a conspicuous specimen taken from tubercular sputum during a time of improvement.† In the sediment of masses of sputa, emulsified with potash, we find not only little rods and filaments, but even numerous tufts of this parasite. I remember also having often found in sputa of every kind freshly stained fragments which I had not known before, and which were portions of ears spoiled in the preparation.

The case which gave the conspicuous specimen, delineated in Fig. 16, was the following :—

* Cornil and Babes, *op. cit.*, p. 135.

† Leyden and Jaffé, *Deutsches Archiv für Med.*, II., p. 488 ; Hunter Mackenzie, *Le Crachat*, etc., trans., Paris, 1888, Fig. 22.

Mr. G. F., 41 years of age, having already been twice attacked by obstinate miasmatic affections, and once by an acute affection of the air-passages (the nature of which he does not remember), fell ill last March with very high fever (from 40·6° to 41·2° in the evening hours, falling about a degree in the morning); with serious dyspnæa (from 45 to 50 breathings a minute), and very rapid pulse (135—145). On the chest it was only noticed that the right apex was less sonorous, giving an indistinct respiratory murmur; little coughing, with scanty and infrequent expectoration; tumour of the spleen by percussion, but probably chronic. The sensorium sound; a little insignificant epistaxis; remarkable bilious diarrhœa. Albumen, one and a-half grammes to every pint of urine, with a few hyaline and granular urinary casts, and increase of ordinary pigments.

In the first days there was a doubtful expectoration, which was thought to proceed from the nose. An eruption of *herpes labialis* appeared (which I remember having deceived me in a case of ileo-typhus). At any rate, the diagnosis of ileo-typhus was given, and three baths were prescribed which gave great relief. On the seventh day the fever suddenly subsided, as generally happens in pneumonia. Dyspnœa remained, only diminished by the quota due to fever; and in thirty-six hours I obtained a small quantity of rusty sputa, a portion of which, nearly free from saliva, I examined, to ascertain whether it proceeded from the air-passages.

The diagnosis was made even clearer when, later on, crackling wheezings and murmurs of friction in unison with the beatings of the heart (pleuro-pericardiac friction of Wintrich) became manifest; these continued up to the sixteenth day of the illness. I have met with this phenomenon on other occasions, in pleurisy or in pleuritic granuloma, but on the left side; here, on the contrary, it was heard on the right upon the line of the sternum, and below towards the mammillary line. Consequently, the grave dyspnœa had been the means of spreading the process to the pleuritic coating of the pericardium, and perhaps to a limited part of the pericardium itself. At any rate, the examples of a great disproportion between the extension of the local and the magnitude of the general facts are not rare, especially when the apices of the lungs are affected with pneumonia.

The following were the elements found in the sputum : a fair number of red blood corpuscles, both isolated and in small clusters ; here and there a great gathering of ellipsoidal epithelia (so-called alveolar), mostly with a coat of granules of myelin, and a few free particles of the same ; many cylindrical and vibratile epithelia of the air passages ; the usual buccal epithelia ; a few slender and short spires of Curschmann ; myriads of uncapsulated pneumococci, generally in vast cumuli ; small well-defined groups of capsulated pneumococci, after type in Fig. 4, *f*; various other forms of bacteria and bacilli ; filaments, and even tufts, of *Leptothrix* ; groups and large cumuli of spores and sprouts of oidium (*Oidium lactis* ?).

As it is my custom to keep the sputa for whole weeks in order to repeat my investigations, on the seventh day after collecting this sputum I detected in it fragments of fructifications, which I could perfectly well recognise. 'I then thought of removing a particle of the patina of the sputum, which was adhering to the internal wall of the tube, to that side on which I had always inclined it, in collecting the specimens. There the layer of the mucus was very thin and wet, the tube being accurately closed. This mucous patina, having been adhesive and undisturbed for several days, ought to have preserved nearly the same conditions favourable to the fructification of *Leptothrix*, as seen on the patina of the teeth. That layer appeared to the naked eye opaque and greyish, like a very delicate mould.

The result of that research could not have been more satisfactory, as Fig. 16 shows. It extends over two visual fields (magnified 850 diameters), and the colouring of *ears* with gentian violet is very striking, the particle of the sputum having been immersed for two hours in the liquid stain, then washed and mounted in distilled water. It will be observed at a glance that the bundle of ears has been severed in two groups : in the superior one, *A*, all the stalks are preserved; in the inferior, *B*, only five are visible. The fructifications or part of them, as well as the missing stalks, were spread here and there over the surrounding visual fields. The inferior group, *B*, has been pressed sideways by the cover-glass, and lies in profile with its base down ; the superior one, *A*, has been pressed down vertically by the cover-glass, and is consequently opened and formed into two, so that it shows five upper stems with the base downwards, and two lower stems, f and g, with the base upwards.*

In the largest stalk, a, the gemmules of reserve are distinctly observed, and these are still better seen in a', where a part of that stalk is reproduced (magnified to 2,500 diameters) after a long saturation in glycerine (see later on). In b, the *ear* is partially scattered, so that a few spores remain *in situ*, and the stalk is entirely visible. In d and e, cumuli of sporules are seen in those points where the *ears* are partially scattered. Then from the concussion the sporules are driven towards the top, leaving bare the stalk near the base. In c, the ear is broken, and shows a clear section, but without its superior part.

In the *patina dentaria* I never came across *ears* so long and perfect. In fact, the better kept *ears*, bent down, in f and g, are the longest found hitherto; as that in f measures 166μ, or one sixth of a millimetre, which corresponds to one and a half the thickness of a cover-glass, nine to the millimetre. The number of the sporules implanted in these ears (upon six longitudinal lines) may be calculated, at the lowest, to be 720.

We shall have, here, to consider two points. The first is, that a similar specimen cannot be supposed to have been previously formed on the teeth, then fallen in the expectoration with saliva, to be found there second hand, as the fructifications do not rise so high in the patina dentaria, nor could they have been so well kept after the various manipulations. The circumstances under which they were found incline rather to the belief that they had germinated on the mucous film from filaments there deposited, and were being fed from the materials of the sputum itself.

In the present case, we shall not take into consideration whether the presence of red blood corpuscles may have favoured the development of those *ears*. It is certain that they are found more abundant in the patina dentaria of persons suffering from tenderness of the gums, and whose teeth are, in the morning, somewhat besmeared with blood.

The second point is that the specimen in question is sufficient to clear up any doubt whatever about the nature of the above-

* In the Plate, Fig. 16 B is erroneously marked Fig. 14 B.

described *ears*.   Taking only the specimens obtained from the
patina dentaria, some doubts might remain, as in it the *ears* are,
generally, embedded in the granulous lumps and the entanglements of threads, in consequence of the pressure of the lips or
tongue ; and we might suppose that they are the mechanical result
of that trampling or superposition of minute cocci around the
branching filaments, so as to become incrusted with them.   But
(apart from the consideration that the incrustation cannot explain
the perfect similarity of the sporules, nor their constant disposition in six longitudinal lines) the specimen of Fig. 16 dispels any
suspicion whatever, as it is there manifestly the sign of a genuine and proper germination ; nor, on the other hand, can there
be found a particle of resting place on which the stems might
have become incrusted.

In support of this view there is the fact that the *ears* acquire,
by colouring with aniline, such a solidity that they can hardly be
dissevered.   The above-mentioned specimen, which, through the
accidental crushing of the preparation by a wrong turn of the fine
adjustment, was hard-pressed down, exhibits a proof of it.   The
ear drawn in *g* was abruptly removed towards *f*, and made into a
spire, but no spores were dropped by the shock, and the *ear* has
since gradually resumed its former place.   Now this fact could
not be explained on the hypothesis of a simple mechanical
incrustation.

This preparation has been kept in glycerine, which was afterwards substituted for the aqueous medium.   The tint, especially of
the *pneumococci* without halo, is very much faded, owing to the
diffusion caused by glycerine.   The *ears*, after two months of
saturation, exhibited a partial decolourisation and withered spores,
although the stalks maintained a brilliant colour.   The viscid
matter has become paler and more transparent, so that the stalks
can be detected fully in their whole length ; and inside can be
seen not only the gemmules, but traces of small knobs, shown in
a.   By turning the micrometer screw I could even see, although
interruptedly, a few sporules belonging to the posterior series or
those lying on the slide.

In conclusion, having thoroughly investigated this preparation,
it seems to be absolutely impossible to dispute the fructification of

those forms of *Leptothrix* and the fertility of their stems. This specimen is at the disposal of anyone who wishes to examine it.

We shall, by-and by, deal with the fructification obtained from urinary mucus.

With regard to the oidium forms found in this sputum, I will say two words concerning their fructifications. I have dealt with this argument formerly (see previous Memoir), and spoken about cultures upon peels of lemons and various other nutrient media; but I have found it more expedient and satisfactory to let the fungi of the sputum germinate in the sputum itself.

The better soil for cultures is that in which spores and branching filaments carry out their immersed vegetation. In order that aërial vegetation should take place, two conditions are required: the first is that the materials of culture (in this case, the sputum) should be moderately dry on the surface; and the second is that the under layer, or the lowest stratum of the sputum, should be kept wet in order to feed the vegetation. Now, both conditions are fulfilled by placing a part of the sputum, impregnated with spores and sprouts, on the bottom of a wine-glass slightly hollow, so that the sputum should sufficiently spread and rise to four or five millimetres. The wine-glass and its cover should be previously cleansed with sulphuric acid, and washed out with alcohol. Through the wide opening of the glass the fructifications can be easily observed and scraped out for microscopical examination. Generally, fructification is completed on the fourth day, by keeping the glass well sheltered in the dark.

The sputum, impregnated with oidium forms, exhibited on the fourth day a simple fructification of *Penicillium glaucum*. Naturally, we do not intend by this to discuss the question brought forward by Hallier—*i.e.*, whether the oidium forms be one phase of the *Penicillium* and *Mucor* kinds.

DISSEMINATION OF THE ABOVE-DESCRIBED MICROBES ON THE POSTERIOR ORGANS.

Considering now the extended vegetation and fructification of *Leptothrix* upon the dental surface and superposed layers; its abundant germination on the tongue, its constant presence in the epithelia of the cavities of the mouth and pharynx, its very active multiplication in saliva, the countless swarms of analogical forms, quiescent or reproductive, in the whole wide surface of the nasal cavities, and perhaps of its appendages ; considering also the extraordinary fertility of this parasite, and its multiplication not only through the fission of cocci and bacteria, but through fructification and budding within the filaments, we may form an approximate idea of the mass of germs or scattered elements which are invading the digestive, aërial, and lachrymal passages.

According to my calculations, not less than from two to three hundred trillions of germs or separated elements are generally present in the mouth and nose, and liable to disseminate the species, at every-minute, into the other parts.

In fact, comparing the single sporules of the *ears* with the dumb-bell bacteria of types *a, b, c, d* (Fig. 2), we should have to place nearly eight sporules in two rows upon the surface of a single bacterium in order to cover it. The bacterium is twice as thick as the sporule ; consequently, the volume of one bacterium will be, at least, equivalent to that of sixteen minute sporules. Now, in our first Memoir, we have demonstrated that about 25 milliards of bacteria of types *a, b, c, d*, go to form a centigramme in weight ; then, for every centigramme of sporules, we shall have to multiply the 25 milliards by 16, which will give 400 milliards of sporules for each centigramme. Calculating now the considerable number of the dropped sporules, that of the gemmules of reserve in the filaments, that of even the most minute granules which constitute the bed of the branching threads and of the clods of *Leptothrix ;* calculating the very small lineal germs grafted on the young productions by *points,* we can safely maintain that their extreme minuteness compensates for the larger volume of bacteria enclosed in the older filaments, that of spindle-like, the comma, and the serpentine bacilli, as well as of spirilli and cocci disseminated in every part. Thus, holding the volume of the sporule as the unity of measure, taking the whole patina of the thirty-two teeth as equivalent, at least, to a gramme, we shall have the number of the elements and germs living on the dental patina equal to the product of 400 milliards (in about a centigramme), × 100, viz., 40 trillions. To these figures is to be added the huge mass of germs and

elements spread in the saliva, of which a third part may be considered to be formed of microbes : *ut tota aqua vivere videatur*, as,
in his time, Leeuwenhoek wrote. The quantity of saliva which
continually moistens the mouth is not less than eight or ten
grammes ; therefore, we have three more grammes of germs and
elements, bringing the total up to *120 trillions*.

But we will not further trouble the reader by calculating the
other germs and elements lodged on the tongue, in the mucus of
the mouth and pharynx, and in the nasal cavities ; from which it
would result that the figures given above are not at all exaggerated.

It is to be noted that we started with an average of 25 milliards
of bacteria per centigramme ; a figure twelve times under the calculations of Naegeli—*i.e.*, 30 milliards per milligramme, which
would make the mass of germs and elements of nose and mouth
upwards of *2 or 3 quadrillions*.

Besides, we must consider that *Leptothrix* lodges, as we have
demonstrated, even in the mouth of domestic animals. Now, the
mucus, saliva, breath, urine, fæces of these animals, jointly with
those of men, constitute, in their whole, an immense preserve. I
am led to believe that a great many germs and elements of *Leptothrix* pass into the fæcal matters alive. There is also incalculable
diffusion in the air, waters, and soil, especially in populous towns.
What wonder, then, that Perroncito has found such a large number
of bacteria in the dust of the streets ? We may, rather, contend
that such a dissemination of buccal microbes constitutes a sort of
cloud, which hardly permits external germs to penetrate.

But let us return to the dissemination of *Leptothrix* in the
internal organs of the human economy.

With regard to the digestive passages, the number of elements
carried there by the saliva must be fabulous, considering that in
the smallest drop myriads of them are to be found. Nor is it
presumable that their descent into the stomach should prove
inactive.

The view that they are partakers in the transformation of ali-
ments is not new ; it goes as far back as Hallier, who considered
them *necessary* to the transformation, especially of starchy substances, not only in the stomach, but even in the mouth itself.*

* Richter, Monograph quoted in the Bibliography, Part I., Section B 2.

Béchamp compared the transformation of saccharine in the presence of its *microzymas* with that of the fæcula through diastasis.* Warkmann held the same action even for starch, and he conceived that from the bacteria was secreted a true diastatic yeast, destined to transform starch into glucose, although unapt to peptonise the albumenoids.† But, later on, others recognised in buccal bacteria even such a peptonising action, and Miller holds it to be equal to that of the pepsine itself, even without the co-operation of acids.‡

A considerable number of the same germs and elements of the mouth, as well of the nose, is cast into the air-passages at every breath, as the daily observation of the normal sputa shows (see preceding Memoir). Those germs and elements thrive and multiply in the mucus of the bronchi, trachea, and larynx, as the fructification in the pulmonitic sputum, just described, shows; it is evidently a soil fit for the vegetation of *Leptothrix*.

Shingleton-Smith,§ in the quoted paper on *Contagium Vivum*, considers, first, the property that the moistened mucous membrane has of holding bacteria. It does not let them out with the breath, but through the ciliary action of the epithelia, which throws them out with the mucus. This happens with pathogenic bacteria, but, *a fortiori*, it must be so with normal bacteria. The same author quotes the experiments of Tyndal, who found the residual air from the lungs absolutely free from particles which would reflect the electric beam. He quotes as well the observations of Lister and others, who maintained that there were no bacteria in the pus of Empyena with Pneumothorax, the opening on the surface of the lung being very small. This must indicate that the microbes

---

* Béchamp, *Les Microzymas et les Zymases*, Arch. de Physiol., 1883. The same, *La Salive, la Sïalozymase, et les Organismes Buccaux*, etc. (*vide* Bibliography), 1883.

† Warkmann, *Untersuch. über das diastatische Ferment der Bacterien* (*Zeitschrift für Phys. Chemie*, Bd. VI., 1883).

‡ See Bufalini, *Of the Peptonising Action of Bacteria* (*Giornale internaz. delle Sc. Mediche*, 1883). Miller, *Ueber Gährungsvorgänge im Verdauung-stractus und die dabei betheiligten Spaltpilze* (*Deut. m. Woch.*, 1885, No. 49). The same, *Die Mikroorganismen der Mundhöhle* (see Bibliography), pp. 77—90, where he specially deals with the fermentative action of buccal microbes upon hydro-carbons, albumenoids, and fats.

§ *Journal of Microscopy and Natural Science*, 1890, p. 34.

are kept back by the mucus wetting the larger air-tubes, where the protecting barrier of the epithelia exists ; but they cannot reach the slender bronchi or alveoli where the epithelium is evanescent or at all wanting. Thus, the microbes, having reached the border of the smallest bronchi, by means of the ciliary action, they are thrust back towards the superior passages, and carried at last, by the sputum, into the normal or pathological mucus.

But if this happens in normal conditions, it cannot be so in certain pathologic conditions, as in consumption, pleurisy, and phthisis. The elements or parasitic germs, in these cases, not only reach the most minute bronchi, but from there they diffuse themselves even to the remotest parts ; this may proceed from the insufficiency of the ciliary action, or from an extraordinary concourse of parasitic elements overcoming it, or from the microbes along a denudated tract of mucous membrane.

Förster and Graefe* were the first to write upon the vegetation of filaments or branching threads of *Leptothrix* on the conjunctival or lachrymal ducts ; Bizozzero draws a specimen in Pl. V., Fig. 52, of the quoted work. This has been reproduced in our Fig. 6, *d*, and it is natural to suppose that the relative germs originate from the nose.

If now, from the buccal and nasal microbes, we pass to consider those of the external genito-urinary passages, we shall find the same disseminations ; but with this difference, that they will not be found so frequently and in such abundance. This depends upon two reasons :—The first is the relative scarcity of parasitic vegetation, especially on the balano-preputial mucus (compared with that of the mouth and nose together) ; the second is the want of a propelling force, sufficiently vigorous and repeated, capable of pushing, at least in normal conditions, the elements or the germs of the parasites in the urethra or in the bladder. At any rate, we must admit a penetration, on a small scale, by looking at the epithelia of the -inferior urinary passages. Such epithelia, even in the most normal .and freshly-discharged urine, are found impregnated with bacteria to such a degree that they cannot be attributed to consecutive pollution (viz., after the urine has been

* Förster, *Arch. für Ophtalm.*, t. xv., page 310. Graefe, *Ueber Leptothrix in den Thränenröhrchen (Arch. cit.*, t. xvi., p. 324).

passed) ; but they evidently denote their invasion *in situ* in the bladder or the urethra.   Of course, the invasion has happened by a slow, progressive motion of the relative elements or germs, either along the mucus coating, or from one epithelian scale to another, and so forth.

In certain urine, round the small flakes or spires of mucus of the male urethra, clods of *Leptothrix* are met, at times, so large as to occupy the whole visual field of an objective (Nos. 7 or 8 of Hartnack). Such clods are so closely similar to the young clods of the *patina dentaria*, that the most experienced investigator would be unable to distinguish one from another.

Lately, I observed a case of this kind in the urine of a patient affected with chronic pyelitis, having already been subject to several attacks of Blennorrhœa. -I placed two of these small flakes in a closed tube with a little of the urine, and for six days I kept them adhering to the internal wall, wetting them gently once or twice a day by inclining the tube.   On the sixth day the material of these small flakes, prepared and stained with gentian violet, like the *patina dentaria*, exhibited exactly all the forms of *Leptothrix buccalis*, with the exception of the productions by points (perhaps spoiled in the preparation).   Undoubtedly, it contained countless bacilli, both comma and spindle-like, rather small, and in brisk activity ; even a fair number of small ears was to be found there.   Briefly, all forms of bacteria and bacilli, chain-like filaments, and bundles, described in *Leptothrix buccalis* (except the *Jodococcus vaginatus*) were likewise there ; and the staining also showed that not even the *spirilla* or *spirochœta*, mostly in motion, were wanting.

Between these forms of *Leptothrix* and those of the mouth, the only difference is that the filaments, bundles, and chains are shorter and the spirilla extremely slender, but longer.   In all other particulars they were perfectly identical.*

* Our present remarks about the multiplicity of the microbes of Blennorrhœa and the presence of *spirilla* in urethral secretion have been confirmed by the research of Legrain, *On Blenorrhœa microbes*. That author found in the flow of gonorrhœa *12 different forms of cocci, 3 distinct forms of bacilli, and 1 spirillic form :* total, 16 forms of microbes *(Archiv f. Dermatologie u. Syphilis,* 1890). Stroganoff has recently dealt with the bacteria of the vagina and cervical canal. —(*Modern Medicine and Bacteriological World*, Oct., 1893, p. 258.)

It would be useful to persevere in this kind of research, and I propose to resume it on the first opportunity, in order to prove the identity of *Leptothrix buccalis* with *præputialis*, and then that of bacteria and bacilli of the air-passages with those of the genito-urinary organs.

We have an argument by analogy for the arrival of the elements or germs of *Leptothrix* in the urethra, in the introduction of more bulky germs, like those of superior fungi, probably of the genus *Penicillium*, either in the urethra itself or in the bladder, of which we have already given an instance in our article upon fungi of the male urethra.

Naturally, the removal, insignificant in healthy conditions, increases in the morbific ones, and that is the reason of the gonococci in blennorrhœa; but we have observed beforehand that in blennogenous pus the most common form of bacteria is not that of gonococci or curved diplococci of type *g, l, p* (Fig. 2), but that of common diplococci with round heads, sometimes surrounded by a halo, either within or outside the epithelia or the corpuscles of pus. I have already treated this argument in the other paper upon a case of carcinoma of the bladder, in which the urine showed a considerable number of curved diplococci or gonococci, independently of any blennogenous contagion whatever.

In the Memoir on Whooping-cough, and in the first section of this work, I have touched upon the presence of such curved diplococci or gonococci in the sputa or saliva; as well as upon the singular discovery of true diplococci in a halo, morphologically identical with the so-called pneumococci, in the spermatic fluid recently discharged through illness.

I think that the surprise or incredulity which at first may have been aroused from the present observations will cease, when it is considered that the incentive of all these varied disseminations is always the same, *i.e.*, that the gonococcus as well as the pneumo-coccus or the bacillus of Koch proceed, in great probability, from small seeds of the same genus; from the parasite that lives nor-mally at the entrance to the digestive and air passages, at the egress of the lachrymal as well as the genito-urinary passages. What wonder, then, if the pneumococcus, towed, so to speak, along the urethra, is found in the spermatic fluid; and the gonococcus, in its turn, is found in sputa and saliva ; and that the bacillus of Koch is met with, at times, even in the buccal epithelia of a healthy man ?

For the rest, anyone is in a position to verify the exactness of our observations upon these varied points.* Now from the exhibited facts we must infer that probably the same bacteria or bacilli considered pathogenic (in the affections of the genito-urinary or air passages) are, in reality, only so many disseminations of germs or elements of buccal microbes or balano-prœputialis, from which, morphologically, their varied types do not differ at all ; or, at least, that, before admitting the existence of this or that pathogenic bacterium in the above-mentioned passages, it is neces-sary to demonstrate, with clear and conclusive proofs, that such bacilli, declared pathogenic, are not derivations from the normal preserve. And this objection does hold good, not only in the hypothesis we have set up of the oneness, or duality at most, of the parasitic species of the mouth, but even in the hypothesis now prevailing of their plurality. We contend that the more species there are in the nasal crypts or the cavities of the mouth, the more difficult will it be to exclude their co-operation in generating bacteria reputed pathogenic.

Everything, indeed, leads us to believe that this inexhaustible preserve of normal microbes, placed by nature at the entrance of the digestive and air passages, may have a defensive mission against the intrusion of micro-organisms from without ; may, in other words, constitute a true excluding vegetation ; a barrier to the effect of preventing foreign germs penetrating and thriving in the adjacent organs.

Nature has imparted to *Leptothrix buccalis* such a power of tenacious resistance to foreign agents that the elements of this fungus upon the human teeth are not destroyed for ages, as it is

---

* We have lately noticed that our views are confirmed in an observation of Bordoni Uffreduzzi and of Gradenigo, who found "in the pus of the left ear (in a case of ottorhœa) numerous diplococci of biscuit shape, some isolated, others joined in tetrahedrons, partly free and partly contained in cellules, *in the whole similar to gonococci.*" The authors quoted maintain that the various microbes of ottorhœa (in which the lanceolatus diplococcus predominates) originate even from the saliva. *Sull'etiologia dell'otite media. Archivio per le scienze mediche*, Vol. XIV., 1890, pages 276 and 278.

proved from the dental tartar of the Egyptian mummies, in which its filaments have been found intact by Zopf and Miller, by means of dissolving the calcareous salts with acids.*

We would like to make further considerations upon certain points of the present Bacteriological doctrines ; but, in order to keep as far as possible within the clinical area to which our researches are directed, we come to the final conclusion.

RECAPITULATION.

In summing up what has been exhibited in the Bacteriological part of the preceding Memoir and the present one, we may conclude as follows :—

I.—Amidst all forms or types of bacteria or bacilli to be found in the sputa, normal or pathological, there exist points of transition, manifesting their polymorphism, and the gradual passage from one type to the other.

II.—Of the types in question, there is not one which cannot be found even in the contents of the mouth, or in the nasal mucus, and the balano-præputialis patina. They do not morphologically differ from each other; therefore, it is generally maintained that the bacteria and bacilli found in sputa, in normal conditions, are simply secondary disseminations of the buccal or nasal microbes.

III.—But when we deal with bacteria, rightly or wrongly reputed pathogenic—(as, for instance, the pneumococcus or the bacillus of Koch)—it is another matter. We cannot grant their buccal origin, setting up, instead, the hypothesis of their having a specific origin from without.

Nevertheless, even these types cannot be morphologically distinguished from their corresponding types of the contents of the mouth or of the nasal mucus. We should have, at least, to demonstrate, in a positive manner, that the former do not proceed from the latter, although those same microbes are (like the other buccal microbes) thrown by swarms into the air-passages. But that demonstration has not been given; nay, the only notion upheld

*Zopf, *Die Spaltpilze*, Breslau, 1883, page 80. Miller, *Prehistoric Teeth* (*Independ. Practitioner*, 1884).

from their resisting the decolourising process has, in some instances, been proved wrong.

IV.—Consequently, there only remain, in support of the specificity of such bacteria, the results, more or less controverted, of the methods of culture and inoculation. We shall speak by-and-by about the methods of culture, having already touched on those of inoculation; but leaving, for the present, that question on one side, we wish to deny that the hypothesis of the speciality of bacteria, found in pathological sputa, is at all supported by the general clinical facts, or even by the daily microscopical observation of the sputa.

V.—The congeries of the buccal and nasal microbes, put together, may be summed up to two hundred or three hundred trillions of germs and elements in continual prolification. This vegetation, either through its diffusion—(at least, in many species of domestic animals)—or its constancy and abundance, assumes the character of a true *excluding* vegetation, placed by nature at the ingress of the digestive and air passages, to aid the former in the digestion, and to defend the latter against the micro-organisms of the external world. An identical vegetation adorns the egress of the genito-urinary passages. Thence, it is not surprising if bacteria reputed specific of the air-passages are found even in the products of the genito-urinary passages, and *vice versa;* if, for instance, incapsulated diplococci *(pneumococci)* are found in the urethral mucus or in the spermatic fluid, and curved diplococci (*gonococci* of Neisser, reputed specific forms of blennogenous virus) are found, in their turn, in the sputa or in the middle ear.

VI.—Whatever may be the physiological importance of *Leptothrix buccalis*, it results from our observations that its degree of organisation is far superior to what has been reputed hitherto. *Leptothrix* does not only live as bacterium, bacillus, or filament; but it possesses real organs of reproduction by which it would resemble fungi and *diœcious algæ*, with distinct sexes upon different filaments or individuals. Its fertile filaments are at times engrafted, with two or three roots, upon clods or firm substrata, and end in a fructification. The *ears* constituting these fructifications, as long (in pulmonitic sputum) as 1/6th of a millimetre, are formed of many very minute sporules (so much so that 400 milliards of them hardly would weigh a centigramme); and the small sporules, taking a bright colour with aniline, are in some instances disposed in six longitudinal series, making a total of 720 for each *ear*. They are linked together and fastened to the stalk by means of an amorphous substance difficult to be coloured.* However, other filaments, less numerous than these, at times multiply; and lastly, branching off, bear certain productions by *points*, or pseudo inflorescences, formed of spindle-like, snake-like, or comma bacilli *(Spirillum sputigenum)*, destined from all appearance, through their lively activity, to the function of conjugation. Finally, there are gemmules in reserve, which (together with the multiplication of the proper sporules) are destined to diffuse the species in the unstable substrata or in the products and in the liquid secretions.

The fructification by *ears* can be reproduced, even in the sputa and in certain small flakes of the urethral mucus.

VII.—Of the six primary species of fungi of the mouth, lately described by Miller, there would, in fact, exist only one—the *Leptothrix buccalis* of Robin (*Leptothrix innominata* of Miller), or, at most, a second one—the Spirillum (*Spirochæte dentium* of Miller). The other four types would represent, if we are not mistaken, only phases or disintegrated particles of the microphyte—viz., *Bacillus buccalis maximus* and *Leptothrix buccalis maxima*, fragments of the stumps that form the inferior layer of vegetation; the *Jodococcus vaginatus* series of special sheaths of bacteria proceeding from certain gemmules of reserve enclosed in the filaments; the *Spirillum sputigenum* (comma bacilli) with our spindle-like and serpentine; appendages detached from the pseudo inflorescences, and probably male organs.

All these particles or articulations cut from the mother plant (except the last—viz., copulative filaments) multiply by themselves, in various ways, according to the condition of the nutrient substratum, in the liquid menstrua or on firm soil.

VIII.—The study of such vegetable forms, in the contents of the mouth as well as in sputa (especially of the fructification by *ears*), requires special rules and care and proper optical means,

* Owing to the connecting of the sporules to the central stalk, by means of peduncles or engrafting threads, visible with a new objective (1/25th in. objective). See former Note.

without which it would be difficult to verify the facts, even for the most experienced investigators. The productions by *points* are fairly well detected, even with the ordinary objectives; but special care must be taken in collecting and preparing the *patina dentaria.*

We do not know whether the specific oneness of the above forms (by *ears* and by *points*) will be well received by competent observers, as well as all the buccal microbes reunited in a single plant. But, were even two or more of the vegetable species in question, the successive dissemination of their germs and elements in the air-passages would not at all be invalidated.

IX.—The most minute form of cocci, scattered or in a line, found in the contents of the mouth, in the omonimous epithelia, or in the sputa, are, in our opinion, nothing but sporules dropped from the fructifications by *ears*, being first disseminated in the mouth and then thrown into the respiratory passages. The bacilli of Koch (we are speaking of *bead-like* bacilli) might be only these same sporules disposed in series, and thus germinating in the tubercular products and elsewhere. Undoubtedly, the sporules, still attached to the fructifications, or fallen near them, fix strongly the aniline colours ; but it remains to be seen whether, in the sputa of consumptives, their resistance to decolourising means is original or simply acquired in the fresh nutrient substratum received from the tubercular lesions. At any rate, that resistance is always relative, and is of doubtful value in the diagnosis, and is sometimes inferior to that of other bacteria, as we have demonstrated before.

The forms of the bacilli of Koch being, on the contrary, in small rods containing granules or vacuoles, with double staining (gentian violet and solution of iodine) behave like the analogous articulations of *Leptothrix.* But upon these and other not less important points, about the clinical study of the tubercular sputa, we shall have to deal on another occasion.

X.—However, if our observations on the morphology and biology of *Leptothrix* are correct ; if all, or nearly all, forms of bacteria and bacilli to be found in the sputa, are nothing but *particles* or various organs of a single plant, everyone can see the extent of the actual methods of culture (at least of the bacteria of the air-passages).

And, to be sure, one person may identify or qualify a fungus on the ground of its fructification ; another, on the ground of the form of cultures, immersed or creeping, of its single particles, or the modifications brought about by these in the various nutrient substrata.   One bacterium or bacillus will fluidify gelatine or elaborate certain principles (even poisons) in a manner totally different from another bacterium or bacillus, without, however, considering the two forms as two organisms or different species, when such forms are for us simply the result of organs or particles of the same plant, destined to attain dissimilar aims, and which may possess the most different qualities.

XI.—Under such conditions, the name of bacterium or bacillus can no longer be considered as synonymous of micro-organism. We may speak of bacteria and bacilli, if by those names we mean particles or articulations detached from the mother plant ; and in such a case the generic noun of *microbes* may be even applied. But the word micro-organism applies to the whole microphyte of which bacteria and bacilli are only single scattered particles, and which do not constitute by themselves complete organisms, although they mostly possess the faculty of multiplying on their own account. Such name does not suit at all the single particles, nor, of course, the single bacteria or bacilli.*

I cannot conclude this paper without a short statement.

Perhaps some of my views will appear too bold, but my only intention was to submit them to the judgment of scientific observ-

---

* In confirmation of our views, we shall quote the opinion of Klein, who lately presented to the Royal Society of London the photographs of a culture of the tubercular bacillus, which appeared with distinct branches like a mycelium fungus.   He inferred that *at least a few schizophytes* are really only *forms of development or transitory phases of superior organisms.* The opinion of Dowdeswell is even more explicit.   " It is clear " (writes the latter, on the subject of comma bacilli) "that either these microbes are not normal schizophytes, or if so they do not represent an independent group (as it has been observed by the last writers on Bacteriology), but are *simple phases of evolution* of some superior organisms, or perhaps they are only *simple organs of other organisms* (*Lancet*, 1890, Vol. I., *loc. cit.*, p. 1422).   Lately, Sheridan Delepine, studying the development of bacteria in its *cultures in interlamellar films* (between the cover-glass and the slide) has come to that conclusion—viz., of the branching off in many bacilli, by means of defined filaments (A New Method, Interlamellar Films, of studying the development of Micro-Organisms, etc., in the *International Journal of Microscopy*, Nov., 1891, p. 343).

ers. I shall only notice that my opposition to certain points of the
present Bacteriological doctrines is more apparent than real. I am
of opinion that, by pursuing the present method of specifying, in
the classification of bacteria, we must more and more multiply their
species to such an extent as to hinder the further progress of these
studies. On the other hand, should my observations be correct,
if the forms of bacteria of the air and genito-urinary passages are
reduced to a single species, or if, consequently, even the other very
varied types of bacteria (either not at all pathogenic or pathogenic
of other parts) might be gradually reduced to defined species of
micro-organisms, *according to the natural phases of their full development* ; this work upon their arrangement and simplification may
aid in the researches of experimental pathology. In other words,
we shall be able to recognise the difference between the botanical
and the pathogenic entity of bacteria ; that two or more bacteria,
totally different in their own pathogenic action, may proceed from
a single micro-organism, and that a bacterium, not pathogenic,
may belong, in its turn, to the same micro-organism from which a
pathogenic bacterium proceeds.

F. VICENTINI,

Chieti, June, 1890.   Corresponding Member.

BIBLIOGRAPHY.

1.—Arndt, Beobactungen an Spirochaete denticola (Arch. für
   path. Anat. und Phys. und d. klinische Med. t. LXXIV. 1880).

2.—Baume, Odontologische Forschungen, II. of the same, see
   also the "Elements of Dentistry" (Lehrbuch der Zahn-
   heilkunde, 3. Auflage, Leipzig, 1890).

3.—Beale, The Microscope in Medicine, London, 1878, p. 490.

4.—Béchamp. La salive la sialozymase et les organismes buccaux
   chez l'homme (Archives de physiologie, 3 série, I., 1883,
   p. 47).

5.—Bizzozero, Manuale di Microscopia Clinica, Milan, 1882,
   pp. 103—104 and p. 153.

6.—Chevalier. L'étudiant micrographe, Traité théorique et
   pratique du microscope, etc. Paris, 1882, pp. 393—94.

7.—Cornil et Babes. Les bactéries et leur rôle dans l'anat. et l'hist. patholog. Paris, 1885, pp. 28 and 135.

8. Decker und Seifert. Ueber Mycosis leptothrica pharyngis (München medicin. Wochenschrift. xxxv. 4., p. 67, 1888, and Sitz. Ber. der physik. med. Ges. zu Würzburg, 2, p. 26.

9.—Dowdeswell. Note on the Morphology of the Cholera comma bacillus (Lancet, 1890, Vol. I., pp. 1419—23).

10.—Ewart and Geddes, Life History of Spirillum (Proceedings of the Royal Society, xxvii., 1878, p. 484.)

11.—Flügge. I microorganismi ed etiologia delle mallattie infettive. It. trans. Naples, 1889.

12.—Förster. Arch. für Ophtalm., xv., p. 318.

13.—Frey. Manuale di tecnica Microscopica. It. trans. Naples, 1873, p. 278.

14.—Graefe. Ueber Leptothrix in der Tränenröhrchen (Arch. für Ophtalm., t. xvi.).

15.—Hallier. Die pflanzlichen Parasiten, etc., Leipzig, 1866.

16.—Jacksch. Manuel de diagnostic des malad. intér. par les méthodes bactériologiques, etc. (translated from the German), Paris, 1888, pp. 50—57, and p. 71.

17.—Kützing. Linnæa, 1883.

18.—Leber. Quarterly Journal of Microscopical Science, 1874.

19.—Leber und Rottenstein. Untersuchungen über die Caries der Zähne, Berlin, 1867.

20.—Lewis. A Memorandum on the Comma-shaped Bacillus alleged to be the cause of Cholera (Lancet, 1884, Vol. II., p. 513).

21.—Miller. Der Einfluss der Mikroorganismen auf die Caries der Menschlichen Zähne (Arch. für exp. Pathologie, xvi., 1882).

22.— ———— Ueber einen Zahnspaltpilz, Leptothrix gigantea (Berichte der Deutschen Botanischen Gesellschaft, Berlin, 1883, Heft 5, p. 221).

23.— ———— Zur Kenntniss der Bacterien der Mundhöhle, 1884.

24.—Miller.　Biological Studies on the Fungi of the Human Mouth (Indep. Practitioner, 1885, pp. 227—83).

25.— ——— Beiträge zur Kenntniss der Mundpilze (Deutsche Medicin. Wochenschrift, XIV. 30, 1888. '

26.— ——— Die Mikroorganismen der Mundhöhle. Die örtlichen und allgemeinen Erkrankungen, etc., Leipzig, 1889.

27.—Peroncito.　I parassiti dell' uomo e degli animali utili. Milan, 1882, p. 56.

28.— ——— Il carbonchio, 1885.

29.—Rappin.　Les bactéries de la bouche à l'état normal et dans la fièvre typhoide.　Paris, 1881.

30.—Rasmussen.　Ueber die Cultur von Mikroorganismen vom Speichel gesunder Menschen, Kopenhagen, 1883.

31.—Richter.　Die Neuern Kenntnisse von den krankmachenden Schmarotzerpilzen, nebst phytophysiologischen Vorbegriffen (Schmidt's Jahrbücher, 1867, 135, pp. 81, 101, and 140).

32.—Robin.　Des Végétaux qui croissent sur les animaux vivants, Paris, 1847, p. 42.

33.— ——— Histoire naturelle des végétaux parassites qui croissent sur l'homme et sur les animaux vivants, Paris, 1853.

34.—Sternberg.　American Journal of Medic. Science, 1884—5.

35.—Vignal.　Réchérches sur les microorganismes de la bouche (Arch. de physiol. norm. et pathol., 1886, n. 8).

36.—Zopf.　Zur Morphologie der Spaltpflanzen, Leipzig, 1882.

37.— — Die Spaltpilze, Breslau, 1883, p. 80, etc.

38.—Zurn.　Die Schmarotzer, etc., Part II., Weimar, 1887—89.

For the works of David and Billet, see note at the end of 3rd paragraph.

EXPLANATION OF PLATE VI. (SEE APRIL, 1895).

FIG. 4.

Other forms of Bacteria (Gentian Violet).

a.—Catenula of type *p, p* (Fig. 1), incapsulated (*Jodococcus vaginatus* of Miller, ?), from a sputum of Whooping-cough, × 690.

b.—Diplococci with and without halo, and small cocci from the *patina dentaria* mixed in the same heap.—*b'*, Very minute diplococcus with halo from the *patina dentaria* mixed with saliva, × 1750.

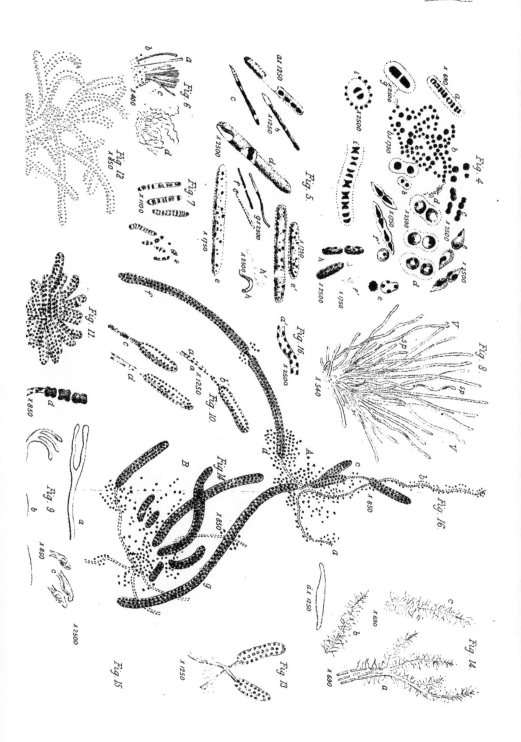

c.—Dumb-bell bacteria, in couples, from the condensed nasal mucus, in normal conditions, × 2500.

d.—Diplococci with halo (pneumococci), partly coloured, × 2500.

e.—Diplococci without halo, also partly coloured, from pulmonitic sputum, × 2500.

f.—Pneumonocci with distinct halos of acuminated shape. *f'*, Another pale one, without halo, in increase, from a pulmonitic sputum, × 1750.

g.—Large diplococcus in a halo, from pulmonitic sputum, × 2500.

h.—Diplococci, without halo, in increase, containing internal granules, from pulmonitic sputum, × 2500.

i.—Large roundish diplococcus, with halo, surrounded by very minute cocci.—*i'*, Incapsulated catenula, formed of analogical diplococci, from pulmonitic sputum, × 2500.

Fig. 5.

Other forms of Bacilli, from a to h, with gentian violet;
h', with solution of iodine.

a.—Bacillus of three types, partly with internal gemmules, partly clear, partly opaque, from sputum of *influenza*, × 1250.

b.—Bacillus, partly opaque, partly clear, from saliva, × 1250.

c.—Bacillus of four types, from saliva, × 1250.

d.—Bacillus partially coloured, containing a transverse dumb-bell bacterium (*Bacillus buccalis maximus* of Miller, ?), from *patina dentaria*, × 2500.

e.—Spindle-like bacillus, internally granulous, from sputum of bronchitis.—*e'*, Analogical bacilli, but cylindrical, from pulmonitic sputum, × 1750.

f.—Two very slender bacilli, one mono-articulated, the other granulous, from pulmonitic sputum, × 2500.

g.—Very slender bacillus with three knots, from the same sputum, × 2500.

h.—Comma bacillus (*Spirillum sputigenum* of Miller) from pulmonitic sputum.—*h'*, Serpentine bacillus, shaped like a spermatozoid (coloured with solution of iodine), from the *patina dentaria*, × 2500.

Fig. 6.

Tufts of Leptothrix, in normal condition, from Bizzozero, × 400.

a.—Bundles of filaments of Leptothrix.—*b*, Articulated filaments. *c*, Spirilla from *patina dentaria*.

d.—Tuft of Leptothrix from a concretion of the lachrymal bag.

Fig. 7.

Jodococcus vaginatus, Miller, × 1100.

Fig. 8.

Tuft of *Leptothrix gigantea* of a sheep in natural state, from Miller, × 540.

Sp., Spirillic filaments.   *V.*, Vibrio-filaments.

Fig. 9.

Roots, Swellings, and a few contracted fertile Threads of Leptothrix, from the patina dentaria, × 850.

a.—Fragments of fertile filament, with two roots, in natural state (*haustoria, ?*).

b.—Fertile filament with three roots, and a swelling at the apex, in natural state.

c.—Various forms of swelling at the apex, or small heads of young filaments, with gentian violet.

d.—A large, somewhat woody filament, ending in a small chain (*Leptothrix buccalis maxima* of Miller), with a contraction, from which springs forth, at last, a slenderer fertile filament, with gentian violet.

Fig. 10.

Three fructifications by Ears, isolated, with gentian violet, from the patina dentaria, × 1250.

a, a.—Gemmules of reserve, within the stalk.—*b*, Intact sporules.— *b′*, Residual sporules on the point of dropping from the *ear*.

c.—*Ear* grafted upon an older stalk, with knots.

d.—*Ear* grafted upon a younger stalk, also with knots.

Fig. 11.

Tuft of Ears seen from above, with weak methyl violet, from *patina dentaria,* × 1250.

Fig. 12.

Tuft of Ears seen in profile with weak fuchsine, from *patina dentaria,* × 850.

Fig. 13.

Two Ears jutting out from a tiny island of *materia alba*, with unacidulated solution of iodine, × 1250.

Fig. 14.

Productions by *points* (pseudo inflorescences), in natural state, from *patina dentaria.*

a.—Branched pseudo-inflorescence, bearing small spindle-like bacilli (male organs, ?), × 690.

b.—Isolated pseudo inflorescence, bearing analogical bacilli, × 690.

c.—Fragment of pseudo inflorescence, bearing spindle-like and comma bacilli together (male organs, ?), × 690.

d.—A dropped spindle-like bacillus, more enlarged, in quick motion (*Bacillus tremulus* of Rappin), × 1250.

FIG. 15.

Articulations and comma bacilli, with vacuoles, in apparent conjugation, with unacidulated solution of iodine, from *patina denturia*, × 2500.

FIG. 16.

Specimen of fructification by *ears*, filling up two visual fields (coloured for two hours with gentian violet) from a very slender layer of pulmonitic sputum, on the seventh day after the emission, × 850 and × 2500.

A.—A principal tuft. *B.*—A part of it drawn down to the bottom from the pressure of the cover-glass, × 850. (This specimen is erroneously marked Fig. 14 in the Plate.)

a.—Denudated Stalk, containing gemmules of reserve, × 850.—*a′*, A part of it more enlarged, after a long saturation with glycerine, with a trace of knot, × 2500.

b.—A long ear, mostly scattered, × 850.

c.—A broken ear, with clear section, × 850.

d, e.—Cumuli of sporules dropped from the adjacent *ears*, × 850.

f, g.—Points of two *ears* bent down from the pressure of the cover-glass, × 850.

N.B.—Other fragments of *ears* were seen scattered about the visual fields, around the specimen here delineated.

SATURN'S RINGS. —It has long been the accepted theory that Saturn's rings consist of swarms of minute satellites crowded together, but hitherto the evidence in support of this view has been chiefly of a negative character, it having been proved by mathematicians that the rings could not maintain their shape if they were solid or fluid masses. A recent telegram from Pittsburgh announces that Prof. Keeler, of the Alleghany Observatory, has proved from spectroscopic observations that the rings *are* actually formed of satellites which do not all revolve at the same rate, thus affording an interesting confirmation of the present theory.

The Microscopes of 1894.

DURING the past year considerable activity has been shown by opticians in the manufacture of new forms of microscopes; indeed, we cannot remember any recent year in which so many radical innovations have been adopted. Of course, it is needless to say that that which is novel may not always be good; but we venture to predict, regarding the new microscopes and apparatus of last year which are both novel and good, that, in popular parlance, they have come to stay, and as with certain articles of universal consumption, no microscopist will be complete without some of them. Another further source of congratulation is that most of the instruments and apparatus in question are within the reach of those of moderate means, and we are afraid that that includes the majority of us.

Messrs. J. Swift & Son's Four-Legged Microscope.

The first instrument to which we would direct attention is Messrs. Swift's four-legged tripod. One by one our cherished ideals are being exploded, and now Messrs. Swift have killed another by proving that a tripod may have four legs.

The microscope referred to (Figs. 1 and 2) has a diagonal rack and pinion coarse adjustment, the fine adjustment being by micrometer screw; at the back of the fine adjustment is a small milled-head screw, by means of which any wear in the adjustment may be taken up. The length of the body-tube from the ocular to the nose-piece is 6 inches, and can be extended to 9 inches by means of the draw-tube, which has a millimetre graduation. The stage, of the horse-shoe shape, is provided with spring clips, or a movable object-stage can be attached. The sub-stage can be had in two forms, one being an ordinary fitting taking the condenser, the other being the regular rack-and-pinion sub-stage. The body of the instrument is supported on an horse-shoe platform, from which the four legs spring; the two front legs are fixed, but the hind legs are pivoted to the platform. This arrangement of pivoting the hind legs enables the microscope to adapt itself to any uneven surface, thus keeping the instrument very steady; it also reduces the likelihood of the instrument being upset by any lateral

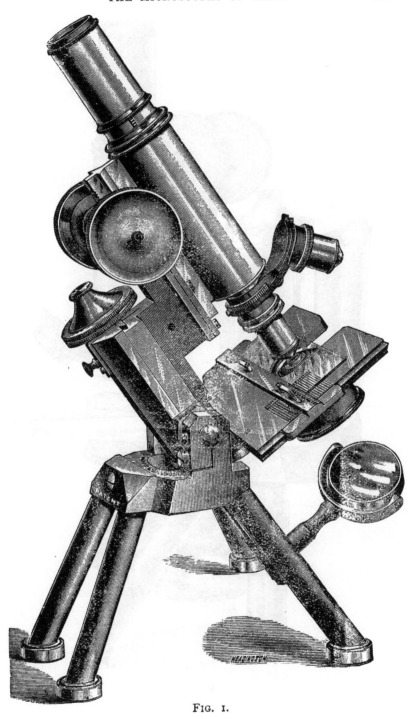

Fig. 1.

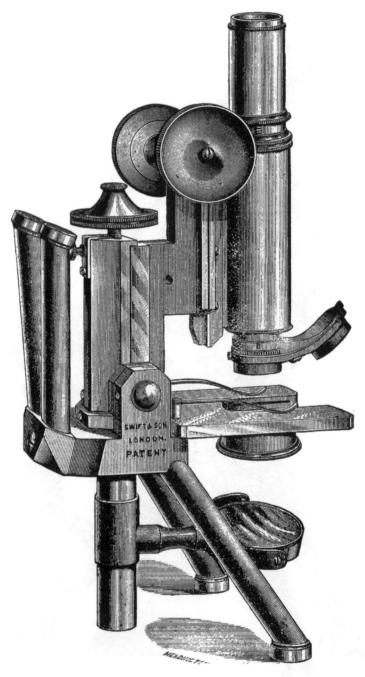

FIG. 2.

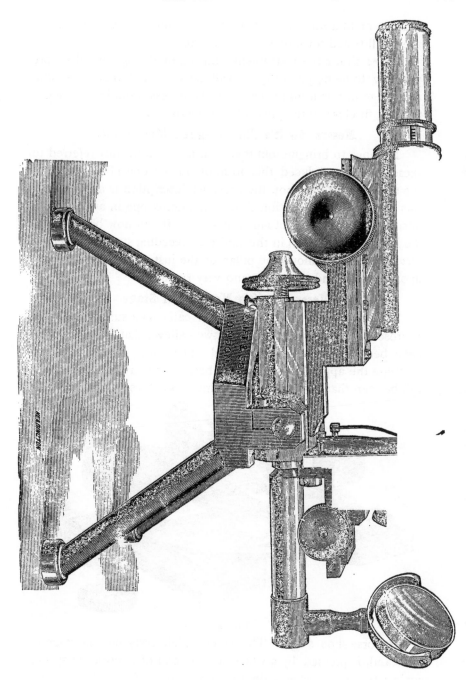

FIG. 3.

movement to a minimum, getting rid of the only difficulty which the small tripod form of foot is open to.

Fig. 2 shows the instrument with the hind legs turned up preparatory to being put away; it will be seen that there is another advantage in this form in that it takes up less room in packing, as the legs fit close to the pillar of the microscope.

Messrs. Swift's Three-Legged Microscope.

Previous to bringing-out their four-legged form just referred to, these makers produced this form of microscope (Fig. 3). With the exception of the foot, the previous description is applicable to this stand. The illustration shows the microscope in an horizontal position, as used in photo-micrography. It will now be seen that the horse-shoe platform in this and the preceding stand is extremely serviceable, as it allows the pillar of the instrument to rest firmly upon it, thus rendering the stand very rigid.

Messrs. Swift's Mechanical Stage.

This stage is simply unique, and came as a surprise to most microscopists. It is the first time, we believe, that friction wheels have been used for microscopical apparatus. The figure will illustrate the apparatus in question better than any description. It will be seen that the object-stage runs in grooves let into the side

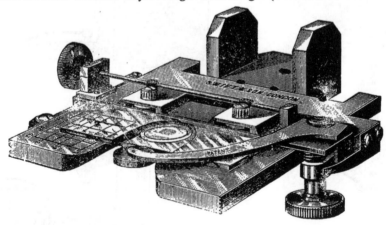

FIG. 4.

of the horse-shoe stage. The slide, which rests on the stage of the stand, is pressed by a wheel at the end of a powerful spring against two small wheels on the inner edge of the object-stage. Transverse motion is imparted to the slide by a turning milled head attached to a rod, which acts on two screw-heads, or rather two endless screws, turning the wheels in question. The movement in this direction is over $1\frac{1}{2}$ inches. The longitudinal movement is by a screw, which runs in grooves on the outer edge of the main stage. We believe that Messrs. Swift have since placed the milled heads of the two movements on the same side, so that an observer needs only to use one hand to the stage, leaving him free to use his other hand to manipulate the adjustments. They have also further improved this stage by placing the two wheels closer together, thus giving a wider transverse motion—*i.e.*, two inches.

Messrs. Ross & Co.'s Eclipse Microscope.

When a simple idea is brought forward, we often feel inclined to kick ourselves for not having thought of it before. Messrs. Ross and Co.'s form of stand with the ring foot is a case in point. It can be seen at a glance how steady this form of stand is, the wonder being that it has not been thought of previously. But, however, better late than never.

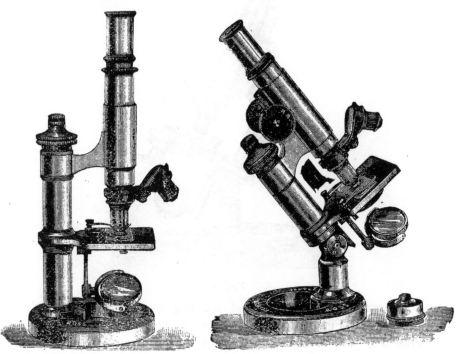

Fig. 5.   Fig. 6.

Figs. 5 and 6 show the two simplest forms of this class of microscope; Fig. 5 being a rigid stand with a sliding coarse adjustment, the fine adjustment being by micrometer screw. The body tube is of the Continental tube-length (160 mm.), extending by means of the draw-tube to 200 mm. (8 inches). The eye-pieces are also of the Continental size. The makers call especial attention to the manner in which the fine adjustment is capped, thus preserving it from disturbance or injury by dust. Fig. 6 shows the stand inclined; it will be seen that the instrument is provided with a knee-joint below the stage. The pillar is so arranged that it can be rotated on its base, reversing the position of the stand, and thus securing the stability of the instrument. When the instrument is used vertically, it will be seen that the centre of gravity is so placed that there is an equal steadiness in every direction.

FIG. 7.

Messrs. Ross have adapted the ring foot to a petrological microscope (Fig. 7). This stand has a revolving circular stage graded to 360°. The analyser, which can be drawn out when not needed, is fitted to the lower end of the body-tube, in which there is also a slot, cut at the angle of 45°, for the insertion of the quartz wedge. The polariser is pivoted to swing in and out of the field ; it has a circle divided into eight and clicked at 0° and 180° to indicate where the nicols are crossed. The eyepiece is furnished with crossed webs.

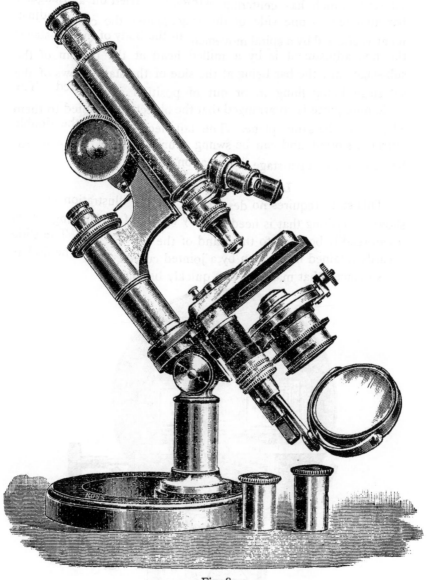

Fig. 8.

Fig. 8 shows the highest class of the "Eclipse" stand, made especially for bacteriological and other high-power work. The adjustments are the same as in the preceding forms and the body-tube is graded to millimetres. The stage is of horse-shoe form, thus allowing free access to the sub-stage when focussing. The sub-stage, which has centering screws, is carried on a triangular bar attached to one side of the stage-plate; the coarse adjustment is effected by a spiral movement in the body of the sub-stage; the fine adjustment is by a milled head at the bottom of the sub-stage bar; the bar being at the side of the stage allows of the sub-stage being flung in or out of position as required. The triple nose-piece is so arranged that the objectives adapted to them all focus in the same plane. The tail-piece carrying the double mirror is slotted, and can be swung round to allow of the mirror being used for super-stage illumination.

Leitz' Mechanical Stage.

This stage requires no description, as our illustration (Fig. 9) shows everything that is needed. By means of a clamping screw, it can readily be fixed to the stand of the microscope. The slide is easily retained in position by a jointed curved arm. The finders are so placed that markings can quickly be found.

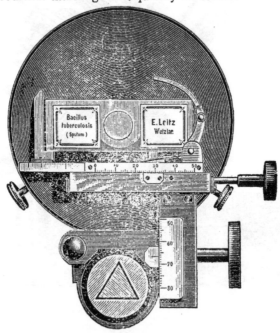

Fig. 9.

Fig. 10.

Leitz' Microscope.

In the January number of this Journal (page 106) for last year, we called attention to the fact that at last one of the continental makers had adopted the tripod form of foot. Continental microscopists have to thank Herr E. Leitz, of Wetzlar, for this improvement in their instruments. By the courtesy of Messrs. C. Baker, we are enabled to give an illustration (Fig. 10) of the stand in question. The only feature that we need point out in this instrument is the sub-stage, which has an excentric rack-and-pinion movement, and can also be moved in or out of position as required. A horse-shoe stage can be supplied instead of the circular one as depicted.

Watson & Son's "Grand Model Van Heurck" Microscope.

This instrument (Fig. 11), which we have reserved till last, as all the preceding forms are of the student's or low-price class, fully answers up to its title; it is a superb microscope. The stand is mounted on a massive tripod foot, which has a spread of 10 inches in each direction; when in an horizontal position, the instrument has a 10-inch optical centre. The coarse adjustment is by rack-and-pinion. The fine adjustment is by a finely threaded screw, which bears on the end of a powerful lever; this adjustment was first adopted by Zentmayer, of Philadelphia, but after a short time was found to work unsteadily; to Messrs. Watson belongs the credit of rectifying the error in the design and in practically making the adjustment their own : it is now one of the most exquisitely delicate and reliable forms of fine adjustment. Rack-and-pinion and sliding movements are also attached to the draw-tube. The stage, which has rectangular rack-and-pinion movements, is provided with finders, and can be completely rotated when required. The sub-stage has centring screws, a coarse adjustment by rack-and-pinion, and a fine adjustment by micrometer screw.

In microscopy it is often said that, given a good instrument, it all depends on the man at the other end. We can now say, Here is a good instrument; now bring your man.

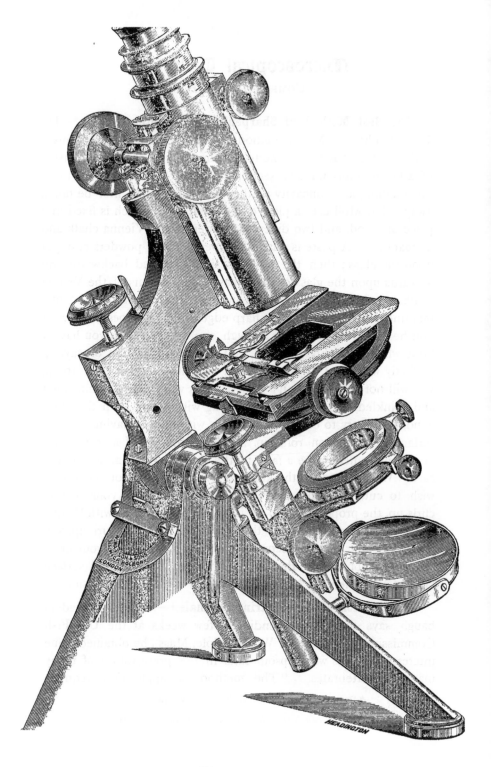

Fig. 11.

Microscopical Technique.

COMPILED BY W.H.B.

The Best Method of Sharpening a Microtome Knife.*—Dr. Lotsy thinks that Moll's method of sharpening microtome knives is the best, as "it allows one to put the knife in good shape inside of a few minutes for any section he wants to cut. Before using this method, any concavity of the knife-blade must first be taken away. Dr. Moll uses a plate of polished glass, which is fixed in a piece of wood, and two different powders, viz.—Vienna chalk and diamantine. A paste is made of one of these powders and put upon the glass; then the knife is simply moved backwards and forwards upon the glass over the paste. By means of the Vienna chalk you can polish your knife in a very few minutes. The diamantine allows you to put a sharp edge on it, but does not give a polished surface, but rather a rough one. Now, when you have a knife which is highly polished, you can cut a section of, say, 5μ perfectly well; but if you try to cut with it a section of 1 to 2μ, you will not succeed at all; your sections will become compressed and wrinkled, and you can do nothing with them. On the other hand, if you try to cut a section of 5μ with a knife having a rough surface, your section rolls up. This rolling up of a section has been represented to be a fault in the paraffin, but that is not the case. We must adapt the knife to every thickness of section we wish to cut. Starting out with a certain knife, if your section curls up, the proper thing to do is to polish your knife with Vienna chalk, and your section for that thickness will not curl up, any more. If your section becomes too much compressed, your knife should be rubbed over the diamantine and the polished surface taken away, when the sections will be cut without compression."

Preservation of some Marine Animals.†—Mr. W. A. Reden-baugh says that while spending a few weeks at the U.S. Fish Commission Laboratory at Wood's Hole, Mass., he obtained some interesting results with Epsom salts in the preservation of many marine invertebrates. "The method of application requires

---

* *Johns Hopkins Hospital Bulletin*, v., pp. 136—7.

† *American Naturalist*, XXIX., 1895, pp. 399—401.

modification in individual cases, but a few experiments will usually enable one to obtain the desired results. . . Complete stupification of the organism must be produced, so that when it is removed to a killing fluid no contraction will take place. Care should be exercised, however, not to carry on the process too slowly, as maceration may ensue.

Cœlenterates.—The most beautiful results were obtained with sea-anemones, which ordinarily are so difficult to preserve in a well expanded condition. These were allowed to expand in a dish with as little water as possible. Then crystals of magnesium sulphate were placed in the bottom of the dish and allowed to dissolve slowly until a saturated solution was obtained. The process of dissolving may be hastened, if necessary, by stirring up the water gently from time to time with a pipette. Several hours were required to completely stupefy large specimens. When narcotisation was complete, a few crystals placed in the mouth of the sea-anemone had no effect; but if the process had not gone far enough, the lips of the animal would slowly spread open, and then would follow sometimes a violent contraction of the whole animal. This method was tried upon *Metridium marginatum, Sagartia leucolena,* and *Halocampa producta* with excellent results, the tentacles remaining perfectly expanded after the animals had been transferred to Perenyi's fluid, picro-sulphuric acid, or formalin. The same method applied to *Astrangea, Scyphistoma,* and various hydroids did not give as good results as those obtained with the sea-anemones. The polyps were not equally affected, so that only portions of the colonies were perfectly expanded. A large Physalia treated in this way was preserved in 4 per cent. formalin, with all the tentacles and polyps fully extended.

Echinoderms.—Star-fishes and sea-urchins were killed with the ambulacral feet and pedicellaria well extended, by placing them upon the aboral surface for a short time in a saturated solution of Epsom salts, and then transferring them to 4 per cent. formalin. The epidermis of the star-fishes, however, was rendered soft, and was subsequently easily rubbed off, but this was probably due to the formalin.

Specimens of *Synapta* were readily preserved without any constriction by very slowly and intermittently adding to the water, in which they had been allowed to expand, a saturated solution of MgSO$_4$ (Epsom salts).

Vermes.—Most annelids, when placed in saturated solution of Epsom salts, in a very short time became perfectly limp, and were easily extended upon a glass plate and treated with a fixing reagent. *Balanoglossus*, when taken soon after being collected, was preserved in this manner in nearly a perfect state. It was necessary, however, to keep it in position between the edges of two glass slides when the fixing fluid was applied. Good results were obtained with *Cirratulus, Amphitrite, Nereis, Rhyncobolus, Clymenella*, and *Phascolosoma. Phascolosoma*, in most cases, was killed with tentacles protruded. Nemertean worms when transferred to a killing fluid before being completely narcotised, sometimes protruded their proboscies.

Ascidians.—*Molgula* and *Cynthia* were readily killed with siphons open after anæsthetisation with magnesium sulphate. In this case it is best to add the saturated solution of sulphate intermittently with a pipette.

Ctenophores.—After considerable experimentation, a method for preserving these delicate creatures in a nearly life-like appearance was devised. Formalin alone in solutions of varying strength had been tried without success. It was found necessary to treat the animals with some hardening re-agent before placing them in the formalin, and the following method seems to be the most successful:—To a solution of equal parts of 2 per cent. formalin and Perenyi's fluid was added enough common salt (NaCl) to increase the density of the mixture to that of sea-water—*i.e.*, until a Ctenophore placed in it barely floated. This adjustment of the density of the surrounding medium prevented the Ctenophores from collapsing of their own weight. After remaining for about half-an-hour in this fluid, they were transferred to 4 per cent. formalin, the density of which had been increased by the addition of either Epsom salts or common salt, so that the Ctenophores again barely floated. Epsom salts is probably better than common salt for increasing the density of the fluid. Some specimens which were preserved in formalin + NaCl began to shrink after a few days; while some (*Mnemiopsis*) which have been preserved for nearly six months in formalin + MgSO$_4$ are still in excellent condition.

After the Ctenophores have been properly preserved, precaution must be taken in transporting them, for they are easily torn to pieces. If they are placed in bottles filled with fluid of the proper density and the cork so inserted as to leave no air-bubbles, this danger is reduced to a minimum."

Preservation of Sea-Weeds.*--Dr. J. P. Lotsy recommends the following method of preserving specimens of *Floridea*, which prevents swelling of the cell-walls or contraction of the protoplasm, and preserves the chromatophores uninjured. The specimen is first laid in a 1 per cent. solution of chrome-alum in sea-water, and kept there for a period varying from one to twenty-four hours, according to the size and texture of the species. The chrome-alum is then completely washed out, and the specimen placed in a mixture of 5 ccm. of 96 per cent. alcohol in 100 ccm. water, and vigorously stirred. The amount of alcohol is then increased by increments of 5 ccm. every quarter of an hour until it amounts to 50 ccm. The specimen is then removed, and placed in a mixture of 25 per cent. alcohol in distilled water, and the quantity of alcohol again increased in the same way till it amounts to 50 ccm. alcohol to 100 ccm. of water. The same process is again repeated with 50, 60, 70, 80, and 90 per cent. solutions of alcohol in distilled water, the specimens being finally preserved in the last.

Staining and Fixing Diatoms.†—Dr. P. Miquel finds the staining reagent best adapted for demonstrating the gelatinous envelope of diatoms to be an aqueous or boric solution of methylin-blue, which is not taken up so readily by the gelatinous stipe. The same reagent, especially in a slightly ammoniacal solution, may be used for demonstrating the nucleus, which is stained blue, while other substances contained in the cell take from it a dark blue violet stain. For fixing, the author uses a solution of 65 gr. corrosive sublimate and 15 gr. sodium chloride in 100 ccm. of water.

* *Bot. Centralbl.*, LX., *vide Journ. R.M.S.*, 1895. p. 130.

† *Le Diatomiste*, II., 1894, *vide Journ. R.M.S.*, 1895, p. 127.

Reviews.

LABORATORY GUIDE for the Bacteriologist. By Langdon Frothingham, M.D.V. 8vo, pp. 61, with 2 plates. (London : Henry Kempton. 1895.) Price 4/- net.

This is a most useful book of Bacteriological Technique, giving full instructions for the Preparation of Specimens, Staining Methods, Preparation of Nutrient Media. and of Embedding Tissues for Cutting Sections. Each alternate page is left blank for MS. notes, etc.

OBJECT LESSONS IN BOTANY from Forest, Field, Wayside, and Garden. Vols. I. and II. By Edward Snelgrove, B.A. Cr. 8vo, pp. viii.—109 and xii.—297. (London : Jarrold and Son. 1895.) Price 2/6 and 3/6.

Two very capital books. Vol. I. is intended to meet the requirements of Standards 1 and 2, and Vol: II. for Standards 3, 4, and 5. The first is thoroughly elementary and treats of subjects easily understood by the youngest children, as apples, oranges, plums, garden vegetables, etc. The lessons in Vol. II. are intended for children from nine to eleven years of age, and are therefore restricted to those subjects of botany which lie within the scope of their capacities. It contains 100 lessons, divided into the following sections :— I., Leaves, Stems, and Roots ; II., Flowers ; III., Fruits and Seeds ; IV., Classification. Both books are thoroughly illustrated.

LABORATORY EXERCISES IN BOTANY. By Edson S. Basten, A.M. 8vo, pp. 540. (Philadelphia, U.S.A.: W. B. Saunders. 1895.)

A fine work, designed especially for the use of colleges and schools, in which Botany is taught by Laboratory methods. It aims to inculcate in the student, by the study of properly selected examples, a knowledge of the elementary principles of botany, to develop his observing faculties, to stimulate in him the spirit of investigation, and to lead him to take delight in this beautiful science. The book is divided into two parts:—Part I., ORGANOGRAPHY, which deals with the grosser structure of flowering plants ; and Part II., VEGETABLE HISTOLOGY, or the microscopic structure of plants. It is illustrated with seven figures in the text and 87 full-page plates from original drawings, comprising upwards of 250 figures.

ACROSS THE COMMON after Wild Flowers. By Uncle Matt.

DOWN THE LANE AND BACK in Search of Wild Flowers. By Uncle Matt.

A STROLL IN A MARSH in Search of Wild Flowers. By Uncle Matt.

AROUND A CORNFIELD in a Ramble after Wild Flowers. By Uncle Matt.

THROUGH THE COPSE : Another Ramble after Wild Flowers. By Uncle Matt.

Cr. 8vo, each about 100 pages. (London: T. Nelson & Son. 1895.) 1/6.

Five charming little books by M. C. Cooke, M.A., LL.D., which will be read with interest by every intelligent boy or girl of twelve years of age, giving instruction in the structure and phenomena of plants. Besides a coloured illustration on the cover, each volume contains a beautifully coloured plate, and some 25 or 30 other illustrations.

WAYSIDE AND WOODLAND BLOSSOMS : A Pocket Guide to British Wild Flowers for the Country Rambler. By Edward Step. 12mo, pp. vii.—173. (London : F. Warne and Co. 1895.) Price 7/6.

A most useful book for the country rambler. It contains 128 coloured plates, giving figures of 156 species, 22 full-page illustrations, and clearly written descriptions of 400 species of plants. Persons taking country walks and who are unacquainted with botany will find this book a welcome pocket companion.

———

DIE NATURLICHEN PFLANZENFAMILIEN. By A. Engler and K. Prantl. Nos. 113—116. (London : Williams and Norgate. Leipzig : W. Engelmann.)

These four numbers contain the conclusion of the Guttiferæ, by A. Engler ; the Dipterocarpaceæ, by D. Brandis and E. Gilg ; Ancistrocladaceæ, by E. Gilg ; Elatinaceæ and Frankenlaceæ, by F. Niedenz ; Borraginaceæ (concluded), by M. Gúrke ; Verbenaceæ, by Briquet ; Bignoniaceæ (concluded), by K. Schumann ; Pedaliaceæ and Martyniaceæ, by O. Stapf ; Globulariaceæ, by R. v. Wettstein ; and Acanthaceæ, by G. Lindau. These four numbers contain 75 illustrations, composed of 673 figures.

———

A HANDBOOK TO THE CARNIVORA. Part I., Cats, Civets, and Mungooses. By Richard Lydekker, B.A., F.R.S. Cr. 8vo, pp. viii.—312. (London : W. H. Allen and Co. 1895.) Price 6/-

This volume of *Allen's Naturalist's Library* is devoted in a great measure to the family FELIDÆ, followed by the various genera of the family VIVERIDÆ. Attention is also given to extinct Carnivora. There are 32 coloured plates in the volume.

———

BRITISH FRESH-WATER FISHES. By the Rev. W. Houghton, M.A., F.L.S. Second edition. Roy. 8vo, pp. xxviii.—231. (Hull & York : A. Brown & Sons. London : Simpkin, Marshall, & Co. 1895.) Price 10/6.

This very handsome volume treats of the natural history of the various species of fishes that are known to occur in the rivers, lakes, and ponds of the British Isles, a description and an accurate engraving of each being given. Sixty-five species of fishes are described and illustrated. There are besides twenty-four prettily tinted plates of river scenery. The anatomy of the fish is exhaustively described in the preface.

———

A MANUAL FOR THE STUDY OF INSECTS. By John Henry Comstock and Anna Botsford Comstock. 8vo, pp. x.—701. (Ithaca, N.Y., U.S.A. : The Comstock Publishing Co. 1895.) $3 net or $4·09 post free.

This is a handsomely got up volume, in which the authors have endeavoured to give to entomologists a handbook, by means of which the names and relative affinities of insects may be determined in some such way as plants are classified by the aid of the well-known manuals of botany. They have taken much pains to render easy the classification of specimens ; groups of insects have been fully characterised, and much space given to accounts of the habits and transformations of the forms described. Special attention is given to wing veins of insects. There are one coloured and five plain plates and nearly 800 wood-cut illustrations.

———

PRACTICAL MICROSCOPY. By G. E. Davis, F.R.M.S., F.I.C., etc. etc. 8vo, pp. viii.—436. (London : W. H. Allen and Co. 1895.)

This is a third edition of this well-known work on the microscope, and contains much information which will prove useful to the microscopist.

THE THEORY OF LIGHT. By Thomas Preston, M.A. (Dub.).
Second edition. 8vo, pp. xvii.—574. (London : Macmillan and Co. 1895.)
Price 15/- net.

At the present time there are many text-books on light. The majority of them fall short of the requirements of those who wish to know how far investigation has been carried, or in what direction it remains to be pursued, and which of these are the most urgent and most likely to be attacked with success. The author's aim in this book is to furnish the student with an accurate and connected account of the most important optical researches from the earliest times up to the most recent date. Although a large portion of the book will be found suited to the reading of junior students, yet it will be found sufficiently full to meet the requirements of those who desire a more special acquaintance with the subject. This second edition has been thoroughly revised and more than 100 new pages added. There are about 250 wood-cut illustrations in the text.

THE EYE IN ITS RELATION TO HEALTH. By Chalmer Prentice, M.D., Chicago. 8vo, pp. 214. (Bristol : John Wright and Co. London : Simpkin, Marshall, and Co. 1895.) Price 6/6.

Many of the opinions advanced by the author are evidently original, and some are even of a surprising nature ; nevertheless, the general tenour of the book is scientific, and will doubtless repay a careful perusal.

MURCHE'S SCIENCE READERS. Books 1, 2, 3. By Vincent T. Murché. Cr. 8vo, pp. 127, 128, 176. (London : Macmillan and Co. 1895.) Price 1/-, 1/-, 1/4.

Three very interesting books of Object Lessons in Elementary Science, dealing in a simple way with the commonest properties of bodies ; the nature, growth, and structure of plants in general ; and some of the leading types of the animal creation. Each book is nicely illustrated.

LESSONS IN ELEMENTARY PHYSICS. By Balfour Stewart, M.A., LL.D., F.R.S. Fscap. 8vo, pp. xii.—475. (London : Macmillan and Co. 1895.) Price 4/6.

A new edition of this work, in which many additions have been made, is now before us. It brings before the student, in an elementary manner, the most important of those laws which regulate the phenomena of nature. It treats of Laws of Motion ; The Forces of Nature ; Energy ; Sound ; Heat ; Radiant Energy ; Electrical Separation ; Magnetism, etc. There are 157 illustrations.

AN INTRODUCTION to the Science and Practice of Photography. By Chapman Jones, F.I.C., F.C.S., etc. Third edition. Cr. 8vo, pp. 326. (London : Iliffe and Son. 1895.) Price 2/6.

In this work the various cameras, lenses, etc., are fully described, as are also the numerous processes. Owing to the rapid advances made in the science of photography, several new chapters have been added, and others carefully revised. The formulæ for solutions are given in both the English and metric systems.

NATURE IN ACADIE. By H. K. Swann. Cr. 8vo, pp. viii.—74. (London : John Bale and Sons, Oxford House, Great Titchfield Street. 1895.) Price 3/6.

We have here very pleasingly told a narrative of a Nature-lover's first voyage westward, with an attempt at the word-picturing of what he saw during his sojourn in Nova Scotia, or, as he poetically calls it, *Acadia*.

THE STORY OF PRIMITIVE MAN. By Edward Clodd. Fscap.
8vo, pp. 206. (London: George Newnes. 1895.) Price 1/-
In this little book we have told in pleasant language—I.—The place of
Man in the Earth's Life-History; II.—The place of Man in the Earth's Time-
history; III.—The Ancient Stone Age; IV.—The Newer Stone Age; and
V.—The Age of Metals. There are a number of illustrations, principally
relating to the Stone Age.

––––––

SHORT STUDIES IN NATURE KNOWLEDGE : An Introduction to
the Science of Physiography. By William Gee. Cr. 8vo. pp. xiv.—313.
(London: Macmillan and Co. 1895.) Price 3/6.
A thoroughly interesting little book, which, whilst making the reader
acquainted with some phases of the natural world, may also serve as a reader,
and companion to the text-books used in the upper forms at schools, etc.
There are 117 illustrations.

––––––

A NEW ENGLISH DICTIONARY on Historical Principles.
Edited by Dr. James A. H. Murray. Vol. IV., FANGED—FEE. By Henry
Bradley, Hon. M.A. Oxon. (Oxford: The Clarendon Press. London: H.
Frowde. April, 1895.) Price 2/6.
We have received the April part of this great work. This section, which
covers the words between FANGED and FEE, contains 897 main words, 179
combinations explained under these, and 187 subordinate words, making a
total of 1263. ––––––

THE STANDARD DICTIONARY OF THE ENGLISH LANGUAGE. By
Isaac K. Funk, D.D., Editor-in-Chief; Francis A. March, LL.D., L.H.D..
Consulting Editor; Daniel S. Gregory, Managing Editor; J. D. Champlin,
M.A., A. E. Bostwick, Ph.D., and R. Johnson, Ph.D., LL.D., Associate
Editors. 4to, pp. xx.—2318. (London and New York : Funk and Wagnalls
Co. 1895.)
We have the greatest pleasure in introducing this grand work to the notice
of our readers. It embodies many new principles in Lexicography and contains
5,000 illustrations made expressly for the work, besides a number of full-sized
coloured and plain plates; 301,865 Vocabulary Terms. Two hundred and forty-
seven editors and specialists and 500 readers for quotations were engaged on it,
and we are told nearly five years were required to complete this work in. Where
the whole work is excellent, it appears useless to mention any special portion.
We wish, however, to point to a few of the special features of this work.
Whilst all the trades and arts have been searched for new terms, in Electricity
alone something like 4,000 new terms have been entered and described. As a
proof of the thoroughness in which the work has been carried out, we will
quote the explanatory note at MYTHOLOGY :—
" Mythology among the Greeks took the form of idealisation of the beauti-
ful and esthetic (see list of gods at OLYMPIAN); as developed by the Romans,
it deified virility, war, and the principles of law and order (see list of gods at
PANTHEON); in India, it deified the forces of tropical nature (see ADITI, AGNI,
ASURI, BRAHMA, DEVA, DYANS, INDRA, KAMA, KRISHNA, NIRVANA,
PURANA, SIVA, TRIPITAKA, VEDA, VISHNU); in Egypt, it centred about the
Nile and its denizens (see ANURIS, APIS, ISIS, OSIRIS, PTAH, RA, SEB,
SERAPIS, SET, TYPHON); in Scandinavia, it idealised the struggle with the
Arctic forces of nature (see ÆSIR, ASGARD, MUSPEL, RAGNAROK, VALHALLA,
VAX). See also ANTHROPOLOGY. We think the editors have attained as near
as possible to perfection. The price of the Dictionary is—
  Single Vol. Edition, Half Russia, £2 8s.   Two-Vol. Edition, £3 0s.
  Full Russia (with Patent Index), £2 16s.   ,,   ,,   £3 8s.
  Full Morocco   ,,   ,,   £3 12s.   ,,   £4 8s.

THE ELEMENTS OF HEALTH : An Introduction to the Study of Hygiene. By Louis C. Parker, M.D., D.P.H. Lon. Cr. 8vo, pp. xii.—246. (London : J. and A. Churchill. 1895.) Price 3/6.

The author tells us his main idea has been to give some simple, yet practical information and instruction on the preservation of individual or personal health in the ordinary routine of domestic life. The book deserves careful reading.

IN SUNNY FRANCE: Present-Day Life in the French Republic. By Henry Tuckley. Cr. 8vo, pp. 249. (Cincinnati, U.S.A. : Cranston and Curts. 1894.) Price 4/-

An interesting series of sketches of people and things in France, the result of the author's personal observation and study, all very pleasantly told.

HOLIDAY TRAMPS through Picturesque England and Wales. By H. H. Warner. Fscap. 8vo pp. 318. (London : Iliffe and Son.) 1/-.

Just the book for the pocket when on our holiday excursions. We notice the most interesting places in England and Wales are visited during the nine tramps covered by this little book. Several maps and illustrations are added.

THE ·AUTOBIOGRAPHY of an English Gamekeeper (John Wilkins, of Stanstead, Essex). Edited by Arthur H. Byng and Stephen M. Stephens. Third edition, post 8vo, pp. 441. (London : T. F. Unwin.) 2/6.

Knowing Stanstead and the country round about well, we have read this book with peculiar interest. Mr. Wilkins has written his autobiography in a plain, straightforward manner, evidently in keeping with his general character. We wish that more keepers were like him.

THE ADVENTURES OF SHERLOCK HOLMES. By A. Conan Doyle. Cr. 8vo, pp. 343. (London : Geo. Newnes. 1895.) Price 3/6.

We are not surprised to learn that this thoroughly entertaining book is now in its thirty-second thousand. This volume contains twelve well-told adventures, with many good illustrations.

CASSELL'S NEW TECHNICAL EDUCATOR. Sixpence monthly.

We have received Nos. 30, 31, and 32 of this work, containing papers on a great number of technical subjects. In No. 31 is a coloured frontispiece, showing 18 specimens of cross-weaving. These numbers are well illustrated.

EUROPEAN BUTTERFLIES AND MOTHS By W. F. Kirby, F.L.S., F.E.S., etc. Parts 11, 12, 13. (London : Cassell & Co.) 6d. monthly.

These numbers describe the Methods of rearing and managing the larvæ and pupæ ; Preparing Lepidoptera for the Cabinet ; and the arrangement and management of a collection. A beautiful coloured plate accompanies each No.

BATTLES OF THE NINETEENTH CENTURY. Nos. 2, 3, 4. (Cassell and Co.) 7d. monthly.

These numbers contain exciting and well illustrated accounts of the Battles of Waterloo, Koniggratz, Ayacucho, Bull Run, the July Battles before Plevna, the Shanghai Patrol, the Siege and Storming of Delhi, Gislikon, Isandhlwana, Lissa, MacMahon at Magenta, Alma, Austerlitz, and several others.

THE BOOK OF THE FAIR. Part 10. (Chicago : The Bancroft Co. 1895.) Price $1.

This beautifully illustrated part continues the description of the World's Fair, recently held at Chicago. The conclusion of the section on Agriculture occupies the greater portion of this part, and Chapter 14, Electricity, is commenced. No expense appears to be spared in the production of this fine work.

Aphaniptera.

By Lt.-Col. L. Blathwayt, F.L.S., F.E.S.

Plate XVI.

I HAVE been interested, for some time past, in studying the anatomy of Fleas, and having accumulated a considerable amount of miscellaneous information regarding them, I thought that these fragmentary notes might be put togther, and woven into something which, if of no great value from a scientific point of view, would nevertheless prove of some interest to the Members of the Bath Microscopical Society, or, at all events, afford a subject for discussion.

The title of my paper is "Aphaniptera,"* a name given by Kirby from the Greek αφανης = inconspicuous, because, although fleas have no wings, they were believed to possess rudiments of wings. Macleay,† writing in 1819, says, "Vestiges of wings are to be discovered even in the flea." And Westwood,‡ in 1840, adopting Kirby's order Aphaniptera, gives as characters—"Wings, 4; minute scaly plates applied to the sides of the body, those of the metathorax being the largest. Huxley, in his *Manual of Invertebrated Animals* (p. 367), says " the two hinder somites of the thorax have lamellar appendages, which possibly represent wings "; and Nicholson, in his *Manual of Zoology* (p. 353), says of the Aphaniptera, "wings rudimentary in the form of scales situated on the mesothorax and metathorax." Theobald, in *British Flies* (Vol. I., p. 21, now in course of publication), defines the Aphaniptera as "parasitic, with scale-like rudimentary wings, the metathoracic scales being the largest." Although I have taken as the title of this paper the word "Aphaniptera," on account of its very general adoption, I confess I prefer the term "Suctoria" of De Geer, and for the following reason: *I do not believe that fleas*

* Kirby and Spence, *Introduction to Entomology*, Vol. IV., p. 382.

† Macleay, *Horæ Entomologicæ*, p. 357.

‡ *Modern Classification of Insects*, Vol. II., p. 488.

International Journal of Microscopy and Natural Science.
Third Series. Vol. V.

z

possess the least rudiment of a wing, and this I hope to explain presently.

Westwood considered that the fleas formed a distinct order, and he placed them between the Hemiptera and the Diptera, being allied to the former in the structure of the mouth organs, and to the latter in the metamorphoses they undergo ; and it is now, I believe, generally considered that their true place is among the two-winged flies. In the latest list of British Diptera, published by Verrall in 1888, the order Aphaniptera is done away with, and becomes a Family of Diptera, the *Pulicidæ*.

Fleas have long been the subject of very various researches ; they have been considered from philological, from historical, and from satirical points of view ; they have been celebrated for their strength and for their jumping powers, and still more frequently cursed for their bloodthirsty proclivities. They have been educated, and proved of considerable pecuniary profit to their instructors. But of their real structure, of their life-histories, and their correct classification, a great deal more knowledge remains to be worked out. For a long time it was thought that the fleas of different animals belonged only to a single species,* and, consequently, that the human flea was not different from that of a cat or a dog. So accurate an observer as Gilbert White writes in his *History of Selborne* (p. 200), " The sand-martin is strangely annoyed with fleas ; we have seen fleas, bed-fleas (*Pulex irritans*), swarming at the mouths of their holes, like bees on the stools of their hives." I need scarcely tell you that the flea found on the sand-martin is as distinct from *Pulex irritans* as a thrush is from a blackbird. Daniel Scholten,† of Amsterdam, showed in 1815, by his microscopical observations, that fleas differ from each other; and in 1832, Dugès, of Montpelier, investigated the distinctive marks of the various species.

Regarding the number of species, however, we are still left in the greatest uncertainty. Walker‡ describes fourteen, adopting the names of the animals upon which, as a rule, they are found ; as *Pulex canis*, *P. felis*, *P. gallinæ*, *P. talpæ*, and so forth. Verrall in his list, though adding several new species not mentioned by

* † Van Beneden, *Animal Parasites*, p. 127.
‡ *British Diptera*, Vol. III., p. 1.

Walker, reduces the number to thirteen; and Theobald also has a total of thirteen, though he includes one omitted by Verrall.

Dale, in his *History of Glanvilles Wooton*, published in 1878, writes (p. 290), "Although only fourteen species of the Aphaniptera have hitherto been described as British, thirty-eight have been taken in this parish." And he then gives a list of names, as *Pulex furoris*, on ferrets; *P. mustelæ*, on weasels; *Ceratophyllus elongatus*, on the great bat; *C. vespertilionis*, on the common bat; *C. musculi*, on the common mouse; *C. muris*, on the field mouse; and so on; considering as a distinct species, and giving a specific name to a flea if found on some animal from which he had not before taken it. His descriptions are remarkably concise; some are as follows :—*Ceratopsyllus merulæ*, in blackbirds' nests, *pallida picio-fusca*, length 1 line; *C. arvensis*, in skylarks' nests, *picio-fusca et obesa*, length 1 line; *C. trochili*, in willow-wrens' nests, *picio-fusca et oblonga*; and so on. All these and a number of others I believe to be merely *Pulex avium*, varying a little in size and colour. The colour of a flea, however, depends a good deal upon its age, and also on the contents of the stomach; while "*obesa*" was probably the result of a plentiful meal, or that the capture was a female containing eggs.

Before entering into an anatomical description of the flea, I shall refer to it from an historical and from a literary point of view. The earliest mention of a flea that I have come across occurs in the Book of Samuel,* cir. B.C. 1060: "for the King of Israel is come out to seek a flea." But about seven hundred years later, Aristotle (born B.C. 364), in his *History of Animals*, shows that he knew the flea underwent a metamorphosis, for he noticed, not only that it had distinct sexes, but also that they produced σκωλης ωοειδεις = egg-shaped worms. I suppose he must be here referring to the pupa, which is egg shaped, for the larva is greatly elongated. He did not, however, trace the changes of the insect far enough, and fancied that the perfect insect was generated spontaneously in the earth†; and Isodorus, a Bishop of Seville, who lived about the beginning of the seventh century, stated that the Latin name *Pulex* was derived from *pulvis* = dust, "*quasi pulveris filius*";

---

* I. Samuel xxvi. 20.

† *Class. of Ins.*, Vol. II., p. 492.

but Skeat considers it a modification of the Sanscrit *plu* = to
jump, from which we have the Dutch *vloo*, the German *floh*, and
English *flea*.   Scaliger thought that they were produced from
humours amongst the hairs of dogs.   Dr. Thomas Mouffett, an
English physician of the time of Queen Elizabeth, who wrote a
very curious Latin treatise of Zoology,* held an opinion similar to
that of Aristotle.   I have searched Pliny's *Natural History*, but,
although the whole of his eleventh book is devoted to insects, I
can find no mention of the flea.

    Latreille,† in his *Natural History of Insects*, says, "the Indians,
on account of their belief in metempsychosis, take the greatest
care of these creatures, as well as of all kinds of vermin that suck
human blood.   At Surat a Hospital has been established for them,
some one being hired for the night who allows the creatures to
make a meal from him."   He gives also an account of the per-
forming fleas which, in his time, were being exhibited in Paris.

    In their *Introduction to Entomology*, Kirby and Spence‡ say,
" Aristophanes, in order to make the great and good Athenian
philosopher, Socrates, appear ridiculous, represents him as having
measured the leap of a flea." §   I think, however, that the ridicule
was directed at the *method* of calculation, not at the calculation
itself.   The account runs, " Socrates and Chærephon tried to
measure how many times its own length a flea jumped.   They
took in wax the size of a flea's foot, then on the principle of " *ex
pede Herculem* " calculated the length of its body.   Having found
this, and measured the distance of the flea's jump from the hand
of Socrates to Chærephon, the problem was resolved by simple
multiplication.

    A great deal has been written about the habits of fleas, but it
is difficult to say how much of this can be relied on.   M. Defrance
wrote, "Small, shining black grains are nearly always found mixed
with the fleas' eggs, and these consist of dried blood.   It is a
provision which the thoughtful mother flea has prepared at our
expense for the nourishment of its posterity.   Figuier writes,||

* *Insectorum sive minimorum animalium Theatrum.*

† *His. Nat.*, Vol. XIV., p. 404.

‡ Vol. II., p. 310.

§ *Nubes*, Act. I., Sc. 2. || *Les Insectes*, p. 33.

" There is something very remarkable about the manners of fleas ; the mother flea disgorges into the mouth of the larva the blood with which she is filled " ; and Van Beneden, in Vol. xix. of the *International Science Series*, 1889, says, " The larvæ of fleas live only on what the full-grown insect brings them ; the mother flea sucks for herself first, and then divides the spoil with her larvæ." The stories are copied from one book to another, but as the author rarely or never informs us that he writes from personal observation, I think these statements require confirmation.   According to Van Beneden it is probable that the dog flea is the intermediate host of the tape-worm common to that animal.

The fleas are all comprised in the two families *Sarchopsyllidæ* and *Pulicidæ*. Passing over for the present the first of these, which contains only a few species, and none of them being found in this country, and adopting the classification in Verrall's list, we have in Britain three genera—*Pulex, Hystrichopsylla*, and *Typhlopsylla*.

The common flea, *P. irritans*, is sometimes regarded as the only species of the genus *Pulex*, and the rest are ranked under other genera; but this seems rather in honour of man as being the host of the first-named flea than from any real difference.

Fleas may be classed as follows :—

   a. Eyes distinct—Genus *Pulex* (in Britain about seven species).

   aa. Eyes wanting or rudimentary—*Hystricopsylla* (one species) and *Typhlopsylla* (four species).   The chief genus *Pulex* can be sub-divided.

   b. No spines on back or head—*P. irritans* (man).

   bb. Spines on back only—Comb with 18 teeth : *P. fasciatus* (rat), *P. melis* (badger), *P. sciurorum* (squirrel).   Comb 24 to 26 teeth—*P. avium* (birds).

   Under side of head with spines : 7 to 9 spines—*P. serraticeps* (dog, cat) ; 2 spines on cheek—*P. erinacei* (hedgehog) ; 5 to 6 on each cheek—*P. goniocephalus* (rabbit).

   Eyes wanting *Typhlopsylla*: 3 spines on cheek —*Assimilis* (mole).

The slides and photo-micrographs that I have brought to illustrate this paper are chiefly of *P. goniocephalus*, which I found in great numbers inside the ear of a rabbit.   This species was named *goniocephalus* by Taschenberg, on account of an angular projection on the forehead, which, however, is not always very apparent.   It is probably the *Ceratophyllus leporis* of Curtis.

The mouth-organs of the flea consist of a pair of mandibles with toothed edges, a tongue, an underlip with two palpi, which, when joined, form a hollow tube in which the tongue lies ; their structure is well shown in *P. avium.*   In this the tongue has been removed, and the mandibles are two long delicate bristles, which lie close together at the top.   They are so transparent that they are difficult to photograph, and in the female of *P. avium* they are scarcely visible ; they are the two faint lines on the left centre.   In *P. goniocephalus,* on the other hand, the mandibles are large, strong blades, doubly serrated along the edges.   There are also two maxillæ, generally triangular plates, each with a four-jointed palpus. Regarding the labial palpi, Westwood* says, "these are three-jointed. Latreille described them as three-jointed ; Curtis described them as four-jointed; while Dujès figured them as five-jointed. Theobald says they are four-jointed ; while Taschenberg, whose descriptions of fleas are by far the most accurate of any I have yet seen, says they are four-jointed, and adds, " Bouché has quite wrongly described them as five-jointed, and Kolenati has fallen into a similar error."   It is impossible to tell, by the examination of an entire flea mounted in balsam, what is the real number of joints in these labial palpi, though in the maxillary palpi the joints are distinct enough.   I did not try to decide the question until a few months ago, and then I found it almost impossible to get any living fleas, except *goniocephalus* from the rabbit, and *avium* from poultry ; but from these two species I found that in the latter they are five-jointed, as described by Bouché, Dujès, and Kolenati ; while in the former they are only two-jointed, a number of joints which, so far as I am aware, has not been mentioned by anyone.   In fact, I suspect it to be the old story of the Chameleon, " You all are right, you all are wrong," and that the naturalists who thus differ have dissected and described different species of flea.

The eye, when present, lies generally about the centre of the head, and is a simple ocellus ; sometimes, however, it is very near the lower margin.   Close behind it lies the antennæ, in a kind of

* *Class. Ins.*

groove, and are sometimes more or less covered by a chitinous plate. I shall return to the structure of the antennæ later on.

The thorax consists of three distinct segments, each bearing a pair of legs; and each segment is composed of one dorsal plate and two side-pieces or pleuræ. There is a very small spiracle, difficult to see, just at the posterior lower corner of the pleura of the prothorax. Then holding a very similar position on the meso-thorax there is the small, round scale, which has hitherto been called a rudimentary wing, but which is really a spiracle. Then at the back of the metathorax is the large flat scale, which is said to represent the hind wing by all except Taschenberg, who, how-ever, while denying that this is a rudimentary wing, asserts that it is a portion of the metathorax. For my part I am inclined to believe that it represents the first ventral plate.

There are eight clearly defined abdominal segments; some authors say nine and some ten; but then they probably include the genital plates. Each segment consists of one dorsal and one ventral plate, the former bending over the back and down on each side, and contains two spiracles, and usually bears one or more rows of bristles. The first dorsal plate, however, is much smaller than the others, and does not contain any spiracle, the missing spiracle being situated quite at the top of the so-called rudimentary wing, or, according to Taschenberg, metathoracic appendage, or, as I think, first ventral plate. The fact that it bears a spiracle, two rows of stiff bristles, that there is no plate beneath it, it alone forming a covering for the soft parts of the body below, is quite conclusive that it cannot be a wing. Taschenberg considers that the posses-sion of bristles and a stigma prevent it being a ventral segment, for, he says, " the other ventral segments have no spiracle and no such bristles. But then, if this is the case, I ask, " What has become of the first ventral segment?" It would only tire you if I went through all my reasons for my opinion; but very likely after all I may be wrong, and I shall certainly not forget the concluding lines of the *Chameleon* :—

> " When next you talk of what you view,
>   Think others see as well as you :
>   Nor wonder when you find that none
>   Prefers your eyesight to his own."

In addition to the presence of eyes and combs, I have found the following very useful distinctions in defining species :—

1.—The relative length of the joints of the tarsi.

2.—The number of joints in the labial palpi.

3.—The structure of the mandibles—in some a large, flat, double-toothed saw ; in others a long delicate bristle.

4.—Their very various antennæ.

Van Heurck, in his work on *The Microscope*, published two years ago, gives an illustration of the pygidium of a flea, and states that it contains from thirty-two to thirty-eight areolæ. As the pygidium of *P. goniocephalus* contains only twenty-eight, I think it probable that this also may be of use as a specific distinction.

I now come to the antennæ. The older naturalists quite overlooked these, which, as a rule, lie hidden in grooves close behind the eyes. Kirby and Spence, for instance, have a very good drawing of the parts of a flea's mouth ; but, like many previous observers, they do not appear to have noticed the real antennæ, but mistook the maxillary palpi for them, and consequently the names they gave to the other parts of the mouth are inaccurate. Latreille, also, although he overlooked the real antennæ, appears to have been in doubt, for he speaks of these maxillary palpi as "*Antennæ, potius palpi.*" Louis Figuier, in one of his popular works, published not more than twenty years ago, still states that these palpi are the flea's antennæ.*

Regarding the true antennæ, Westwood writes :—" The antennæ are minute articulated organs, varying in form in the different species, composed apparently of four joints, the third of which is very minute, and forms the cup-shaped base of the terminal joint, which in some species is furnished with numerous transverse incisions, which have been considered by Curtis as so many distinct articulations." There is, I think, no doubt that Curtis was quite right, for in this photo they are plainly separate joints. I do not think, however, that the so-called third cup-shaped joint is a joint at all, but consider it as merely a thickening of the top of the second joint, which forms a support for the base of the third. To show how greatly these organs vary both in size

* *Les Insectes,* p. 30.

and structure, according to the species and even the sex of the flea, I have prepared photographs of the antennæ of ♂ and ♀ of *Pulex avium*, one of *P. goniocephalus*, and one of *P. erinacei*, all magnified to the same extent—viz., 200 diameters.

Many species of flea have a number of spines on the back of the prothorax, forming a sort of comb, and some have similar spines on the sides of the face. These combs have the teeth always directed backwards, and must greatly assist the fleas in maintaining their hold among the hairs of the animals they infest, and the probability that this is their use is strengthened by the fact that the human flea alone is entirely devoid of them.

At the extreme end of the flea's body is the pygidium, a circular plate with a division down the middle, in which are situated a number of areolæ, each containing several wedge-shaped elevations placed in a circle, and with a rather long hair or bristle implanted in the centre. The spaces between these areolæ appear to be covered with small spines on the top of small tubercles. These show rather indistinctly in *P. goniocephalus*, which is magnified 750 diameters, taken with ½-in. oil-immersion without ocular. The addition of a projection ocular, which has increased the magnification to 1,500 diameters, has, however, failed to disclose any further detail. The pygidium of a flea is an excellent test to judge of the defining power of an objective.

The second genus, *Hystrichopsylla*, contains only one British species—the *Pulex talpæ* of Bouché, *Ceratophyllus talpæ* of Curtis, regarding which, however, Dale says, writing in the *Entomologist's Monthly Magazine*, in June, 1890, " It is quite certain that *Talpæ* is a misnomer, as it is not found on the mole." As I have not been successful in obtaining a specimen, I pass on to the third genus, *Typhlopsylla*, which contains four or five species, some of them infesting bats. One species, *T. assimilis*, is very common on the mole.

Not having as yet bred any fleas, I have no personal knowledge of their metamorphoses. It is said that each female lays about a dozen eggs at a time. These are deposited in various places, such as the cracks in boards, in carpets and rugs, on the hairs of dogs and other animals, and in various other places according to the species. These eggs soon hatch, and from them come worm-like maggots. These larvæ have no feet, but the last segment has two strong curved hooks. In a fortnight or less some species spin a silken cocoon. (It was this, I fancy, which made Aristotle speak of them as "egg-shaped" worms.) In some species the larvæ are said to live through the winter and not pupate until the spring. Certainly, during the past winter, I have had the greatest difficulty in obtaining any fleas.

Regarding the parasites of fleas I know very little; but only a few days ago, on some fleas I took from a rat, I found some minute white acari, which were to a wonderful degree tenacious of life. I put the fleas into liquor potassæ, which appeared to kill them in about ten minutes; but twenty-four hours later I found an acarus walking about on the body of one of them to all appearance none the worse; the remaining acari had all disappeared.

I have not been able to find any record of fossil fleas. The geographical distribution of fleas seems to be world-wide; they are found from the Arctic regions to the equator. Darwin, in the *Voyage of the Beagle*,* writes of the mountains near Coquimbo :—
"I enjoyed my night's rest here from a reason which will not be fully appreciated in England ; namely, the absence of fleas. The rooms in Coquimbo swarm with them; but they will not live here at the height of only three or four thousand feet." This, he states, cannot be on account of the trifling diminution of temperature at that height. They occur just as plentifully in the mountain châlets of Switzerland as in the warmer valleys.

Before bidding adieu to the fleas, I must just mention the other family of *Aphaniptera*—the *Sarcophsyllidæ*, of which fortunately we have no representatives in this country.

According to Taschenberg there appear to be only two or three species, but it contains the famous, or rather the infamous, "Jigger" (*Sarcopsylla penetrans*), an American pest, especially abundant in the West Indies and in the north of South America. It is often called the sand-flea. The males and the immature females are no worse than other fleas; it is the impregnated female which is the chief trouble. She burrows under the skin of animals, in the feet or under the toe-nails of man. There her body swells to the size of a pea under the pressure of the eggs within her,

* *Voyage of Beagle*, p. 344.

which eggs are ready to hatch in about a week. In the meantime the flea herself has undergone a marked change, the internal organs being atrophied by the growth of the eggs.

According to some authorities, if the female is not extracted mortification of the toe or foot may set in. The late Rev. J. G. Wood disbelieved the stories of the injuries caused by the Chigoe. He gives his brother as his authority, and his words are:—" A great deal of rubbish has been written about the Chigoe. It is true our friend is a great nuisance in his way, but in six years I have never known, or ever heard, of anyone being much the worse for the Chigoe, though I have seen some people too lazy to extract them until their feet were full of their nests. . . There is a slight itching, and then, if they are extracted with any reasonable amount of care, the nest of eggs comes away all correct. If it should be broken, which will happen sometimes, a pinch of snuff is put into the hole, and there is an end of the matter."* Other authors, however, who had as good or better opportunities of finding out the true state of the case, have come to a quite different opinion. Until about twenty years ago the Jigger was believed to be confined to America, and Mr. Newton, Vice-Consul at Loanda, states that before 1872 the Jigger was not known on the West Coast of Africa; but in that year a ship, the *Thomas Mitchell*, arrived from Rio Janeiro, the crew of which were suffering from Jiggers. These were quickly communicated to the crews of the boats and introduced on shore, and they have since gradually spread along the coast. Mr. Newton says that he has seen many natives without toes, and in a dreadful state from allowing the eggs to remain and hatch and the wound to fester.

Walton, in his *History of St. Domingo*, records that a Capuchin friar, desirous of settling the dispute as to what genus the Jigger belonged, brought away with him from that island a colony of these animals, which he permitted to establish themselves in one of his feet; but unfortunately for himself and for science, the foot intrusted with the precious deposit mortified, was obliged to be amputated, and, with all its inhabitants, committed to the waves.†

I think that more credit must be given to the affirmative than

---

* *Insects Abroad*, p. 772.
† Kirby and Spence, Vol. I., p. 100.

to the negative evidence, and the Jigger be put down as one of the worst of insect pests.

Although I shall be very glad if some other member of this Society will take up this subject, I must point out that it has its drawbacks. During the winter I could obtain very few specimens, but with the arrival of the warm weather the fleas arrived, mostly the dog-flea, by hand and by Parcel Post. I now look on any unknown small box with considerable suspicion, and open it with as much care as a policeman examining a box which he thinks may contain an infernal machine.

———

Besides a number of slides, the following photographs were shown to the members :—

Pulex irritans, man.

P. fasciatus, rat.

P. sciurorum, squirrel.

P. avium, sand martin.

P. serraticeps, cat.

P. erinacei, hedgehog and dog.

P. goniocephalus, rabbit.

Typhlopsylla assimilis, mole.

Ditto   ditto   turkey.

P. avium, ♀, mouth organs.   Mandibles like long fine bristles. × 100.

Ditto, Mandibles, labium, with 5-jointed palpi; Maxilla, with palpus, × 100.

P. erinacei (from dog), mouth-organs complete, tongue serrated as well as mandible, × 100.

P. goniocephalus, ♀, mandible serrated, × 200.

Ditto, antenna in groove behind eye, × 100.

Ditto, 1st ventral plate, or rudimentary wing, showing spiracle, × 50.

Ditto, one of the dorsal plates, × 50.

P. avium, portion of head with antennæ, showing articulations, × 200.

P. goniocephalus, ♀, antenna, × 200.

P. avium, ♀, portion of head with antenna, × 200.

P. erinacei, antenna, × 200.

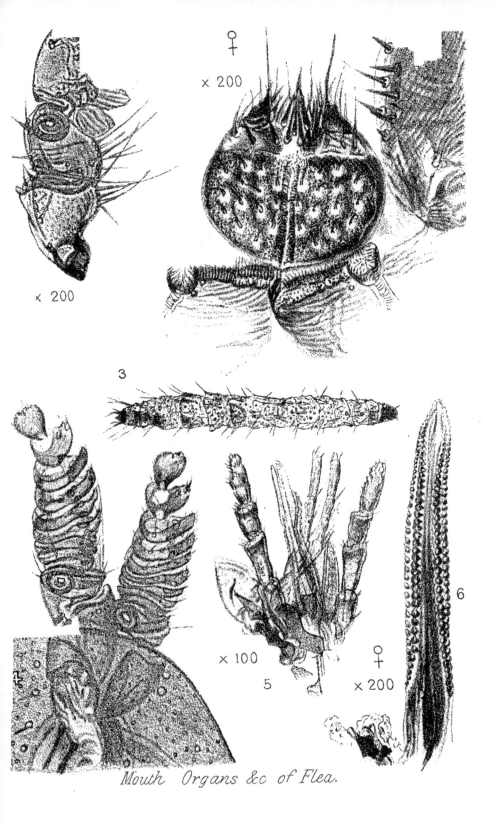

× 200

♀
× 200

3

× 100

5

♀
× 200

6

Mouth Organs &c of Flea.

P. goniocephalus, ♀ , pro-thor. comb with 15 teeth, × 100.

Ditto, maxilla, with 5 genal spines, × 100.

Ditto, pygidium, with last abdominal segment, × 100.

Ditto, ditto, showing hairs which spring from areolæ, × 750.

Ditto, portion of pygidium, showing areolæ, × 750.

Ditto, areolæ of pygidium, × 1500.

EXPLANATION OF PLATE XVI.

Fig. 1.—Antenna of Flea from dog, *Pulex erinacei.*

,, 2.—Pygidium of *Pulex goniocephalus*, showing hairs which spring from areolæ.

,, 3.—Larva of Flea from Cat.

,, 4.—*Pulex avium*, ♂ . Portion of head, with antennæ.

,, 5.—Complete mouth-organs of *Pulex erinacei*, showing serrated tongue as well as mandible, × 100.

,, 6.—Serrated mandible of *Pulex goniocephalus*, × 200.

𝕽𝖊𝖑𝖆𝖝𝖎𝖓𝖌 𝕴𝖓𝖘𝖊𝖈𝖙𝖘 𝖋𝖔𝖗 𝕮𝖆𝖇𝖎𝖓𝖊𝖙.

By G. H. BRYAN, M.A.

I HAVE not tried the effect of naphtha as recommended in the extract from the *Entomologist's Record*, quoted on p. 226 of this Journal, but I have tried shellac, and I have also tried removing the insects from the setting board, both at the end of twenty-four hours and after a few days or a week, and I can therefore state from experience that there is no surer way of rendering valuable insects worthless as cabinet specimens. Shellac disfigures the bases of the wings, and is at the same time no safeguard against springing, and the only relaxed insects which I have had to destroy in my collection on account of the wings having "sprung" so hopelessly as to ruin them were some that I fixed with shellac many years ago. There is no "royal road" to relaxing insects.

As far as my experience goes, I find the best plan is to pour a layer of plaster of Paris into the bottom of a biscuit-tin and keep the plaster well moistened, and place the insects over this. It is better to run the plaster into a separate tablet instead of pouring it straight into the tin, as the tin is less likely to rust, being less closely in contact with the plaster. For butterflies in envelopes whose wings are folded, I stand a number of pins upright in rows in the plaster just as it is setting, and stand each insect between two pins. I have been more successful with plaster than with sand, but a great deal depends on keeping the relaxing box at a sufficiently warm temperature (yet not so hot as to soak the wings with steam). A hot-water cistern is very good for this, but I have even had recourse to a small spirit-lamp on emergency.

A little wet, if it condenses on the wings, can be absorbed with blotting-paper, and the insects will stand a fair heat, which will generally render them more pliable in a day than if they had been kept in the cold for an indefinite length of time. A little care about heating is of far more value than all the nostrums which various writers recommend on the experience of setting half-a-dozen insects. As regards preventing springing, if the insects have been on the setting boards for a *month*, one may *begin* to think of taking them off. Three weeks is decidedly risky, unless the boards are kept in a particularly warm, dry place—*e.g.*, hanging on a wall, with a kitchen chimney behind it. If on being placed in the cabinet, the wings still show signs of springing upwards, they may often be kept down with card braces transfixed with pins for weeks until all tendency to spring has been cured.

Finally, if you do not wish to have your specimens disfigured with unsightly rubs and gashes across the wings, such as are seen in nine collections out of ten, you cannot be too particular about the mounted setting needles used for drawing the wings into place. Curved needles are particularly useful, as they enable the wings to be moved forward into place from the underside, where the needle has a better grip on account of the projecting veins, and rubs do not show above. I have nearly thirty different kinds of setting needles, each adopted to its own special purpose, some with only 1/32nd of an inch projecting beyond the wood, some with the eye end projecting, besides a large number of curved ones. A "needle holder," as used for dissecting purposes, is also invaluable. Without a good supply of setting needles ready at hand, a considerable percentage of insects will be damaged in setting.

Questions bearing on Specific Stability.

By Francis Galton, D.C.L., F.R.S.

[The following paper, which has just appeared in the *Transactions of the Ento-mological Society* (1895, pp. 155—7), is, we think, of such importance that we obtained the kind permission of Dr. Galton to reprint in this Journal, in the hope that some of our readers may help to elucidate the many problems connected with Specific Stability. Communications relat-ing to the subject should be addressed to Dr. Galton, 42 Rutland Gate, London, S.W.]

"THE questions are more especially addressed to those who have had experience in breeding, but by no means to breeders only; nor are they addressed only to entomolo-gists, being equally appropriate to the followers of every other branch of Natural History.

"I should be grateful for replies relating to any species of animal or plant, whether based on personal observation or referring to such observations by others as are still scattered through the wide range of periodical literature, not having yet found a place in standard works.

"The questions are for information on the following subjects:

"(1) Instances of such strongly-marked peculiarities, whether in form, in colour, or in habit, as have occasionally appeared in a single or· in a few individuals among a brood; but no record is wanted of monstrosities or of such other characteristics as are clearly inconsistent with health and vigour.

"(2) Instances in which any one of the above peculiarities has appeared in the broods of different parents. [In replying to this question, it will be hardly worth while to record the sudden appearance of either albinism or melanism, as both are well known to be of frequent occurrence.]

"*Note.*—The question is *not* asked now whether such pecu-liarities, or 'sports,' may be accounted for by atavism or other hypothetical causes.

"(3) Instances in which any of this peculiarly characterised individuals have transmitted their peculiarities, hereditarily, to one or more generations. Especial mention should be made whether the peculiarity was in any way transmitted in all its original intensity, and numerical data would be particularly acceptable that shows the frequency of its transmission (*a*) in an undiluted form, (*b*) in one that was more or less diluted, and (*c*) of its non-transmission in any perceptible degree.

"It is impossible to explain . . . the precise way in which the derived facts would be utilised. An explanation that would be sufficiently brief for the purpose could not be rendered intelligible except to those few who are already familiar with the evidence, and the technical treatment of it, by which the law of Regression is established, and with the consequences and requirements of that law. Regressiveness and Stability are contrasted conditions, and neither of them can be fully understood apart from the other.

"I may as well take this opportunity of appending a list of my various memoirs on these subjects. The most important of these are Nos. 1, 3, part of 6, 7, and 8, in the following list. Nos. 1 to 5 refer to regression only.

"List of Memoirs by the Author on Regression and Stability.

1.—'Typical Laws of Heredity.' *Journ. R. Institution*, 1877. This was the first statement of the law of Regression, as founded on a series of experiments on sweet peas.

2.—'Presidential Address, Anthropol. Section, Brit. Assoc., 1885. Here the law of Regression was confirmed by anthropological observations.

3.—'Regression towards Mediocrity in Family Stature,' *Journ. Anthrop. Inst.*, 1885. A revised and illustrated reprint of No. 2.

4.—'Family Likeness in Stature.' *Proc. Roy. Soc.*, 1886.

5.—'Family Likeness in Eye-Colour.' *Proc. Roy. Soc.*, 1886.

6.—'Natural Inheritance.' Macmillan and Co., 1889. This vol. summarises the results of the previous work.

7.—'Patterns in Thumb and Finger Marks, and the resemblance of their classes to ordinary genera. *Phil. Trans. Roy. Soc.*, 1891.

8.—'Discontinuity in Evolution.' *Mind*, 1894. This is an article on Mr. Bateson's volume."

On the Study of Micro-Fungi.

By W. Thomson.   Plate XVII.

———

THE study of fungi, while being regarded by many as perhaps one of the most uninviting which the world of nature affords, is, nevertheless, regarded by others as one of the most fascinating of the many branches of Natural Science. One man smites down with his stick every fungoid growth that appears before him, whereas another man would almost as soon think of smiting down a friend. If those of us, however, who are regardless of the fungi world were privileged to enter the *sanctum sanctorum* of some mycologist who has devoted years to the study and collection of fungi, our indifference would be supplanted by deep interest. As he displays the treasures of his collection, drawer after drawer filled with boxes, and shelf after shelf laden with bottles, all containing specimens, and describes with enthusiasm some particular points in their life-history, or some exciting hunt he had in quest of them, it will begin to dawn upon us that there must be something in it after all. I have been in the rooms of a certain mycologist (lately deceased), where every available resting place was occupied either with specimens or with literature on the subject, and where even the dining-table was literally covered, save a small portion reserved at one end for its legitimate use, with fragments of bark, broken pieces of branches, withered leaves, and miscellaneous piles of specimens, which had just been brought home for examination. In the midst of such disorder, however, my friend was in his element. He seemed thoroughly possessed with his favourite hobby, and dwelt with manifest pride upon his specimens. He ate many of the species, talked continually about them, and indeed appeared to live entirely for fungi.

If we had followed our friend into his happy hunting grounds, we should have found him prowling about old stick-yards, creeping along dry ditches, or laboriously examining one by one the fir cones, leaves, or broken twigs, which covered the floor of some plantation. Indeed, so enthusiastic and diligent was he that it was often said that he had fungus on the brain; but he heeded not the remarks nor the opinions of those who, as he would say, knew nothing about fungi.   His perseverance was rewarded by the discovery of many species which had not previously been found. He was ever ready to have a chat with kindred spirits, and to try to influence others to take an active interest in a study which had afforded so much pleasure and enjoyment to himself.   No stranger who visited his rooms and evinced any interest in the specimens would depart without first having seen his collection of microfungi, and more particularly his collection of those micro-fungi which grow upon living plants, and we know that those who took his advice to bestow a little attention on this comparatively neglected study have received an ample reward for their trouble.

Now, it is to this latter-named branch of mycology that we wish specially to draw attention in this paper, and we will try to show that the domain of micro-fungi is not such an uninviting and desert land as it might at first sight appear.   The honey-bee finds in the modest-looking clover more of the sweetness of which it is in search than in the brilliantly painted foxglove ; and many insects and animals find a more congenial abode on plants of a very humble and unattractive character than on many of those which are most highly esteemed by man.   In the same way, therefore, as an insect may prefer to live upon a common nettle rather than upon the Rose of Sharon, so there are men who, in making a choice of some branch of study as their own speciality, prefer to pass over the more prominent and fashionable fields of research, and to settle down in a comparatively obscure corner.   There is doubtless a certain novelty and charm in a choice which leads one out of the hard-beaten track, and away from the busy crowd of observers, into a less-populated country, where there are yet many things to be discovered.

Thomas Carlyle remarks that the Torch of Science has been brandished and borne about with more or less effect for 5,000 years and upwards, and that " in these times not only the Torch still burns, and perhaps more fiercely than ever ; but nnumerable Rushlights and Sulphur Matches kindled thereat are also glancing in every direction, so that not the smallest cranny or doghole in Nature or Art can remain unilluminated."   Though almost every cranny now obtains a share in the general illumination, still many of them are yet dim enough to render acceptable the smallest light, and around the subject of our present study there is still sufficient darkness to give our sulphur matches an opportunity of letting their light be seen.

There is a certain fascination in poking about hedgebanks and odd nooks and corners in search of these fungi, knowing that at any moment some new specimen may present itself, or the missing link in the life-history of some species be found. This opportunity of turning up fresh soil without very much difficulty should prove a strong recommendation to anyone who wishes to find out something new. The study of micro-fungi should also commend itself to those who desire to take up a subject which will require the use of a microscope, for in it they will find that their favourite instrument is in continual demand ; indeed, that it is an absolute necessity.

It is, of course, a much pleasanter occupation to search for the fungi which are parasitic upon the leaves and stems of living plants than for those which are to be found only by grovelling among rotting wood and decaying vegetation. A beginner may probably experience some little difficulty at first in finding specimens, and will be sure to overlook many which would be seen at once by a more experienced eye. As the fungi are small and many of them inconspicuous, they require to be very carefully searched for, or else they will never be found at all. A hasty observer may cover a great deal of ground in an afternoon's excursion, but, in all probability, he will not secure half so many specimens as the collector who travels no further than a stone's throw from the starting-point, and who carries out the search in a complete and exhaustive manner.

Most of the fungi we are considering grow on the under-surface of leaves, and therefore, until some experience has been gained, a considerable amount of time will be spent in diligently turning over the leaves without any specimens being found. Very soon, however, the fungus-hunter will learn to recognise, from certain signs—such as the sickly appearance of the plant, the discoloration of the leaf, or the abnormal growth—those plants which are most likely to have a fungus established upon them. Continual practice will soon develop a sharp eye for the most minute growths, so that in course of time it will be found that even when walking quickly along a road or over a field, the suspicious-like appearance of a plant or a small coloured spot upon a leaf will often arrest the attention of one who is familiar with such signs.

We remember that upon one occasion, when hastily crossing a field, a slight yellow coloured spot on a leaf of the common daisy was sufficient to cause a halt in order to see what was the reason for the discoloration   It turned out to be the Cluster Cup Fungus (*Æcidium bellidis*), and although a considerable search was made, no more than three very small diseased leaves could be found. This habit of always looking for diseased plants grows upon one, so that no matter how beautiful the flowers and surrounding vegetation may be, the eye is continually wandering about in search of any sign which betokens the presence of a fungus; and, although it may sound strangely to the uninitiated, the joy is much greater when the plant is found to be attacked by one of these parasites than when it is in the full vigour of health.

It is not always easy, by a cursory examination, to distinguish those marks which are caused by fungi from those which are due to insects or the natural decay of the tissues of the plant; indeed, until the microscope has been brought to bear upon them, it is sometimes almost impossible to settle the question with certainty. The common dock by the wayside is a prominent offender in this respect, as the leaves are very frequently marked by coloured spots, which prove an abundant source of disappointment, until the collector is rendered wary and taught not to expect too much from them. Mistakes of this nature, however, become fewer as the observer becomes more familiar with his subject.

The majority of the species must be looked for on the under-surface of the leaves, and there will be found to be a great diversity, not only as regards the microscopical appearance of the different spores, but also as regards the appearance which they present to the naked eye, both in colour and in the manner of their growth on the host plant. The beautiful cluster cups, to which we shall refer more particularly further on, will, on some plants, be seen in clusters, on others in circles, and on others scattered about in no apparent order or running down the leaf-stem, as is frequently seen on the Violet; while on the Willow

Herb, for instance, they will cover the whole under surface of the leaf. In some cases the fungi appear as black, brown, orange, or reddish patches or spots, and in other cases as a fine black or golden powder peppered over the leaf. Then we have species which adorn the backs of leaves like spores on the fronds of ferns, species which give the plants the appearance of having been well sooted, and species which invest the stems and leaves with a covering as though enamelled with a coat of white paint. We shall find fungi which follow the lines of venation, and fungi which spread themselves quite independently. On the Dog's Mercury we shall see the parasite *(Uredo confluens)* slowly destroying the leaves on which it is established ; and on the Nettle we shall often notice how the Cluster Cup *(Æcidium urticæ)* irritates the tissues and causes the stem to swell and to become much bent and distorted.

Of course, there are species—such as *Puccinia calthæ* on the Marsh Marigold—which are rarely met with, and species—such as the *Æcidium* on the Coltsfoot—which are very common. In one locality a certain fungus may be exceedingly plentiful, whereas in another place, although the host plant is common enough, it may not be possible to find a single specimen of it. These parasites make a host of all manner of trees and plants—the ōak tree, the rose bush, and the tender moschatel alike forming a habitation for them. But it does not by any means follow, that because a fungus has been recorded as growing upon the young leaves of the oak in one part of the country, that therefore we are likely to find specimens of it wherever we find an oak tree ; nor should it always be taken for granted that because we have found a certain fungus in one locality, that all the host plants of the same species which are growing even comparatively near at hand will also be infected with the fungus. The case of the *Æcidium* on the Daisy before mentioned illustrates this, as, although there were plenty of the same plants growing around the spot where the three diseased leaves were found, yet all of them appeared to be perfectly free from the fungus.

What causes affect the distribution of the fungus and lead to its appearance in one place and not in another, and how it comes to pass that the spores which grow upon, say, the Primrose will not develop upon, say, the Dandelion nor on any other of the many plants which may be flourishing around, will probably become clearer when we consider their life-history.

In commencing to make a collection of these minute forms, it would undoubtedly be a great advantage to have the advice of someone already acquainted with the locality, who could point out where particular specimens might be looked for, and which were the most prolific places in the district; but, though such assistance might often save a good deal of needless searching, it might, on the other hand, not be very conducive to the acquirement of much fresh information as to their habitats, nor leave very much scope for the discovery of new specimens. Neither does it seem an altogether satisfactory plan for the collector to make out a list of host plants from a book, and then, searching out these plants, to examine them for the parasite which has already been recorded as occurring upon them. This method would no doubt expedite the making of a collection, but it is manifest that it would curtail the chances of adding to the number of host plants. It appears to me that the best plan is to go forth unbiassed towards any particular plants or district, and to search first one plant and then another, totally regardless as to whether a fungus has ever been found upon them or not. By so doing we should, at any rate, put ourselves into the path of original observations, and have an opportunity of being successful.

The specimens when collected should be pressed, and then either affixed to paper or preserved in an envelope, care of course being taken to keep a record of the place where obtained, and of the date of finding. When the name of the host plant has been noted, and the spores of the fungus examined with the microscope, we shall very probably be able to make out the species without much trouble, but when this is done we should not let our interest in it cease, for each fungus has a life history of its own, and it ought to be our duty to learn something regarding it. There is certainly considerable pleasure to be derived from collecting the specimens and ascertaining the species; but if we content ourselves merely with gathering together a collection, we do not differ very much from the man who stocks his library shelves with books and chooses to remain in ignorance as to what is written within them.   If this man were to try to arrange his library, he might easily put the parts of a three volume work in different places in his bookcase, instead of keeping them together as one work, and this is exactly what was done with these fungi, for many of them were classified as being distinct species until their life histories were studied, and it was shown that one species might have several stages, and that, though the spores in each stage were different, they required to be reckoned only as one species.

If we briefly trace the growth of one of these fungi and follow it through the different stages, we shall, perhaps, be enabled more easily to understand what is here referred to, and for this purpose we propose to take as a sort of typical example the species which grows upon the Nipplewort (*Lapsana communis*). We have chosen this species for our illustration because we have found it, in addition to being one of the commonest, to be one of the earliest to appear after the winter is over, and because the host plant can so readily be obtained for purposes of experiment by those who wish to try to cultivate the fungus.   If we were to examine one of these plants which ultimately produces a fungus, we should at first see no evidence of the presence of the fungus, although it had already taken up a lodgment in the tissues of the plant; but though nothing is outwardly visible, the parasite is nevertheless gradually gaining a footing.   Within the tissues there is growing and developing that part of the fungus, the mycelium, which in due season will make itself manifest and produce the spores which appear on the surface of the leaf.   This mycelium consists of an assemblage of hyaline tubes, which ramify chiefly between the plant-cells, and these mycelial tubes branch off and unite with one another until a perfect network is formed within the tissues.

When we remember that the fungus is a parasite, it is scarcely necessary to say that consequently it lives at the expense of the plant upon which it has taken up its abode, and it will at once be evident to us that the mycelium is for the purpose of supplying the fungus with the necessary nutrition by abstracting the material which the plant has elaborated for its own use.   The fungus cannot, therefore, be considered as a very welcome guest, seeing that the host must, *nolens volens*, provide it with sustenance quite independent of any question of comfort or convenience.   When the mycelium has developed to a certain extent, the areas in which it is present will be distinguished by the green of the leaf being changed into a reddish or purplish colour. Not only is the chlorophyll affected, but, as might be expected from the presence of such a growth in the tissues, the parts will very probably soon become swollen and bulged out, giving a deformed appearance to the leaf.

When the mycelium is sufficiently far advanced, large numbers of very fine hyphæ or branches are given off from it. These hyphæ grow towards the epidermis, and their apices incline to a central point, so as ultimately to form a flask-shaped body, termed a spermogonium, immediately beneath the epidermis. The apex of this body, spermogonium, then pierces the cuticle and becomes visible on the surface of the leaf as a minute elevation. From the ends of the hyphæ, which project into the interior of the spermogonium, are budded off a large quantity of very small, round bodies, to which the name spermatia has been given. These spermatia continue to increase in numbers till the whole cavity is filled with them.

The hyphæ which form the apex of the spermogonium then separate and make, as it were, a mouth to the flask, thus providing a free passage from the interior of the body. Through this passage the spermatia, which are held together by a sticky substance, gradually ooze out to the surface of the leaf, and it is stated that this propulsion of the contents is due to the action of moisture in causing the gelatinous matter which binds the spermatia to swell. Plate XVII., Fig. 1, shows a section of a spermogonium. When the spermatia are placed in water containing sugar, it has been found that they multiply almost in the same manner as the well-known yeast spores. Many experiments have been made by those who make a special study of these objects with a view of ascertaining definite information regarding their functions, and of finding out what part they play in the life of the fungus; but it does not appear that any of the theories which have been advanced can be considered as having given a satisfactory explanation.

The exit from the spermogonium is surrounded by the ends of the hyphæ, and they are supposed to act as a sort of hedge to prevent the spermatia from being washed off. Whilst the viscid substance in which they are embedded restricts their distribution by preventing the wind from blowing them away, it has been observed, on the other hand, that this investing material forms an attraction for insects on account of the saccharine substance it contains, and that when the insects depart after their feast of good things they carry away, adhering to their feet, small portions of this viscid substance, with, of course, the spermatia that are enclosed in it. The spermogonia are very small, and appear immediately before, or at the same time, as the spores about to be mentioned, and, unless they are specially looked for, may very easily pass unnoticed.; but if a beginner experiences any difficulty in making out the spermogonia to his satisfaction, he will have no trouble in recognising the development we are now to consider : the æcidiospore stage.

As soon as ever the Nipplewort begins to put forth its young leaves in spring, just so soon should we look out for the first stage of the fungus. We have found it bursting through the leaf as early as the 11th February, but it will, as a rule, be most plentiful about the month of April. By, say, the end of May, the first stage will have disappeared entirely. If, then, we examine the plants almost as soon as they appear above the ground, we shall, if the mycelium is present, very soon see evidence of its existence in the change of colour which begins to take place in the affected areas. Before long there will appear, shining through the epidermis, the yellow spots which show that the æcidium is being formed beneath, and if we look carefully we shall perhaps be able to find the spermogonia. In a short time the æcidium will burst through, and we shall have the pleasure of looking upon a number of little cups filled with yellow spores.

We say pleasure, because a cluster of these cups, each fringed with a margin of white teeth and brimming over with the golden grains, is a sight which never fails to please, and, especially when examined with a pocket-lens or low power, it forms an object of great beauty. When the fungus is further advanced, it will be noticed that the peridia, or cups, are abundant on both sides of the leaves, and, as it has been stated already that the æcidium on many plants appears only on the under surface, it will be noted that *Æcidium lapsanæ* is an exception in that respect.

The peridia grow in very irregular patches or clusters, and often cover more or less all the leaves of an infected plant. Indeed, this stage of the fungus affects the host plant much more prominently than do the succeeding stages, and it is consequently more easily found. These cups arise from the same mycelium which produced the spermogonia, and are formed at centres where the mycelial tubes have become branched and gathered into small, round masses. These almost spherical bodies increase in size, and, when about ready to break through the cuticle, it is found that in the centre of each there has been developed a number of upright hyphæ or branches of the mycelium, each supporting a column of spores, and that this forest of hyphæ and multitude of spores are enclosed within a covering of barren cells. The spores become filled with a yellow colouring matter, those at the tops of the column coming to maturity first, but the enveloping cells— though in other respects similar to the cells from which the spores have been formed—remain empty.

As soon as the æcidiospores—that is, the spores of the *Æcidium*—are ripe, the epidermis is ruptured and the fungus presents itself to the world. The covering of barren cells gives way and becomes curved backwards around the margin, thus turning the body into a small cup, and laying bare the æcidio-spores which have been produced within. Pl. XVII., Fig. 2, represents a magnified section of an æcidium, and makes clear its general structure. The spores at the surface of the cups being ripe, and having been detached from the tube in which they were developed, are at the mercy of every wind that blows, and it will easily be observed how very readily they are scattered abroad. On examination with the microscope, these æcidiospores will be seen to be smooth and nearly spherical. In diameter they are about 18 micromillimetres (one micromillimetre = 1/25,400ths of an inch). As we look into one of these cups and see the quantity of spores which it contains, and then glance at the plant and see the number of cups flourishing all over the leaves, we cannot but be struck by the prodigious number of spores which there must be altogether.

Dr. M. C. Cooke, in his book, *Rust, Smut, Mildew, and Mould*, writes, in reference to the *Æcidium* on the Goatsbeard, that if we compute 2,000 cups as occurring on each leaf, and suppose each cup to contain 250,000 spores, then we shall have not less than five hundred millions of spores on one leaf of the Goatsbeard.* Whether *Æcidium lapsanæ* can boast of such a wealth as that, we shall not pause to consider, but the figures will give some idea as to the sort of numbers attained when the spores are actually counted.

If, now, we obtain some fresh specimens, and shake or brush off the ripe spores at the tops of the cups into a little water on a glass slip, we shall, without difficulty, be able to observe the germination of the spores. In order to confine a small quantity of water for this purpose, it will be found very convenient to affix a small indiarubber band to an ordinary glass slide with balsam or some proper adhesive substance that will not readily wash off. I have used cells made in this way for a considerable time and have found them very satisfactory. When the spores are sown in the water, the slide should be kept under a glass shade, in which a dish of water is also placed to keep the chamber moist, so that the water which is necessary to the germination of the spores does not evaporate.

In reference to these rubber-band wells perhaps I should mention, by the way, that they answer very well for the examination of pond water with low powers. They are easily made, and being much stronger than many types of life slide, can be freely cleaned without danger of breaking. The cover-glasses adhere closely to the indiarubber, and do not slip off. When some active organism is under examination everyone knows how difficult it is to keep the creature under view in the usual sized well, as it gives such ample scope for demonstration of agility. If, however, a portion be cut out of an indiarubber circle, and a small hole be punched in the segment, it will, when fastened on a slide, provide a small well, the whole area of which will be covered by the lens, and consequently the creature will always be under view (see Pl. XVII., Fig. 7).

But to return to the æcidiospores which we have placed in water. They should be examined every now and then to see what progress is being made. In two or three hours it will be seen that

* Dr. M. C. Cooke, *Microscopic Fungi*, 1870, p. 8.

a germ tube has begun to issue from each of the ripe æcidiospores, and this tube will continue to grow until many times the length of the spore from which it arises.   Some of the tubes will become much bent and curved about from side to side, and the coloured protoplasm will be seen to leave the spore and flow slowly along the tube until it reaches the end.   Soon all the protoplasm will have travelled along the germ tube, and the spore will be left quite empty.   The tube may become branched, and the protoplasm will flow into the branches.   Plate XVII., Fig. 3, represents a germinating æcidiospore.

Now that which takes place on the glass slide takes place also on the Nipplewort.   We have already stated that the spores become scattered about, not only on the host plant itself, but also on the surrounding vegetation ; when, therefore, there is a shower of rain or fall of dew, these spores begin to germinate precisely in the same way as we saw them do in the experiment ; but the process goes further, for the ends of the germ tubes, or of the branches of those spores which happen to be upon a leaf of the Nipplewort, enter through the stomata into the tissues of the leaf, and there develop, finally producing a fresh mycelium.   The empty spores and the portions of the tubes remaining outside the leaves then fall off and disappear, as their mission is ended.

It must, of course, be borne in mind that, although all the ripe æcidiospores germinate when moisture is applied, it is only those which are in touch with the proper host plant that enter the stomata and give rise to a mycelium.   The spores may have germinated in hundreds on some neighbouring plant, but without producing the slightest effect.   If we secure a few specimens of *Lapsana communis,* and also of some other plants, all of which should be free from any traces of fungus, we can easily test the accuracy of what has just been stated.   One or more leaves of the Lapsana should be well moistened and some freshly gathered æcidiospores brushed on to the wet surfaces.   The plant ought then to be covered with a globe for a day or two, so as to keep the leaves damp and afford the spores an opportunity to germinate.   If at the same time that this is done we also inoculate some other plants, say the Daisy or the Nettle, and in addition preserve a Lapsana from all contact with any æcidiospores, they will help us to understand the process, and will serve also as a check.

When these different plants are examined in about a fortnight, no change will be observed in the Lapsana which was not inoculated, nor will there be any signs of fungus on either the Daisy or the Nettle ; but on the Lapsana, on which were placed the æcidiospores, there will be seen a quantity of very small light-brown spots. These spots on examination with the microscope will resolve themselves into sori or clusters of spores. They form, indeed, the next stage in the life-history of the fungus we are considering, and are termed uredospores. These uredospores are the product of the mycelium formed by the development of the germ tubes of the æcidiospores. The mycelial tubes accumulate in special centres, and from these parts there are branched off large numbers of upright hyphæ. At the apices of these hyphæ the uredospores are formed, which when ripe become separated from the hyphæ, in order that others may be developed in the same way. The growth of the fungus at length breaks through the leaf, and the ruptured epidermis can often be plainly seen around the sori.

When the uredospores are separated from the beds on which they have been produced, they usually become detached, not exactly at the junctions between the spores and the hyphæ, but at a short distance below them. The spores are then free and ready to be dispersed with the first breath of wind. The sori or spore clusters, which are scattered over both sides of the leaves, are not nearly so prominent and easily discovered as the previous stage of the fungus, neither are the leaves distorted by the presence of the mycelium as we saw they were when the æcidium was being developed. The spores themselves are ovate, of a brownish colour, about 18 micro-mm. in diameter, and are covered with short, sharp points or spines. They will, of course, be found on the Lapsana immediately after the æcidiospores, say, during the months of April, May, and June, and if we obtain, as we did with the previous spores, some of the ripe uredospores, we shall find that, when placed in water, they will germinate in precisely the same way.

In a few hours after immersion the germ tube will have been extended through one of the germ pores in each ripe spore, and the protoplasm and colouring matter transferred from the interior of the spore to the distant end of the tube.   The tube will become branched, and, if the spore be upon a leaf of the Lapsana, it will find its way through one of the stomata into the tissues, and there grow and give rise to another mycelium.   These germinating uredospores may be placed upon other plants than the Lapsana— as, *e.g.*, the Daisy and Nettle again—but they will produce no mycelium in them.   In this new mycelium fresh spore beds will soon be formed, and hyphæ given off as before in the direction of the surface of the leaf.   These hyphæ become expanded at the upper ends into oblong spores, or rather compound spores, for a transverse division in the centre of each of these oblong spores divides it into two chambers, thus practically making two spores, one above the other.   These are called the teleutospores ; that is, the last spores which are produced.   The teleutospores in many species have only one cell or compartment, but in the present species they have, as already stated, two cells, and are known by the distinguishing name of *Puccinia*.   They will be found probably from May to August, and will appear sprinkled over both surfaces of the leaves in black or dark brown little spots.

We have seen that as soon as the æcidiospores and uredospores were ripe, they germinated at once on being placed in water ; but the teleutospores that we are now considering do not germinate until spring of the following year.   Throughout the winter months they remain unchanged and exhibit no signs of life ; but after this period of rest is òver, and when the season once more comes round for the Lapsana to put forth its young leaves, they are ready to germinate.   The process is somewhat different from the germination of the previous spores.   Soon after moisture is applied to a teleutospore two tubes are emitted, called promycelial tubes, one from each compartment.   The protoplasm passes out of the spore to the far end of the promycelial tubes, where the tubes presently become divided off into several compartments.   From each compartment a fine short branch arises, along which the protoplasm flows and accumulates at the end of it in a small, round body, which is there formed, called a promycelial spore (see Pl. XVII., Fig. 4).   In a short time this promycelial spore drops off, and then germinates by emitting a short germ-tube (see Fig. 5).   This germ-tube, when upon a leaf of the Lapsana, pierces through the epidermis with its sharply-pointed end, and thus conveys into the tissues of the plant the protoplasm from the promycelial spore, which very shortly gives rise to the mycelium that produces the spermogonia and æcidiospores.

Thus have we arrived at the point where we started, having followed the fungus through its various stages and described the cycle of its life-history. The æcidiospores germinated and produced the uredospores, the uredospores germinated and produced the teleutospores, and the teleutospores, after resting for several months, germinated and produced the æcidiospores. From this it may easily be understood that, until their life-history had been made out, these several different spore-forms might very well have been treated as so many different species of fungi ; but now it is clearly established that they are but so many stages in the life of one fungus, *Puccinia lapsanæ.*

Perhaps the following sketch of how a new species was found and its life-history worked out may not be without interest to those who are strangers to this branch of work. In searching any district for these fungi it generally happens that some particular locality at length becomes singled out as one which provides more chances of success than others. One place which the writer found particularly rich in specimens was a small, open, boggy spot, perhaps only about twenty yards in circumference, in the midst of a small wood. Very few persons ever found their way thither, so that the vegetation was permitted to grow undisturbed and in a wild luxuriance calculated to delight the heart of anyone who loved to see things in a true state of nature. Part of an old tree-trunk lay rotting amid the herbage, and near at hand trickled a tiny stream. Around the margin grew various trees and shrubs, and everywhere there was an abundance of flowers. In course of time this delightful little spot became, in imagination, my own peculiar property and a place across which no vulgar feet had any right to tread. Under such circumstances it was only natural that great disappointment, if not wrath, should be expressed when, on proceeding one day to the old haunt, it was found that the woodman had been hard at work and levelled to the ground the little wood and obliterated the happy hunting ground. *Sic transit gloria mundi.*

However, while this spot existed, I had free scope to prowl about in quest of specimens, being only occasionally startled by the sudden leap of some surprised frog or wondering toad. .One afternoon, when gathering in this place specimens of *Uredo confluens*, which were growing freely on the Dog's Mercury (*Mercurialis perennis*), I noticed among the Mercury plants a single specimen of the Herb Paris (*Paris quadrifolia*), and that on one of the leaves there was a small, lightish-coloured spot which seemed to indicate the presence of a fungus. On closer examination this proved to be the case, for on turning over the leaf an *Æcidium* was seen to have established itself. Further search revealed the pleasing fact that other specimens of the Herb Paris which were growing near at hand were also infested with this parasite. The fungus was growing both on the leaves and on the stems of the plants, and although not present in any great profusion was nevertheless sufficiently plentiful for all purposes. It was further noticed that the area in which infected plants could be found was very circumscribed; indeed, that they were only to be obtained within a circle of some six yards or so in circumference. Beyond this circle a stray specimen or two might be discovered, but the limit of their habitation was pretty well defined.

I was unable to find any reference to a fungus upon the Herb Paris in the list of those which had been recorded as occurring in Britain; but it appeared probable that it might be the same species as the one which grows upon the Lily of the Valley, seeing that that plant is nearly related to the Herb Paris. However, on forwarding some specimens of it to Dr. Plowright, King's Lynn, for his opinion, he stated that in all probability it was quite distinct from that species; but in order to determine the species exactly its life-history would require to be worked out. The æcidium was found in the month of May, and by the end of June it had entirely disappeared. Now, if this fungus went through the same cycle as the *Puccinia lapsanæ*, the æcidiospores should have been followed by the uredospores and then by the teleutospores, but no further spores appeared on the leaves of the Herb Paris. It was therefore evident that a clue must be looked for elsewhere.

Before going further, we should explain that there are a goodly number of these fungi which spend part of their lives upon one host plant and part upon another host plant of a totally different species. For instance, *Puccinia obscura* dwells during the æcidium stage on the common Daisy (*Bellis perennis*), and during the Uredo and Puccinia stages on the Field Woodrush (*Luzula campestris*). Fungi of this nature are termed heterœcious, and they will not develop unless the necessary two host plants are present. For instance, the latter two stages could not be produced on the Daisy, neither could the former stage be produced on the Luzula. When the æcidiospores on the Daisy are ripe, they fall down upon, or are blown on to, any specimens of *Luzula campestris* that may be near at hand, where they germinate and produce in the ordinary way the mycelium, which eventually gives rise to the uredo and teleutospores. These last-named spores develop, in like manner, when they find a resting-place on the Daisy leaves and produce the æcidiospores. This strange and peculiarly interesting feature was therefore the clue which had to be followed in the search for the missing links in the history of *Æcidium paradis*, as it appeared highly probable that the species was heterœcious. I was unable to follow up the quest that same year, but when opportunity served an investigation of the plants growing in proximity to the Herb Paris was made, and though there were several other species of fungi close at hand, yet they were well known and had nothing to do with the one in question. But just over the place where the young Paris plants were raising their heads from the soil, a considerable quantity of dry, withered Canary Grass (*Phalaris arundinacea*) of the last year's growth was lying upon the ground, and this grass was covered with very minute black spots, which on examination proved to be little beds of fungi, *Puccinia*. This discovery at once gave rise to what proved to be a well-founded suspicion: that these spores were the very ones that were required, but the only way in which the truth could be definitely ascertained was by actually taking the spores and endeavouring to produce from them the æcidium on the Herb Paris.

A number of Paris plants, which were just beginning to unfold their leaves, were therefore collected from a place where it was considered they were in all probability free from the disease, and some of the spores from the canary grass were placed upon the leaves of several of them. One or two of the plants, however, were not infected, so they were retained as control plants ; so that if the infected ones developed the æcidium and the control plants did not, then it would be pretty conclusive evidence that the spores on the grass were the cause of the fungus on the Herb Paris. Owing to some difficulties I experienced in getting the plants to grow, my cultures were not successful ; but Dr. Plowright, who very kindly took a deal of trouble in helping to work out its life-history, was more successful, and I shall therefore give the results which he obtained from his experiments.

When infecting with the teleutospores, it is prudent to make sure, by placing them in water and examining with a microscope, that they are actually germinating, otherwise spores might easily be used for the experiments which have already germinated. Such exhausted spores would, of course, be useless for the object in view. Previous, therefore, to the spores being applied to the Paris plants, it was ascertained that the promycelial tubes were actually being developed. In about eleven or twelve days after infection spermogonia appeared on the leaves of the plants which had been infected, whilst the control plants remained perfectly free from the parasite. This was very satisfactory, and showed at once the connection between the spores on the two different plants. The grass in question (*Phalaris arundinacea*), however, forms a host for several species of *Puccinia*, which respectively produce æcidio-spores on *Allium ursinum, Arum maculatum, Convallaria majalis*, and *Rhamnus frangula*, and it was therefore possible that the Puccinia in question might be one of these species which, in addition to producing æcidiospores on the already recorded host, also produced them on *Paris quadrifolia*.

As the Puccinia spores are in many cases so very much alike, it is impossible by microscopical examination alone to identify one species from another, and it is therefore necessary to cultivate the various spores, and thus show which are related. The spores in question on the Phalaris grass were therefore applied to *Allium ursinum, Arum maculatum*, and *Convallaria majalis* ; but on these plants they produced no æcidium. This proved that the Puccinia had nothing to do with the other species referred to, as it was inefficacious in producing the æcidium on all save the Herb Paris, and showed that one more Puccinia had been added to those already recorded as occurring on the Phalaris. It certainly is very remarkable that, although they all find a congenial home on the Phalaris and do not differ microscopically very much from each other, yet each has its own particular host on which to spend the first part of its life.

In order to understand how the Herb Paris becomes infected in a state of nature, it is only necessary to remember that the withered grass bearing the Puccinia was found lying on the ground directly over the area where the Herb Paris was growing. The young plants, therefore, required literally to push aside the fallen grass in order to get space to grow, and multitudes of the teleuto-spores would consequently be encountered both on the ground and on the grass. The spermogonia were soon followed by the æcidia, that is the cups containing the æcidiospores. The cups are developed on the under surfaces of the leaves, hypophyllous, and are arranged generally in the form of small circles. See Pl. XVII., Fig. 6. The circles, which are usually very definite, are formed by a single row of peridiæ or cups, with a vacant or clear space in the centre. The places on which the æcidia grow, and for a short distance round the margin of each circular spot, are lightish coloured, owing to the chlorophyll being destroyed by the fungus. The number of such circles of cups on each leaf does not very often exceed two or three, but a leaf may occasionally be found bearing five, six, or even more of them.

Having thus proved the case so far, the next step was to try the reverse culture, to see if the æcidiospores, when sown on the Phalaris, would produce the uredospores. This was accordingly done and also proved successful, for in about three weeks the desired spores made their appearance on the grass. The sori or spore clusters are small, of a reddish yellow colour, and were developed in fair abundance on both sides of the leaves. The spores are globose, finely echinulate, and measure in diameter 30—35 micro-mm. In a state of nature, the Phalaris becomes infected, of course, from the spores which are blown on to it from the Herb Paris, and if there be moisture on the grass they straightway emit the germ-tubes, which enter into the tissues in the usual way. The uredospores are usually to be found in the month of July. When they are ripe, they also germinate, and very shortly the teleutospores (Puccinia) are formed on the same grass. The sori of the teleutospores are very small, and are scattered about in abundance like minute black specks upon the grass. The spores are brown, smooth, and somewhat variable in form. The constriction between the two compartments is generally very slight, and the measurements are 40—50 × 18—25 micro-m.m. When the teleutospores are fully developed in the month of July or August, no further change takes place, and they remain inactive throughout the winter. The grass which bears them withers and falls to the ground, but when the Herb Paris is again springing up the spores are ready to germinate and produce the mycelium of the æcidiospores in the way we have already described.

We have thus completed the cycle of the life-history of *Puccinia Paridis*, having traced it from the teleutospores on *Phalaris arundinacea* to the æcidiospores on *Paris quadrifolia*, and then back again from the latter plant to the former one, and so proved the parasite to be heterœcious.

When we bear in mind the facts which have here been referred to, the lowly and insignificant fungi will gain a fresh attraction for us, and will become invested with a halo of interest which at our first acquaintance with them we scarcely thought it possible that they could possess. Indeed, it appears to us that because we obtain a treat where we least expected it, that we are therefore the more pleased when it is suddenly presented to us.

What has just been described regarding *Puccinia lapsanæ* and *Puccinia Paradis* no doubt applies in a general way to the development of many other species; but there are, of course, a great number, to which we have not referred here, which have life-histories much different from those just given. (Dr. Plowright's work on *The British Uredineæ and Urtilagineæ* will be found a most useful and deeply interesting book in giving further information on the subject.) Neither have we made any reference to those species which attack the corn, the potato plants, and many other of the fruits of the earth which are so valuable to man, and in the study of which there is more than a mere passing interest. But perhaps we have said sufficient to show that these micro-fungi are highly deserving of our careful and particular attention.

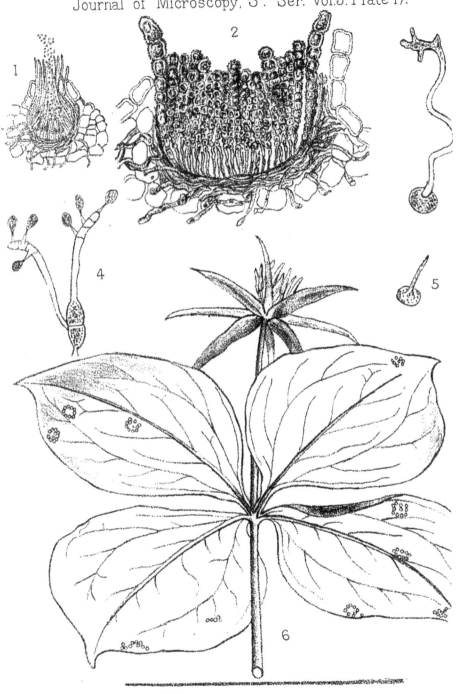

EXPLANATION OF PLATE XVII.

Fig. 1.—Section of a spermogonium, × 200.
,, 2.—Section of an æcidium, × 150. De Bary.
,, 3.—Æcidiospore germinating, × 250.
,, 4.—Teleutospore germinating.
,, 5.—Promycelial spore germinating.
,, 6.—*Paris quadrifolia*, showing æcidium on under surfaces of leaves.
,, 7.—Glass slip, with india rubber cell.

Ookinesis in Limax Maximus.*
By F. L. Washburn.

THE observations here given are confined to early stages of the egg while in the oviduct, and before the expulsion of either polar globule. The article, therefore, deals with stages which, for the most part, precede any discussed by Dr. Mark in his excellent treatise on *L. campestris.*†

Of the following wood-cuts, Fig. 1 is a diagrammatic represen. tation of the oviduct from a laying animal, from which eggs were taken, and studied serially as numbered. The vitellus averaged

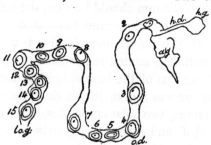

FIG 1.

156·2μ in diameter. Various methods were made use of in fixing —Fols solution: Osmic acid, 1 per cent., followed by Merkel's fluid ; chromic acid 1/3 per cent., etc. ; but the one which gave

* Reprinted from *American Naturalist*, June, 1894.

† "The Maturation, Fecundation, and Segmentation of *Limax Campestris* Binney," by E. L. Mark, *Bulletin of the Museum of Comparative Anatomy*, Vol. VI., parts 11 and 12, Cambridge, Mass., 1881.

the best satisfaction was as follows :—The body cavity of a laying animal was opened with a quick cut of the scissors, and the animal plunged into a boiling hot solution of corrosive sublimate; allowed to remain one minute ; transferred to water, and eggs removed from oviduct and shelled.* Vitellus allowed to remain in distilled water two minutes, then transferred to 35 and 50 per cent. alcohol, remaining three mintes in each grade; then to 70 per cent. alcohol for permanent preservation. I found that if eggs were allowed to remain in distilled water three hours or more, they shelled better, the vitellus coming out clearer and freer. For examination of eggs *in toto*, Czokor's alum cochineal gave, as a rule, good results. Ten minutes' stay in this dye appeared to give the necessary differenti-ation ; but for examination of sections much longer time was necessary, two to three hours or more. Picrocarminate of lithium was also found to be excellent, if anything, better than Czokor, on account of its differentiating nucleus structures. For examination *in toto*, twenty-four hours in this stain, and then washing with distilled water and pure alcohol, gave good results.

Section staining on slide was also found desirable, and Safranin was the stain used—two and a half hours, followed by acidulated ($\frac{1}{2}$ per cent. HCl) alcohol of 90 per cent. grade for seven to ten minutes. The Schällibaum should be new, the sections carefully applied to a well-smeared slide, and kept at 60° C. for exactly fifteen minutes. If Mayer's albumen fixative is used, only warm, and as soon as paraffine is melted remove slide from heat.

A number of sections of the hermaphrodite duct (*h.d.*, Fig. 1) were made. One egg was found, in this duct, near the hermaph-rodite gland, containing two polar corpuscles, each surrounded with a faintly stained Hof, and each showing striæ radiating from cor-puscle through Hof. About eight chromosomes were observed irregularly grouped in the well-defined archoplasm of Boveri.†

From these sections it appears that the centres of attraction

---

* In the upper part of the glandular portion of the oviduct there were a number of eggs in which the outer membrane or shell was barely formed ; in some, egg No. 1 for example, there was no membrane at all, and in others only the inner membranous coat was present.

† *Zellen-Studen* von Dr. Theodor Boveri, Jena, 1887.

which Garnault * says do exist in the ovarian egg of *Arion* and *Helix*, and which were not seen in the hermaphrodite gland of *L. maximus*, do exist in the duct very near the gland. They evidently appear immediately after the egg has left the ovary. This duct was lined, for the most part, with ciliated epithelium, and contained much mucus.

Fig. 2 illustrates an optical section of egg No. 1 from glandular part of the oviduct (see Fig. 1) viewed obliquely to the long axis of the spindle, and showing the two polar corpuscles and chromosomes, there being about twenty of the latter lying in an irregular cluster in the clear space between the corpuscles. This egg was

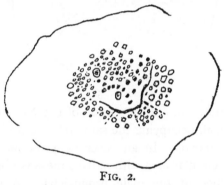

FIG. 2.

stained in picrocarminate of lithium for thirty hours. In its examination a Zeiss Oc. 2 and Obj. E were used. A broken membrane, "membrane rongée," was seen with apparently chromatic thickenings in it. Observations on this egg coincide closely with those of Garnault on *Arion* and *Helix*, and, in a measure, with those of Vejdovsky on *Rhynchelmis*.†

The larger corpuscle is the one nearest the observer. The structural peculiarity of one side of the nucleus should be noted— where cytoplasm and yolk granules are in intimate relation with contents of nucleus. This is Garnault's " prophase ;" it is the stage just previous to formation of nuclear plate leading to the forming of first polar globule. In another egg, No. 9, from the same oviduct, an optical section showed rays of hyalocytoplasm

* " Sur les phénomenes de la fécundation chez l'Helix aspersa et l'Arion empiricorum."—*Zoöl. Anzeiger*, Nos. 297 & 298, Dec., '88, and Jan., '89.

† *Die Entwecklungsgeschicete der oligochaeten (Rhynchelmis)*, 1888.

pushing out from clear area through granules of vitellus. Chromosomes irregularly placed in a hyaline area. Spindle striæ observed in viewing the egg at right angles to spindle axis.

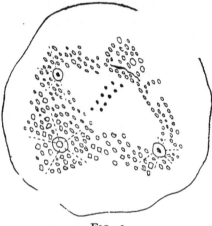

FIG. 3.

Fig. 3 illustrates an optical section of egg No. 11 from oviduct of another animal, occupying the same relative position as No. 11 in the oviduct drawn. In an eccentric position, and near, the surface, a clear circular area with radial striæ was observed, indicating the presence of the male pronucleus. A portion of the membrane of the germinal vesicle still present. Egg No. 10, in the same animal, also showed circular male area in direction of axis of spindle, and chromatin granules within it. In egg No. 9 the head of spermatozoön was seen in optical section, some little distance from periphery, circular with narrow Hof about it, and striæ radiating from Hof. Very fine granules were evident within this pronucleus.

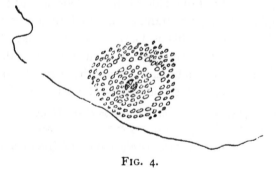

FIG. 4.

Fig. 4 illustrates part of a section of egg shown in Fig. 2, cut in such a plane as to show the sperm nucleus near the periphery. Drawn with Zeiss Oc. 1 and 1/18 oil immersion. Garnault says, in speaking of formation of sperm nucleus in *Arion* and *Helix*, "the spermatozoön enters just before first kinesis, or immediately after. The contracted head does not begin to change until after the expulsion of the second polar globule. The sperm-head first divides into two chromatin spherules, then, by successive divisions, there is formed a greater number of spherules which remain inclosed in a clear areole. This clear areole recalls the hyaline centre of attraction when that has received the half plate for the formation of a vesicular nucleus." *

* The following few notes pertaining to the fixing and staining of *freshly laid* eggs may be of interest :—

Eggs placed for five minutes in Fol 99 (1 per cent. chromic, 25 vol ; 2 per cent. acetic, 50 vol ; H_2O, 25 vol), then shelled in water ; vitellus in same solution for five minutes, H_2O ten minutes., and 35 per cent. and 50 per cent. alcohol five minutes each, 70 per cent. thirty minutes, and 90 per cent. *ad lib.*, gave good results, taking picrocarminate of lithium very well if left long enough in stain. They also took borax carmine very well after the above treatment.

Both of these stains did well after the eggs were immersed in chromic ½ per cent. ten minutes, then shelled in large quantity of water, then vitellus in chromic ½ per cent. four minutes, and H_2O and grades of alcohol as above.

Whole egg in osmic acid 1 per cent. five minutes, followed by Merkel's fluid four hours ; shell, then water, and grades of alcohol two minutes each to 70 per cent. for permanent preservation were quite satisfactory. It gave good results as to nuclei when eggs were left in picrocarminate of lithium for forty-eight hours.

———

THE TORPEDO FISH.—At the last meeting of the Academy of Science, Philadelphia, Prof. D'Arsonval, of the College de France, said :—A fish 30 cm. in diameter could give out a shock of 20 volts. He applied some small electric lamps to the fish, and they were lit by the discharge from its body. In some instances they were so powerful as to carbonise the lamps.—*Sci. American.*

Photographing Minute Objects by Means of the Microscope.

By T. E. FRESHWATER, F.R.M.S., F.R.P.S.

(Read before the North Middlesex Photographic Society.)*

I FEAR I have nothing new to bring before you this evening, but having been pressed into the ranks by your secretary I will endeavour to do my best. The subject is one that has been dealt with by many men much more able than I am, and, if I am not mistaken, you have several good workers in this Society. I do not intend this evening to touch upon high-power work, but treat the subject quite from a popular point of view. The slides which I will show you are a few of the many that I have done to illustrate various subjects, and taken under different conditions to show a few of the methods of illumination.

It is not necessary to spend much money in rigging up an apparatus for ordinary photo-micrographic work, for most objects can be photographed with very inexpensive tools. The simplest method is to use an ordinary microscope. Of course, the better the stand is and the more stage adjustment the better, but it is not necessary. The instrument should be turned down with the body horizontal to the base of your apparatus ; then at the eye-piece end fit up your camera. A half-plate with a long body will give you plenty of extension for low-power work. For light you may use that which is most convenient, daylight, lime, or lamp light. A good large single-wick paraffin lamp will answer for all ordinary purposes. The bull's-eye condenser on a stand should be placed in such a position as to fill the object to be photographed with light evenly all over. This part of the operation must be done very carefully. Too much trouble cannot be taken with this part of the work, as so much depends upon even and proper illumination of the object in the resulting negative. It is well that the whole apparatus should be fitted up on a long, heavy, flat board, and very carefully centred, and made as true and steady as possible. With such simple arrangements as

---

* From *The Amateur Photographer*. We beg to thank the editor for the loan of electros illustrating this article.

this, most of the ordinary objects may be photographed, such as parts of insects, sections of wood, and most of the larger diatoms.

There are many different forms of photo-micrographic apparatus on the market now ; most of the opticians make one or other form a stock article. Of some of the better instruments I will show slides that I have made on purpose, then perhaps it will be better to explain them. The most complete apparatus that is in use, I think, is the one at the Royal Veterinary College. This cost a lot of money, and is quite out of the reach of the amateur; you will see by the slide how it is arranged. There are several vertical instruments made. One by Van Heurck consists of an oblong box, mounted on four legs of such a length that the ocular end of the microscope passes through the bottom of the box ; the box is large enough for the head to pass in for focussing purposes. Another, designed by Mr. Pringle, has many advantages over the solid box, as it is made with a conical bellows, and the screen is made to slide down. For some objects a vertical camera is necessary ; there are objects that you cannot keep flat unless the stage of the microscope is horizontal.

When one starts in this line or branch of photography there are several points to be taken into consideration. First, there is the apparatus—I should advise anyone who is beginning, to buy the best and steadiest microscope stand that they can afford, and with a sub-stage for the various fittings that will be wanted from time to time ; the long bellows camera, with board, is not an expensive part of the apparatus, and can be bought for £3 or £4. Then come the lenses ; these are most important, and, like photographic lenses, rather expensive articles to deal with, if one goes in for high-power work. But I am not going to touch on this line, so will leave it to others who go in for deep scientific research, such as the different forms of bacteria. The Germans have, hitherto, made most of the best lenses for photographic work, and sell lenses with a focussing eye-piece for correction—lenses that have given better definition, flatter field, light, and more even illumination which come about by the use of the new glass from Professor Abbé, and again they are cheaper ; but now the English opticians are beginning to wake up in this direction, and many lenses are made that are quite as good and no more expensive ; in fact, I have some that it would be difficult to beat as regards defining power and the amount of light that they pass.   For micrographic work it is an immense advantage to have more light than you want, as it can always be stopped down by the use of the iris diaphragm, or as some prefer a set of stops, these perhaps, in many cases, have an advantage.   I think it is well never to use a higher power than you can help ; it is better to get the amount of magnification by extending the camera, and put in an ocular or compensating eye-piece ; these are made on purpose, and work with a spiral motion to the eye-lens, so as to focus the diaphragm stop in the tube.   A simple form of apparatus I have designed and will show you a slide of presently.   The camera is of long extension, and has a bellows 30 in. long ; the front part is made to rack back, so as to clear the eye-piece of the microscope, to enable the operator to revolve the instrument on its centre.   The microscope is one of a new series of stands Messrs. Newton have recently brought out, and for all ordinary work will answer every purpose ; it has a simple mechanical substage, fitted with an iris diaphragm and Abbé condenser, with adjustments for centering ; the mirror is made to swing out of the way when not in use.

A very convenient way of seeing how to arrange the object is to place a flat mirror at some distance from the ground glass at the end of the camera ; the image on the screen is reflected on to it.   You can by this means see that the object is in the centre of the field, and easily focus the object and centre the light ; in fact, you can, without any difficulty, get the whole thing ready in a short time, except the final focussing, which has to be done very carefully.   Having now roughly focussed the object, remove the ground-glass screen, insert a plate of plate-glass with lines ruled on it, and it will be found useful to have these lines ruled at a given distance apart, say one-tenth of an inch, and a Ramsden eye-piece used for focussing aerial images ; coloured screens or light filters are very useful.   It is well to be provided with several different ones, such as signal green, bluish grey, and yellow, and a few cells fitted with coloured fluids, for, in many cases, working in monochromatic and coloured light, according to the object to be photographed, is necessary.   I had intended to show you a number of photographs taken with various screens, but have not had time to prepare more than one or two.   One of the advantages of their use is that, if the object is very delicate and likely to be flooded with light, the use of the screen comes in, and for objects that it would be almost impossible to expose quick enough ; the insertion of a suitable screen enables you to make a good exposure and get all detail, notably in the delicate membrane of some of the wings of flies, etc.

With regard to illumination, as I have said before, an oil lamp will do very well ; but if you can get the limelight it is more satisfactory.   The light is more pure in colour, more intense, and more easily under control.

Working from a small spot of light, and the smaller and more intense the better, the ray goes more direct through the centre of the optical system than a large volume scattered about.   A blow-through jet will answer all purposes, and if it is fitted with a Pringle cut-off—that is, an arrangement for lowering the gases between the exposure—you save the gas and also make sure of getting the same amount of light each time.

For time of exposure one can give no fixed rule, so much depends on the subject, its colour, thickness, the amount of density, and the magnification wanted.   With reference to the plate, use a slow, thickly-coated one for most work.   The isochromatic will be found very useful for many subjects, but I have not found all the advantages that are claimed for them, for of the number of slides that I hope to show you very few are taken on those plates. Nearly all my negatives were done on Paget xxx or xxxxx.

Each class of object requires a special study.   By class I mean transparent, high or low power, opaque, objects taken on a dark ground, and objects taken by means of polarised light ; each of these I shall more or less speak upon.   The ordinary objects, such as a blow-fly's tongue, wing of bee, stings, head, internal and external organs, sections of scalp, and thousands of others, can be taken in the way I have explained, and with a 4-10th to 2 in. objective with very little trouble and not much practice.   High-power work, using an immersion lens of, say, 1-12th, requires a very great deal of care and manipulation of the whole apparatus, from the achromatic condenser to the ocular.

Many objects that are opaque are very interesting to photograph. Very little, so far as I can see, has been said about them; I allude more particularly to the eggs of the various parasites, the coarser foraminifera, scales on the wings of beetles *in situ*. Now, these are more difficult to do, and some of the best work in this way I have seen has been done by my friend Mr. Evans, and I think—in fact, I know—they are not to be equalled. I candidly confess I cannot do them nearly so well. A few of my attempts I will show you. As all the light we have to deal with is that which is reflected from the object itself, it is absolutely necessary to concentrate all the light possible on the object. There are several ways of doing this, but first of all your object must be mounted very flat, and in the centre of a small black disc, so as to allow the light to pass round the object from your condenser.

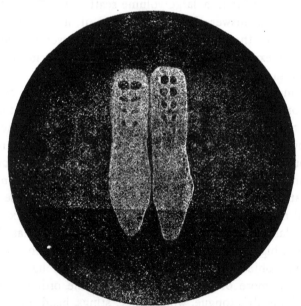

THE STOMACH BONES OF BRITTLE STAR FISH.

This light then falls upon a silvered reflector or Lieberkuhn mounted on the objective, the curve of which must be equal to or a little longer in focus than the objective, so that when the object is in focus, a small amount of collar adjustment enables you to focus down and adjust the rays of light that have been received upon the silvered surface of your reflector. Of course, the more light you can get on your object the better will be the resulting negative.

Such objects as small shells, eggs of butterflies, polycystina, light-coloured seeds, in fact, anything that will reflect and not absorb the light, may be photographed in this way. A parabolic silver-side reflector is sometimes used, but this is not nearly so good, you are apt to get strong shadows on one side. I am sorry I shall not be able to show you many examples of my work in this direction, but the few I have will show you the beauty of their structure, and how easy it is to make *pictures* of such minute things. I have a lovely specimen of the eggs of the parasite of the Reeves pheasant *in situ* clustered on the feather of the bird.

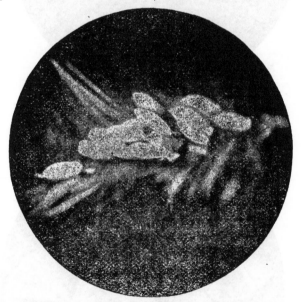

EGGS OF PARASITE OF REEVES PHEASANT.

Objects photographed under polarised light are very pretty; some objects, such as starches, sections of rock, crystals, that are perfectly transparent under ordinary light, when polarised are very beautiful. Now if the two prisms are so turned that the object comes on the dark field, they show up all the beauty of the structure and varied form and composition that, under other circumstances, would not be seen. The polariser, that is the prism near-

EGGS OF HOUSE FLY.

SALICINE.

SELECTED DIATOMS.

DIATOMACEÆ—VARIOUS.

est the light, should be as large as possible, so as to pass as much light as you can through the object.   The analyser follows on at the back of the objective, and should be mounted as close to the back lens of the objective as possible, otherwise it will cut off some of the field, and, to get the best result, both the polariser and analyser should be made to rotate.   A secondary condenser may be placed in front of the Nicol with advantage.

Work done by dark ground illumination has a great charm, more particularly to the student who has a love of pond life, and to see the minute organisms scudding about lit up like small particles of silver on a black ground.   The common hydra, vorticella, and many others may be photographed by one who gives it his careful study.   Diatoms, sponge spicules, polyzoa, and many other objects look grand done in this way.   The few slides that I shall show I hope will bear me out.   I fear that I have taken up a lot of your time, and you are getting tired of this dry stuff; there are several other matters that I might have dealt with, but it would take up too much time in the one evening.

THE RAYS OF THE SOLAR SPECTRUM.—The fact is well known that if we examine the spectra furnished by the light emitted by the various points of the sun, the rays that appear are very variable in number.   There exist but eleven that are constant—that is to say, that we find in the light derived from all the regions.   Among these, five belong to hydrogen, two to calcium, and four to unknown elements.   Mr. Ramsay, however, has identified one of these rays—that of helium, with the ray of a terrestrial element. There remained then but three, corresponding to extra terrestrial substances.   Mr. Deslandres has decomposed clevite by sulphuric acid, and then, on studying the spectrum of the gas disengaged, has ascertained the existence or a ray 447·18, identifiable with one of the three remaining rays.   In consequence of this discovery, there exist but two unknown rays among the permanent ones of the spectrum.—*Sci. American.*

Predacious & Parasitic Enemies of Aphides
(including a Study of Hyper-Parasites).

By H. C. A. Vine.

Part III. Plates XVIII. and XIX.

THE *Hymenoptera* as originally arranged by Linnæus included those four-winged flies whose wing venation, differing wholly from the Hymenopterous type, consists of longitudinal nervures, more or less connected by transverse branches. The value of this character has been strongly disputed by later naturalists, but it seems probable that in this, as in so many other instances, the generalising insight of Linnæus led him to a conclusion which the labours of entomologists will ultimately confirm.

I may be able in the succeeding section to adduce some facts which bear upon this question of the relationship of the *Odonata* to the *Planipennia*, which has a certain interest in connection with the aphidivorous genera, as establishing their relationship to a very destructive carnivorous group, which dates back in geological times to the Palæozoic formations, and which may be continuously traced through the Triassic and Oolitic strata. Indications of the aphis-eating genera are not found until a late date, the remains being found chiefly, I believe, in the Tertiary deposits. This may be accounted for partly by the more delicate and fleshy nature of the *Hemerobiinæ*, and if we accept their relationship with the *Odonata* as established, we may consider the probabilities of their later appearance as arising from their being the product of selection acting upon the more plastic species of the earlier Neuroptera.

The beauty and elegance which characterise so many of the Neuroptera are not wanting in those genera which, being aphidivorous in their larval stage, come within the scope of the present memoir. It is true that in *Hemerobius* one sees little save a dull and often somewhat clumsy brown insect sometimes resembling a small moth. But a slight examination of the wings reveals an intricate venation and often a beautiful mottling which this genus shares with many of the larger dragon flies; while the delicate greens of the *Chrysopidæ*, the elegant reticulations of their ample wings, and the metallic brilliancy of their protruberant eyes, place them among the most beautiful of insects.

It has been already mentioned that the larvæ of these two groups resemble one another very nearly in general characters, but present also many points of difference.

In the last section I described the more important appendages of a larva of *Chrysopa*, which were delineated with some detail on Plate XII. The tapering length of the organs and the specialisation of the parts seems to indicate that the development has progressed considerably beyond that of the *Hemerobiidæ*, and one might almost think that in the *Chrysopidæ* the evolution of the larva had reached its maximum, did we not know that in such things there is no finality. The corresponding organs of the larva of an Hemerobius, shown on Pl. XVIII. at Figs. 1, 2, 3, and 4, will illustrate this difference very clearly. The chitinous external casings are much more fully retained, forming upon the palpi a close series of incomplete rings, or, rather, perhaps, of flat plates tapering to the ends and bent into segments of a circle. These rings or plates are of irregular length, and are regular in their disposition only inasmuch as they are so arranged as to cover the surface equally. In fact, their appearance at once gives rise to the notion that the chitinous covering has originally been continuous, or comparatively so, but that the necessity, and consequent efforts, at flexion have first wrinkled and ultimately separated (on the lines of the wrinkles) the hard surface into a series of irregular plates, which, disposed upon a flexible membrane beneath, present no obstacle to the free movement of the organ. The palpi are comparatively short and thick, while the first and second joints of the organs which are equivalent to labial palpi are encased in a continuous chitinous coat. The maxillary palpi—which I have elsewhere spoken of as antennæ—exhibit in some of the larvæ of *Hemerobius* enlargements as shown in the drawing, which recall at once the characters of the antennæ in some of the less developed (or retrograded) *Aphididæ*. The impression thus formed, that the larva of *Hemerobius* represents a less specialised development than that of *Chrysopa* is confirmed when we observe the short and slight nature of the terminal bristles.

The Mandibles and Maxillæ.

The mandibles of this larva, as shown at Fig. 2, are correspondingly shorter and thicker than those of the *Chrysopa* larva, shown on Pl. XII., and are provided towards the apex with three or four denticulations, well adapted from their shape to prevent the escape of prey, when once pierced. Lower down, the inner edge is finely serrated, but the value of this is far from evident, the mandible being rarely inserted in the body of a victim to any depth.

The hollow-grooved shape of the mandible is well displayed in this specimen. At Fig. 3, on the same Plate, is shown the maxilla withdrawn from the mandibles. The transverse opening at the extremity, through which the juices of the aphis are sucked, is very evident, and a long fold or groove down either side simulates very closely a longitudinal continuation of the slit, but a careful examination reveals its true nature.

It has been stated by some writers of authority that the maxillæ are so shaped as to simply close the groove of the mandibles, so that the two together form a hollow tube, which acts as the channel to the œsophagus at their base. I have been unable to find any evidence to confirm this view, and after frequently watching under the microscope the action of larvæ in abstracting the juices of aphides, I cannot doubt that the latter enter the *tubular* and greatly modified maxillæ by the narrow opening at their extremity, and pass downwards by the suctorial action of the œsophagus into that organ. The action of the mandibles is, in my opinion, limited to the purposes for which their structure adapts them—the piercing the skin of the victim, and affording a safe passage for the action of the maxillæ, which they protect. Anyone who will take the trouble to watch a larva seize and destroy an aphis under an inch or half-inch objective, which may readily be done by confining the insects in a shallow cell, will almost certainly deduce, from the movement of the maxillæ and the downward passage of the ingested oil-globules from the aphis, that the function of the mandibles themselves is merely that of a weapon and a sheath.

The entire absence of any opening at the superior end of the œsophagus, answering to the ordinary nature of a mouth, is one of the most curious features of the Neuropterous group to which the

Aphidivorous larvæ belong. The labium appears to be firmly united to the lower portion of the head, and no representative of the ligula foreshadows the elaborate mouth-organs of the perfect insect. This peculiar construction is the more remarkable, as in other sections of the Neuroptera, in which the larvæ, though of different habits, present many points of resemblance to those of *Hemerobiinæ*, the victim is likewise seized by a pair of formidable forceps, but is conveyed to the *mouth* to be devoured. So far as I have myself observed, the absence of mouth in the larvæ coincides with certain features of the ligula in the imago (which will be illustrated), but I am scarcely able to assert this as a general fact.

THE TARSUS OF THE LARVA.

Another noticeable feature in the species illustrated on Pl. XVIII. is the entire absence of the elongated extension of the tarsus, carrying at its extension the pulvillus, which is so strongly developed in the larva of *Chrysopa*. In the Hemerobius larva the suctorial pad is situated immediately at the base of the claws, but although thus much less obvious, it appears to have very considerable adhesive power, as is evidenced by its ability to climb vertical glass surfaces when fully fed. The larva of *C. Perla* during the latter part of its existence seems quite unable to adhere to such a surface for a much greater height than about twice its own length, when it usually falls back and regains its feet with a struggle. The length of the slender articulation bearing the suctorial organ in the latter larvæ probably has much to do with the difference, as the leverage must greatly weaken the power of adhesion.

THE WINGS.

The *Neuroptera Planipennia* derive their sectional designation from the flat and broad expanse of the wings, which, especially in the typical genus, *Chrysopa*, are of great size, and in many species assume a shape approximating to that of a parallelogram. The hind wings are never larger than the fore wings, but in many of the genera are equal in size and similar in shape and venation. They are not, as in the Hymenoptera, supplied with wing-hooks, but each pair of wings is propelled independently by its own muscles, and, in some species of the nearly related dragon-flies, are provided with means of instantaneously reversing their action.

In the *Chrysopidæ*, as typified by *C. perla*, the venation of the wings is characterised by three or four longitudinal nervures proceeding in determinate lines from the base of the wing. In addition, some more or less zigzag veins connect the numerous transverse veins in such a manner as to form an approximation to a longitudinal nervure; but a slight examination of a few species shows that this appearance is due merely to the coincidence of the junctions of numerous short veins.

The main character of the wing is given by the transverse veins, which, dividing the longitudinal space into more or less numerous cellules, give the net-like effect that has obtained for these flies especially the name of " lace-wing."

Both the anterior and posterior wings in the *Hemerobiinæ* are studded with short, curved, and rather thick hairs along the lines of the veins, becoming more marked in some species around the edges and towards the base, and also about the point where more or less indication of a stigma is sometimes found. No dark mottling is found in the wings of *Chrysopa*, but the green colouring sometimes varies considerably, especially on the nervures.

The wings of the *Hemerobiidæ*—one of which, carefully drawn for comparison, is shown on Pl. XIX., Fig. 6—present less appearance of transverse venation, and the main longitudinal nervures are fewer, or, at any rate, are not so much in evidence. The veins, which in *Chrysopa* are transverse, here become more or less oblique or longitudinal, some of the short connecting veins remaining to form the spaces between into cellules. These wings are often freely mottled with a considerable number of dark patches, affecting the nervures as well as the membrane, and short hairs are fringed pretty thickly along the lines of the veins, as well as sparsely on the surfaces between.

Many years ago, Mr. Bowerbank pointed out in the pages of the *Entomological Magazine* (Vol. IV., p. 179) the beautiful phenomenon of circulation exhibited in the wing venation of *Chrysopa perla*. This interesting illustration of the vital processes of the insect seems to have been but little noticed by microscopic observers, owing, no doubt, to the fly being rarely examined while living under any higher magnifying power than that of an ordinary hand-glass. On Pl. XIX., at Fig. 1, I have reproduced Mr. Bower-bank's excellent drawing, showing by means of arrows the course taken by the currents within the veins. So far as I have been able to observe, it appears that the veins themselves form the circu-latory channels or vessels, the fine tracheæ which pass through them being perpetually bathed in the moving fluid.

An enlargement of portions of the wing veins is given at Figs. 2, 3, and 4 on the same Plate, showing within, the trachea easily visible from its characteristic structure. These tracheal tubes may be easily seen and their course traced throughout the principal veins of the wing by placing it in Canada balsam (fairly thin) under a cover-glass, when, on exerting a slight pressure by means of a needle while the slide rests upon the stage, the balsam may be seen following up the tracheæ and driving the air before it in a very striking manner. The diameter of the tracheal passages, as shown in Fig. 4, is stated by Mr. Bowerbank at about 1/2222nd of an inch, while the circulating passages within which they lie are about 1/408th of an inch.

The Antennæ of the Imago.

The antennæ of the *Hemerobiinæ* have shared the confusion which has affected the descriptions of the family. They have been variously described as moniliform, filiform, setiform, beaded, cylindrical, and thread-like.

The antennæ of *Hemerobiidæ* are always of the monili-form or beaded type, although the shape of the segments varies from that of a shortened and reversed champagne bottle to almost globular. The number of segments is always con-siderable, fifty to sixty being about the general number. The lower segments sometimes present a slight elongation, but the ultimate segment differs in no respect from those pre-ceding it, except that it is terminal. The antennæ are more or less hirsute, and some species—as that figured on Plate XIX., Fig. 8—are characterised by short, stiff bristles, arranged almost or quite at right angles to the axis of the antennæ, on the edge of each segment.

The antennæ of the *Chrysopidæ* are long, consisting of some sixty, seventy, or eighty joints. The segments are perfectly cylindrical throughout, closely jointed, and, excepting the third, which is longer, are of equal length Their appearance is that of a round cord, divided at short intervals by transverse divisions. They are covered with small hairs, laid evenly in the direction of the axis, and I have never observed any deviation in this respect. An antenna of *Chrysopa*, containing an unusually small number of segments, but otherwise typical in structure, is shown on Pl. XVIII., Fig. 7. The antennæ of the *Coniopterygidæ* are moniliform and less hirsute than in *Hemerobius*. The number of joints is also less, although it does not appear to have any definite relation to species.

THE COCOON.

The material from which the silky threads of the cocoon is spun is derived from a gland situated in the posterior part of the abdomen, and which opens by means of a short duct just within the anal orifice. There appears to be no special organ for regulating the outlet, but the muscular bands connected with the surrounding parts seem to control the duct, and probably when the flow of silk-forming secretion is once established the larva has little power of arresting it until the gland has completed its function. Consequently, if a larva commences to spin on an unsuitable surface—such as a glass cell—it does not stop after a few trials, but goes on until it seems unable to make more silk, exhausting its supply in making continual attachments to the glass in the hope of obtaining a sufficient basis for the cocoon. The thread itself, which is shown at Fig. 5, as spun on a cover-glass, is a fine cylindrical cord of glutinous substance, which hardens immediately. It is free from enlargements or irregularities of any kind, and entirely devoid of any structure or duplication. It varies at times in size, and is sufficiently adhesive as it leaves the larva to broaden into a sticky mass, which is attached to convenient positions for the support of the pupa.

The cocoon varies in shape and size, that of *Chrysopa* being round or barrel-shaped and of a very dense texture ; that of *Hemerobius* oval or egg-shaped, and consisting of comparatively few fine and gauzy threads ; while in *Coniopteryx* the cocoon is also oval, but is composed of dense spun silk.

In all the genera of *Hemerobiinæ* the pupæ are far smaller than the larvæ; in *Chrysopa*, about one-half the size; and are so utterly disproportioned to the large insects that emerge from them that it is difficult to believe, after the metamorphosis is complete, that there can be any connection between the very small pupa-case and the handsome fly.   Dr. Fitch remarks that it is as if a full-grown larva had hatched from an ordinary egg.

The Classification of Aphis-Eating Neuroptera.

The classification of the Aphidivorous genera of the Neuroptera has become strangely confused, the names of one or another genus having at different times not only being altered, but transposed.   At present it will not be needful to address ourselves to the vexed question of the transference of certain families of the original order or to the Orthoptera, as has been done by some naturalists, inasmuch as the Aphis-eaters are to be found among those genera which are peculiarly typical of the Linnean order of Neuroptera.   But if it be possible in a future section, it will be extremely interesting to consider the evidence of relationship between those genera and some genera of the *Odonata*.

In order to enable the reader to see at once the views of leading writers on the position of the genera of *Hemerobiinæ*, I shall give the classification adopted by them, commencing with that of Professor Westwood.   He divides the order as follows :—

Order, *NEUROPTERA* (Linneus).

Families.—*Psocidæ* (Leach), Book Mites.
         Perlidæ   ,,   Stone flies.
         Ephemeridæ ,,   Day fly.
         Libellulidæ (Westwood), Dragon flies.
Sub-fam.—*Agrionides* (Westwood).
         Hemerobiidæ (Leach), Lace-wings.
         Sialidæ (Leach), Alder flies.
         Panorpidæ (Leach), Scorpion flies.
Family *Hemerobiidæ :*—
Genus *Osmylus* (Latreille).   Ocelli, three, placed in triangle on forehead.

Genus *Drepanopteryx* (Leach). Ocelli, none; anterior wings very broad; posterior margin, sub-falcate.

Genus *Chrysopa* (Leach). Ocelli, none; wings, entire; antennæ, cylindrical; labrum, notched.

Genus *Hemerobius* (Linn.). Ocelli, none; wings, entire; antennæ, sub-moniliform; labrum, entire.

Genus *Coniopteryx* (Curtis). Tarsi, five-jointed; wings covered with white powder; few nerves; nerves disposed nearly alike in all wings; labial palpi terminated by large ovate joint.

Mr. Moseley, of Huddersfield, arranges the Neuroptera as follows :—

Order, *NEUROPTERA*.

Sub-Orders, TRICHOPTERA (Caddis flies); PLANIPENNIA.

Families.—*Panorpidæ* (Scorpion flies).
    Hemerobiidæ (Lace-wings).
    Raphididæ.(Snake flies).
    Sialidæ (Alder flies).

Sub-Order, PSEUDO-NEUROPTERA.

Families.—*Libellulidæ* (Dragon flies).
    Ephemeridæ (Day flies).
    Perlidæ (Stone fly).
    Psocidæ (Book-mites).

Moseley says :—" This order of arrangement is exactly the reverse of that given by the British authority (McLachlan) on these insects. The *Trichoptera* are so inseparably connected with the *Lepidoptera*, and the *Psocidæ* so much resemble the *Orthoptera*, that under the present disposition of the orders I have been compelled to adopt this course."

Referring to the *Hemerobiidæ* he says :—" The members of the genus *Chrysopa* are of largish size, delicate green with light gauzy wings, and bright brassy eyes. The species of the genus *Hemerobius* are small insects, blackish or buffish, and wings very iridiscent."

Dr. Hagen, whose series of Synopses of the *Neuroptera* are standard, gives thirty-two species of the *Hemerobiidæ* as British, and classifies them as follows :—

Family Hemerobiidæ.

Genus Osmylus (1 species).—O. Chrysops, a pretty brown insect, with the wings spotted with black. It is met with in the month of June, and seems to prefer stony, rapid streams, fringed with alders. The larva lives partly in water. In this genus there are ocelli visible.

Genus Chrysopa (15 species).—In this genus the ocelli are wanting. The larva feeds on aphides.

Genus Sisyra (2 species).—The larva lives in water, and has been described by Westwood under the name of Branchiostoma spongillæ.

Genus Micromus (3 species).

Genus Hemerobius (7 species).—The larvæ of the species of this genus preys on aphides, and clothes itself with the empty skins of its prey.

Genus Drepanopteryx (1 species).

Genus Coniopteryx (3 species).—These are small and covered with a white mealy powder. The larvæ live in fir trees, the aphides frequenting which are their food.

Mr. W. S. Dallas removes the Dragon flies to the Orthoptera, and arranges his Neuroptera as follows:—

ORDER, NEUROPTERA.
Sub-Order, PLANIPENNIA.

Families.—Megaloptera (including sub-family Hemerobiidæ).
Sialidæ (Alder flies).
Panorpidæ (Scorpion flies).
Sub-Order, TRICHOPTERA (Caddis flies).

Mr. Dallas describes the Hemerobiidæ as having "antennæ either thread-like or necklace-like and not clubbed."

Chrysopa he describes as a delicate green insect, about half-an-inch long, with gauzy wings traversed by a most delicate network of green veins. The prominent eyes are of a beautiful golden colour.

Mr. R. McLachlan, in his "Monograph of the Neuroptera," published in the Transactions of the Entomological Society for 1868, classifies the Neuroptera Planipennia as follows:—

Division I.—*Sialina* (Alder flies).

„   II.—*Hemerobiinœ* (Lace-wings).

„   III.—*Panorpidœ* (Scorpion flies).

Hemerobiinœ he divides into three families :—

Family I., *Hemerobiidœ.*—Antennæ moniliform ; wings mostly with numerous transverse veinlets ; margins ciliated.

Family II., *Coniopterygidœ.*—Antennæ moniliform ; wings, scarcely any ; transverse veinlets ; margins not visibly ciliated, minute ; covered with whitish powder.

Family III., *Chrysopidœ.*—Antennæ setiform ; wings with moderate number of transverse veinlets ; margins ciliated ; colour, usually greenish.

The family *Hemerobiidœ*, Mr. McLachlan divides into seven species, of which two, or perhaps three, are aphidivorous, as follows :—

1.—*Osmylus* (1 species), not aphidivorous.

2.—*Sisyra*   „   „   ::

3.—*Psectra* (4 or 5 species)   „

4.—*Drepanopteryx* (1 species), aphidivorous.

5.—*Hemerobius* (13 species)   ..

6.—*Megalomus.*

7.—*Micromus.*   Larva very probably aphidivorous.

The Aphidivorous Genera are divided by him as follows :—

Drepanopteryx.—1 species, *D. Phalœnoides* (Linn.).

Hemerobius.—*H. elegans* (Stephens).

..   *H. pellucidus* (Walker).

H. inconspectus (McL.), Fig. 6.

H. nitudulus (Fabr.), Fig. 7.

H. micans (Olivier), Fig. 8.

H. humuli (Linn.), Fig. 9.

H. marginatus (Stephens), Fig. 10.

H. limbatus (Wesmael), Fig. 11.

H. pini (Stephens).

H. Atrifrons (McL.).

H. subnebulosis (Stephens), Fig. 12.

H. nervosus (Fabr.), Fig. 13.

H. Concinnus (Stephens), Figs. 14, 15.

To assist in recognising these, I have availed myself of the drawings, given by Mr. McLachlan, of the final segments of the abdomen of the male in several species, and of both male and female in one species. These will be found highly characteristic, and entomologists are much indebted to Mr. McLachlan for placing at their disposal such definite features in so clear a manner.

The *Chrysopidæ* the same naturalist has divided into two genera :—

i.—*Chrysopa* (Leach).   2.—*Notochrysa* (McLachlan).

The genus *Chrysopa* comprises the following species :—

Mostly green, often with black markings	-	-	C. *Flava* (Scop.). / C. *vulgaris* (Schneider).
Ditto	do.	do.	C. *Aspersa* (Wesmael).
Ditto	do.	do.	C. *ventralis* (Curtis).
Ditto	do.	do.	C. *septem punctata* (Wesmael).
Ditto	do.	do.	C. *flavifrons* (Brauer).
Ditto	do.	do.	C. *vittata* (Wesmael).
Ditto	do.	do.	C. *alba* (Linn.).
Full green colour	-	-	C. *phyllochroma* (Wesmael).
Ditto	do.	-	C. *abbreviata* (Curtis).
Blue-green	-	-	C. *perla* (Linn.).

The genus *Notochrysa* consists of two species only, but these are remarkable for their colour :—

Dark, reddish orange, black,   N. *fulviceps* (Stephens).
fuscous, and yellow -   - N. *capitata* (Fabr.).

The *Coniopterygidæ* consist of one genus only (*Coniopteryx*), and it includes but three species :—

C. *Psociformis* (Curtis) ;
C. *Tineiformis* (Curtis) ;
C. *Alcyrodiformis* (Stephens).

This is the most complete classification of the Aphis-eating Neuroptera with which I am acquainted, and, save for some confusion in the characters ascribed to the main groups, there cannot well be a more satisfactory arrangement of the species found in Britain. The larval forms are of great interest, and, being comparatively little known in many cases, offer an attractive field for investigation to any observer who may have time and opportunity to study their life-histories.

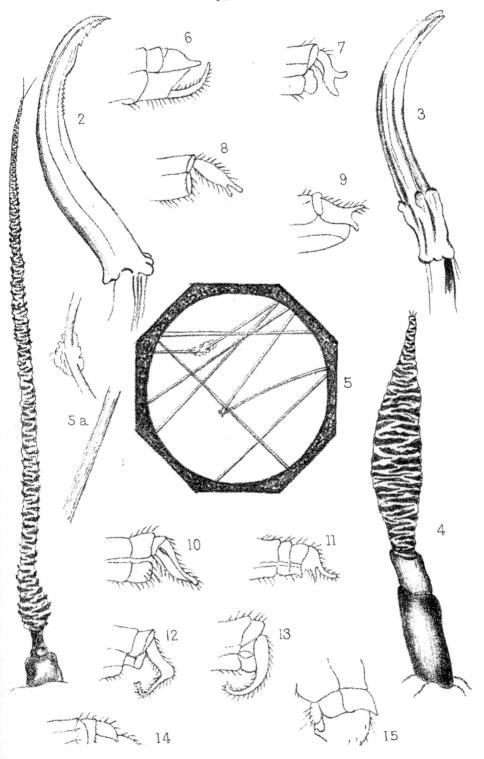

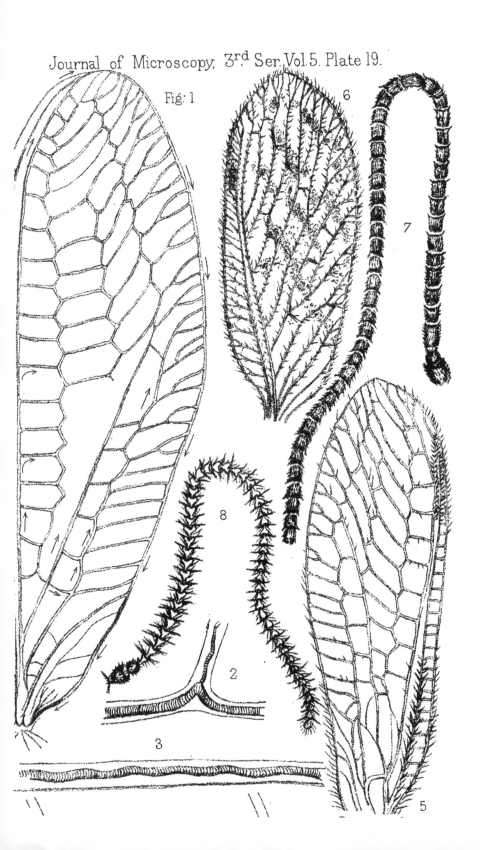

Fig: 1

6

7

8

2

3

5

EXPLANATION OF PLATES XVIII. & XIX.

Plate XVIII.

Details of Aphidivorous *Hemerobiinæ.*

Fig. 1.—Maxillary palpus of Larva of *Hemerobius*, showing the peculiar chitinous incomplete rings.

,, 2.—Mandible of same larva, exhibiting the toothed termination and the fine serrations of the edge.

,, 3.—Maxilla of same. The opening at the extremity through which the juices of the aphis victims are ingested is clear, and also the fold extending longitudinally down the organ.

,, 4.—Labial palpus of same larva, showing in the widened terminal joint the same arrangement of incomplete rings of chitin.

These organs are much smaller than the similar organs of *Chrysopa perla*, shown in Pl. XII., but are magnified in a much greater degree.

,, 5.—Threads of silk-like substance, spun by the same larva to form its cocoon. . These threads were spun on a cover-glass, and the glutinous nature of the secretion is clearly visible at the points of attachment. No indication of duplication or structure of any kind is visible.

,, 5*a*.—Portion of same, more highly magnified.

,, 6.—Extremity of abdomen of ♂ *Hemerobius inconspicuous.*

,, 7.— ,, ,, ♂ *H. nitudulus.*

,, 8.— ,, ,, ♂ *H. Micans.*

,, 9.— ,, .. ♂ *H. humuli.*

,, 10.— ,, ,, ♂ *H. marginatus.*

,, 11.— ,, ,, ♂ *H. limbatus.*

,, 12.— ,, .. ♂ *H. subnebulosis.*

,, 13.— ,, .. ♂ *H. Nervosus.*

,, 14.— ,, ♀ *H. Concinnus.*

,, 15.— ,, ,, ♂ *H. Concinnus.*

The last 10 figures are after those given as characteristic by Mr. R. McLachlan in his *Memoirs on the Neuroptera.*

Plate XIX.

Fig. 1.—The wing venation of *Chrysopa perla*, after Bowerbank, in the *Entomological Magazine*, showing the circulation throughout the nervures. The arrows show the direction taken by the fluids within.

,, 2.—A branching vein, showing the manner in which the included trachea divides.

Figs. 3 and 4.—Veins showing the ringed tracheæ within. The diameter of the tracheal tube is about $\frac{1}{2222}$ in. ; that of the vein
   is about $\frac{1}{408}$ in.

 ,,   5.—Wing of the male of a smaller species of *Chrysopa*, drawn
    directly from the insect.

 ,,   6.—Wing of *Hemerobius*, drawn under the same conditions as the
    last figure, illustrating the differences in structure and
    venation.

 ,,   7 —Antenna of *Chrysopa perla*. consisting of a very large number
    · of similar cylindrical joints.

 ,,   8.—Antenna of *Hemerobius*, exhibiting the moniliform structure.

Change of Air.

THE necessity for change of air not only during convalescence
from illness, but as a means of maintaining the normal
standard of health, is now generally recognised. The father
of a family who fails to make some arrangements for giving his
children an annual holiday, either at the seaside or elsewhere, is not
acting judiciously. Young people who live in large towns and who
have comparatively little opportunity of indulging in outdoor games,
and getting healthful exercise, soon "run down" and become
depressed and debilitated. Many people seem to think that a preliminary illness is requisite as a justification for a holiday, but there
can be no doubt that there would be very much less illness if
change of scene and air were regarded as necessities of life. One
annual outing is not enough for the maintenance of robust health,
but should be supplemented by excursions five or six times a-year.
People are waking up to the absolute necessity of varying their
ordinary daily toil with periods of relaxation. English people are
energetic enough in what concerns business matters, but as a rule
take their pleasures sadly, which may possibly arise from the feeling
that they are doing something unusual. The breadwinner of a
family who cannot take a holiday in the summer should set aside a
weekly sum as a nucleus of a "pleasure hoard" to be devoted to
recreation. A man with an assured income of £300 is justified in
spending 5s. weekly for this object, which would give a family each
week's end invigorating exercise in the fresh air. When a holiday
is taken the programme should be minutely worked out before, or
there will be loss of valuable time. Trips arranged systematically
every week not only afford a vast amount of pleasure, but obviate
resorting to the doctor and the chemist.—"*The Family Physician.*"

The Ascomycetes.*

THE fourth volume of this comprehensive work on *British Fungi* is devoted to the Ascomycetes, and with the kind permission of Messrs. George Bell and Sons, we purpose making one or two short extracts.

The very large number of species of fungi included in the group known as the Ascomycetes are characterised by having their spores produced in *asci*, or mother-cells. In the great majority of species the asci are numerous, closely packed side by side, and form the *disc* or *hymenium*, seated on and protected by a structure called the *ascophore*, which is either *parenchymatous*—that is, composed of a mass of more or less polygonal cells, united to form a tissue—or consists of densely interwoven septate hyphæ.

The Ascomycetes are divided into the following families, viz. : Gymnoascaceæ, Hysteriaceæ, Discomycetes, Pyrenomycetes, and Tuberaceæ.

For our first plate in illustration of this work we have chosen that representing the second family :—

HYSTERIACEÆ (Plate XXI.).

Ascopores erumpent, innate, or superficial ; horizontally elliptical or linear, or vertical and laterally compressed ; texture carbonaceous or membranaceous ; dehiscing by a narrow slit running the entire length of the ascophore, black or blackish-brown ; asci 4—8 spored ; spores hyaline or coloured, continuous or septate; paraphyses usually present. The species are all minute, and mostly gregarious ; all are saprophytes, growing on old wood, bark, and also on dry leaves.

Explanation of Pl. XXI., showing Figures illustrating the *Hysteriaceæ*, etc.—Fig. 1, *Lophium mytillinum*, Fries, a group of plants, nat. size.—Fig. 2, one ascophore; slightly magnified.—Fig. 3, ascus and paraphysis ; highly magnified.—Fig. 4, spores of

---

* BRITISH FUNGUS FLORA: A Classified Text-Book of Mycology. By George Massee, author of " Plant Life," " The Plant World," etc. Vol. IV. Cr. 8vo, pp. viii.—522. (London: Geo. Bell & Sons. 1895.) Price 7/6. We thank the publishers for the loan of the electros, from which the accompanying plates are printed.

same; × 300.—Fig. 5, *Farlowia repanda*, Sacc., group of plants, nat. size.—Fig. 6, one plant of same, slightly magnified.—Fig. 7, Section of same; slightly magnified.—Fig. 8, Ascus and paraphyses of same, highly magnified.—Fig. 9, Spores of same; × 300.— Fig. 10, *Dichæna quercina*, Fries; nat. size.—Fig. 11, Ascus and paraphyses of same; the spores should not be muriform, as represented, but 3–septate; highly magnified.—Fig 11*a*, Spores of same; × 300.—Fig. 12, Spores of *Dichæna faginea*, Fr., var. *capreæ;* × 300.—Fig. 13, *Hysterium pulicare*, Pers.; nat. size.— Fig. 14, One plant of same, seen from above; slightly magnified. —Fig. 15, Ascus and paraphysis of same; highly magnified.— Fig. 16, Spores of same; × 300.—Fig. 17, *Actidium hysterioides*, two plants; slightly magnified.—Fig. 18, Free spores of same; × 300.—Fig. 19, Ascus of same; highly magnified.—Fig. 20, Group of plants of same; nat. size.—Fig. 21, *Schizothyrium ptarmicæ*, plants on a living leaf of *Achillea ptarmica;* nat. size.—Fig. 22, Plants of same on portion of a leaf; slightly magnified.—Fig. 23, Ascus and paraphyses of same; × 300.—Fig. 24, *Schizoxylon Berkeleyanum*, portion of a spore, breaking up into cells at the septa; × 750.—Fig. 25, Ascus and paraphyses of same; × 300. —Fig. 26, One plant of same, slightly magnified.—Fig. 27, Plants of same on dead stem; nat. size.—Fig. 28, *Ephelina radicalis*, Mass., showing the blackened swelling on the stem of *Rhinanthus* caused by the fungus; nat. size.—Fig. 29, Ascus and paraphyses of same; × 300.—Fig. 30, Stylospores of same; × 300.—Fig. 31, *Ostreion Americanum*, Duby, spore; × 300.--Fig. 32, *Hysterographium fraxini*, group of plants; nat. size.--Fig. 33, Spore of same, × 300.—Fig. 34, *Ocellaria aurea*, Tul., group of plants bursting through the bark of a branch, slightly magnified.—Fig. 35, Ascus and paraphyses of same; × 300.—Fig. 36, *Gloniopsis curvata*, Sacc.; nat. size.—Fig. 37, Ascus and paraphyses of same; highly magnified.—Fig. 38, Free spore of same; highly magnified. —Fig. 39, *Nytilidion læviusculum*, Sacc., group; nat. size.—Fig. 40, One ascophore, slightly magnified.—Fig. 41, Free spores of same; × 300.—Fig. 42, *Xylographa parallela*, Fries., three plants seen from above; slightly magnified.—Fig. 43, Sections of same; slightly magnified.—Fig. 44, Ascus and paraphyses of same; × 300.—Fig. 45, *Aulographum vagum*, Desm., plants on portion

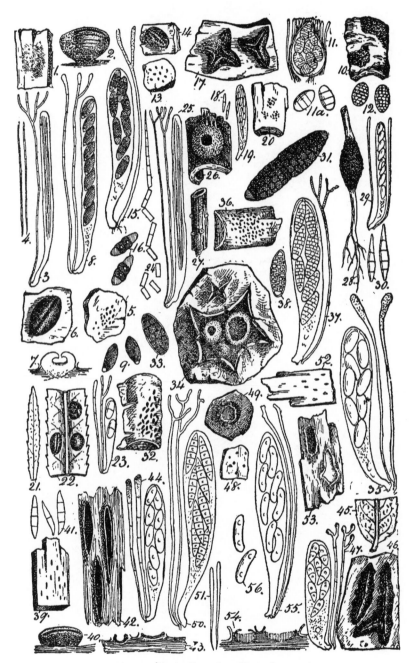

Figures illustrating the *Hysteriaceæ*, etc.

Figures illustrating the *Helvelleæ*, etc.

of a leaf of goat-willow ; nat. size.—Fig. 46, Two plants of same, slightly magnified.—Fig. 47, Ascus and paraphyses of same ; × 300.—Fig. 48, *Pseudographis pinicola*, Rehm., plants ; nat. size. —Fig. 49, One plant of same ; slightly magnified.—Fig. 50, Ascus and paraphyses of same ; × 300.—Fig. 51, *Colpoma degenerans*, Mass. ; spores × 300.—Fig. 52, *Propolis faginea*, Karst., group of plants ; nat. size.—Fig. 53, Two plants of same ; slightly magnified.—Fig. 54, Section of same ; slightly magnified.—Fig. 55, Ascus and paraphyses of same ; highly magnified.—Fig. 56, Free spores of same ; × 300.

Discomycetes.

The most important distinctive feature of the present great group or third family of the Ascomycetes consists in the disc or hymenium being fully exposed at maturity. There is a very wide range in size, form, texture, and colouration, and, as would be expected, there are transitions to allied groups at various points. We have selected for our second plate that illustrating the

HELVELLEÆ (Plate XXII.).

The principal common feature of this group is that the disc or hymenium is fully exposed from the earliest stage. There is an absence of the incurved margin of the ascophore when young, and the gradual exposure of the disc so characteristic of the Pezizæ.

Explanation of Plate XXII., showing Figures illustrating the *Helvelleæ*, etc.

Fig. 1, *Morchella esculenta*, Pers., entire fungus ; about one-third nat. size.—Fig. 2, Ascus and paraphysis of same ; highly magnified.—Fig. 3, Spores of same ; × 330.—Fig. 4, *Rhizina undulata*, Fries, entire fungus about one-half nat. size.—Fig. 5, Section of same, showing the numerous rhizoids ; about one-half nat. size.—Fig. 6, Ascus and paraphyses of the same ; highly magnified.—Fig. 7, Spores of same, × 300.—Fig. 8, *Gleoglossum glutinosum*, Pers., entire fungus ; about two-thirds nat. size.— Fig. 9, Ascus and paraphyses of the same ; highly magnified.— Fig. 10, Spores of same ; × 300.—Fig. 11, *Leptoglossum microsporum*, Sacc., ascus and paraphysis ; highly magnified.—Fig. 12, spores of same ; × 300.—Fig. 13, *Gleoglossum viscosum*, Pers., paraphyses highly magnified.—Fig. 14, *Gyromitra esculenta*, Fries, entire fungus; about one-half nat. size.—Fig. 15, Spore of same, × 300.—Fig. 16, *Gyromitra gigas*, Cooke, spore; × 300.— Fig. 17, *Helvella crispa*, Fries, entire fungus; about one-half nat. size.—Fig. 18, Spores of same; × 300.—Fig. 19, *Verpa digitaliformis*, Pers., entire fungus; about two-thirds nat. size.— Fig. 20, Section of pileus and upper part of hollow stem of same; about two-thirds nat. size.—Fig. 21, Ascus and paraphysis of same; highly magnified.—Fig. 22, *Spathularia flavida*, Pers., entire fungus; about two-thirds nat. size.—Fig. 23, Free spore of same; × 300.—Fig. 24, Ascus and paraphyses of same; highly magnified.—Fig. 25, *Leotia lubrica*, Pers., single plant; about two-thirds nat. size.—Fig. 26, Section of pileus of the same; about two-thirds nat. size.—Fig. 27, Ascus and paraphysis of same; × 300.—Fig. 28, Free spores of same; × 300.—Fig. 29, *Mitrula paludosa*, Fr.; about two-thirds nat. size.—Fig. 30, Ascus and paraphyses of the same; × 300.— Fig. 31, *Morchella conica*, Pers., var. *deliciosa*, Fr., fungus; about two-thirds nat. size.—Fig. 32, *Vibrissea truncorum*, Fries, group of plants; nat. size.—Fig. 33, Section of pileus of the same, showing the spores escaping; slightly magnified.—Fig. 34, Portion of a spore of same; × 750.—Fig. 35, Ascus and paraphyses of same; highly magnified.—Fig. 36, *Peziza ochracea*, Boudier, two plants; nat. size.—Fig. 37, Ascus and paraphyses; highly magnified.

SECRETIONS IN PLANTS.—M. Tschirch announces in the *Botanisches Centralblatt* that in all the normal cases in which he has been able to study the formation of a secretion, he has found that it was a function, not of the protoplasm, but of the cellular wall. In the oil-glands of the labiates, composites, etc., the secretion is due entirely to an internal layer of the cellular wall, and the same is the case with the Papilionaceæ. The secretions, however, are never produced by the metamorphosis of the substance of the cellulose itself. The observation applies likewise to the resins, which M. Tschirch considers as aromatic acid compounds with a particular group of alcohols, which he calls resinols. —*Scientific American.*

Bacteria of the Sputa and Cryptogamic Flora of the Mouth.

THIRD MEMOIR.

By Filandro Vicentini, M.D., Chieti, Italy.

On Leptothrix Racemosa. Plate XX.

Translated by Professor E. Saieghi.

NEW RESEARCHES ON THE FRUCTIFICATION OF THE NORMAL PARASITE OF THE MOUTH
(LEPTOTHRIX RACEMOSA?)

I TRUST it will not appear too pretentious on my part if I propose for such an isolated form—or, better to say, for the tiny plant which thrives (and appears to germinate and fructify)—on our teeth, taken as a whole, the name of *Leptothrix racemosa*, instead of that of *Leptothrix buccalis ;* and this solely on purpose to qualify it better, if it be true that the cryptogamic species must take their name and character from the fructification (if any) rather from the inferior or rudimentary appearances of their single particles, incidentally incomplete, scattered, or isolated.

In describing the various aspects of the fructification in question, we shall endeavour to group them in such a way that, according to our view, there may be a better connecting of the varieties ; but, in doing so, we do not wish to prevent a more matured judgment.

Various Elements and Aspects of the Bunches.

To understand the various aspects which the ears or bunches of *Leptothrix racemosa* present, we must value the different constitutive elements of those clusters, which, according to our observations, we consider to be four, in their natural order of development.

The principal element, from which the other three proceed and upon which they are formed, is the fertile filament, generally slender, pale, containing internal parietal gemmules, invisible in the iodine solution (simple or acidulated), but distinct enough in gentian violet. For this kind of *stalks* we refer to the preceding memoir. Then is it not strange that these fertile filaments should be so slender, whilst the severed or truncated (*Bacillus buccalis maximus* and *Leptothrix buccalis maxima* of Miller) appear much thicker and with a strong iodine reaction. We attributed this thickness, or woody state, to the retrocession of the germinal matter, somewhat analogous to the action of the saps in pruning-plants. Our parasite, being a vegetable, or rather a tiny plant, and having to be studied as such, according to Dallinger,* it is natural that the form of its stem should resemble that of the cone. The stalk being, in fact, the proper organ of vegetation, destined to support the upper organs, it is natural that the lower part should be stronger and that it should get thinner towards the top, which, finally, must bear the fruits. It is identical with the process of other plants.

However, whilst this is the general rule, we shall see, as we proceed, that certain fertile filaments become rather an exception, thinning themselves similarly on the top; but afterwards, on reaching the base of the future fructification, they thicken in the shape of a cylindrical club, colouring brilliantly with aniline, but pale like the rest in iodine solution. We shall speak later on of the probable meaning of this swelling of the stalk, recalling only on this subject the other apical swellings, already mentioned and delineated in the preceding Memoir (Fig. 9, *b*, *c*).

To this first element (the internal stem) follows in order of formation the second, of peduncles or ingrafting threads, destined to bear the spores, like the *sterigmata* of many fungi. Such peduncles are very pale, invisible with less powerful objectives, but very distinctly observable (under certain conditions) with the 1/25th power just mentioned. They are short, funnel-shaped, with a point on the stalk and the opening towards the spore. In the young ears, being yet deficient or scanty in the secretion of the viscid substance, the peduncles are more visible; whilst in the older ones they remain more or less opaque.

* Dallinger, *The Microscopical Organisms and their Relations to Disease* (*Journal R.M.S.*, *1885*).

The spores constitute the third element, and are of globular form ; pale in iodine solution, more or less coloured in the aniline. They are smaller in the younger clusters, on which, for this very reason, the peduncles or sterigmata are more discernible; whilst, the spores becoming afterwards larger and covered with a more viscid substance, the peduncles remain hidden. Besides, the spores are not equally thick in all the fructifications ; but where the spores are thinner, as in Fig. 23, *a*, there the details mentioned are more visible. The little or non-visibility of the peduncles may result, not only from the opacity produced by the abundance and density, or from the heavy colouring of the viscid substance, but even, where this is thin and transparent, from the identity of the index of refraction of the two elements.

The fourth element is the viscid substance or *glair*, which we hold to be the last to form. It proceeds from a sort of oozing or secretion of the stalk or of the spores themselves, for, in the younger ears, with yet small spores, it appears thinner and indistinct.

And these are the fructifications which exhibit a more striking resemblance to real clusters, either in the solution of iodine or when they are very slightly affected by gentian violet. On the other hand, in the ears with an internal swelling of the stalk, the size of the spores, as well as the density and colouring of the viscid substance, reach the highest degree, as we shall see later on. That the viscid substance may proceed from the filament is exhibited by the cited examples, in other species, by Billet, as well as in our parasite, by those filaments of *Leptothrix buccalis maxima* and of *Bacillus buccalis maximus*, which, with the new 1/25th objective, appear thoroughly enveloped in a hyaline sheath. That it may proceed from the spores is shown by the example of the incapsulated diplococci. However, between the exudation of the old filaments and that of the ears there is this difference : that the first has greater affinity for the acidulated solution of iodine, in which it is better discerned, and the second for the gentian violet.

Keeping in view, on one side, these various elements, and, on the other, the degree and different nature of the colouring, we may fully explain the varied aspects of the fructifications.

Concerning the gathering and the preparation of the patina

dentaria and its treatment with the aniline colours, or with the solution of iodine, there is little to add to what has been said in the previous Memoir. We gave there the precise rules, in order to obtain from the patina dentaria the greatest possible number of fructifications, taking it before a meal or early in the morning. We found afterwards that the fructifications are more abundant on teeth with a thin film of tartar, particularly in the superior eye-teeth. About the disintegration and colouring of the clods we refer to what has been said. For the acidulated solution of iodine, we placed on the slide, first, a small drop of lactic acid, disintegrating in it afterwards the patina, and lastly adding to it one or two drops of solution of iodine. It generally takes a quarter of an hour to get a proper colouring.

In the preceding Memoir we suggested that the superfluous liquid should be allowed to trickle down from the sides, after applying the cover-glass ; but afterwards we discovered that it was better to let most of the liquid evaporate before applying the cover-glass, so as to avoid the wave of liquid caused by the pressure, which might remove many bacteria and bacilli and several isolated filaments and clusters set free from the tiny islands that often supply the most instructive specimens ; blotting paper would also take up many of the specimens. To eliminate the air-bubbles keep the cover-glass on edge (straight up) with the two first fingers of the left hand, so as to form an acute angle with the slide, at two centimetres from the preparation ; then with a straight needle, held in the right hand to support the glass, and with a bent needle pushing it on the specimen, lower it down gently on to the edge of the preparation, when the air-bubbles will be set free.

If we wish to institute a comparison between the two stains adopted in this work, we may say that gentian violet colours briskly the little spores, but at the same time attacks and obscures the viscid substance, so that, wherever it fully invests, the peduncles or sterigmata are either not seen at all, or may hardly be distinguished. This happens even by applying very little tint. Nevertheless isolated fructifications slightly affected by the tint may sometimes be found in the preparation, and it is just upon one of these occasions that we first detected the peduncles in. question. For this reason, the whole image, either of the single fructifications, or of their branching filaments, is better seen in the anilines (see in the previous Memoir the Figures 10, 12, 16); the solution of iodine would hardly give the same result. This solution has little or no action upon viscid matter; therefore, the fructifications assume with it a granular aspect, and if the spores are not too thick it shows the peduncles better; also because the solution of iodine acidulated with lactic acid attacks relatively better the peduncles themselves, whilst the gentian violet invades them less than other elements. Consequently it happens that the solution of iodine better exhibits the complex aspect of the clusters or the truncated sterigmata. (See Fig. 23, d.) It has, however, the disadvantage of not satisfactorily allowing the use of powerful eye-pieces, thus limiting the enlargements; whilst in the fructifications stained with gentian violet, the details of structure may be (under favourable conditions) detected even with a No. 6 eye-piece, as is shown in Fig. 24, magnified to 3,100 diameters.

Now from the concourse of these various circumstances, either relative to the age of the single fructifications, and to the more or less thickening of peduncles and spores, or to the quality and degree of the colouring, we are able to obtain images conspicuous in the whole, but with peduncles only partially or not at all visible, or sometimes less conspicuous in the whole, but with quite distinct peduncles. The necessary conditions to the clear vision of the peduncles in question may be summed up in the following series :—a, proper optical instruments; b, clusters still young; c, rather thin spores; d, a weak gentian violet; or e, solution of iodine. We have already said that the best images are obtained from isolated fructifications fallen from clods in a clear field, and it is our intention in this work to consider the two colourings above mentioned, apart from the use of other tints.

The fructifications in question can be observed by axial illumination or by oblique light. The best images of the clustered forms and of the single sterigmata are obtained by axial illumination by properly adjusting the correction collar, and by centering the iris diaphragm and the Abbé condenser along the optical axis. On the other hand, the general relief of the ears and the position of the spores in six longitudinal series, are better detected by oblique light, by pushing aside the diaphragm and letting it after-

wards go round the optical axis, or substituting for it the dark
diaphragm, pushed on one side, and making the stage rotate with
the upper part of the instrument round the same axis. In the
latter case, a better effect is obtained when the ear is horizontally
disposed and is *struck* from behind by the pencil of light, almost
parallel with the stalk, beginning from the root, so that the
luminous rays run along its axis. Then, focussing, the general
effect of the six series of spores becomes striking, the whole ear
takes a beautiful mulberry appearance, of which it is impossible
to give a satisfactory representation.

The figures which we have drawn are, however, sufficient to
give a proper idea of the peduncles in question. In Fig. 23, *a*, we
have represented a short and young ear, as seen stained with the
acidulated solution of iodine, magnified to 1,700 diameters; in it
the spores are thin, and its clustered form is most striking, as we
can even perceive posterior rows of little spores. In *c* (same
figure) is seen an older ear, thicker and longer (same staining,
and magnified 1,170 diameters) in which, however, the peduncles
are sufficiently distinct. In *d*, then (same staining and magnification), is drawn a short fragment of a fertile stalk, found by chance
in one of the numerous preparations; only two spores are seen at
its base, and higher up the peduncles still thick, but without
spores; the rest of the ear is wanting.* In the above Fig. 24
(weakly stained with gentian violet, magnified to 3,100 diameters)
the spores are intact, their funnel-shaped peduncles are visible
through the viscid substance, moderately stained. The stalk
exhibits several gemmules of reserve.

The facts hitherto given (in support of the arguments
expounded or simply suggested in the preceding Memoir), which
everybody can verify for himself, are, in our opinion, sufficiently
conclusive to warrant us in affirming the existence of a really
external sporification or fructification of the normal parasite of the
mouth upon six longitudinal series of peduncles and spores. But
for a more evident proof, take the specimen drawn in Fig. 25.
This specimen, obtained by pure chance, amongst the numberless preparations examined, represents a stalk which, emerging

---

*The attenuated appearance of this stalk probably depended upon a flowing-
out of its sap or germinal matter owing to rupture.

from the *materia alba* or heap, was extending horizontally, and changed abruptly for an upward direction, with the upper part bearing the cluster. It is seen much aslant, but fortunately intact (stained with acidulated solution of iodine, and magnified to 1,700 diameters); it has a round end, with a fine rosette of six rays, formed by six terminal sterigmata, and having the six last spores, probably yet unripe, to their tops.

FRUCTIFICATIONS BY TEMPORARY SPORES (*Sporids?*) AND BY PERSISTENT SPORES (*Teleutospores?*).

The mycetologists and the algologists call temporary spores agamic spores, sporids, and conids the spores which are formed without previous fecundation, and are intended by the multiplication of the species in more favourable and immediate conditions (namely, the conids of the *Peronospora*, destined to diffuse the species during a part of the year); whilst the persistent winterly spores, oöspores, or teleutospores, are produced by the act of conjugation, and help to preserve the species from external injuries, and to strengthen their future shoots (like the hybernating spores of the *Peronospora* in the thickness of the hospitable parenchyma, deputed to reproduce the species in the following year). Of the two processes of sporification, one (says De Bary) seems intent upon preserving *intensively* the species through conjugation, the other only to increase its *extension.**

Of the other processes of propagation through gemmules, sprouts, etc., we have already treated in the previous Memoir, and shall refer to the subject again later on.

All the fructifications of our parasite, hitherto described and drawn, probably belong to the temporary series, with the exception of the specimen given in Fig. 13 of the preceding Memoir, which might be included in the *persistent* series; but from the last researches and other isolated cases, of which we will speak presently, it appears that the same parasite presents also a comparatively scanty number of fructifications of another kind, which, although similar in shape or type, assume special characteristics, so that we are rather inclined to refer them to the persistent series.

---

* De Bary, *Du Développement des champignons parasitaires (Ann. des Scien. Nat. Sér. Bot., t. XX.).*

We have already mentioned the club-shaped stems, two of which, stained with gentian violet, are seen drawn in Fig. 26, magnified to 1700 diameters. The club is generally long enough, as in *a* ; but sometimes we meet short ones, as in *b*, which might be called an incipient phase. In the first named specimen, the club, although complete, is still quite bare, and, to all appearances, represents a hardly-formed expansion, before the sterigmata and the spores have germinated. In fact, in the Fig. 27 (same staining and magnification) may be noticed an ear, on the whole larger and with spores proportionally more conspicuous, exhibiting an internal stem, club-shaped, and brilliantly coloured, quite dissimilar from the pale and slender stems of the first series, but analogous to the bare stem of Fig. 26.

To our knowledge, these ears never reach the length of some others having slender stems (see Figs. 12 and 16), which may depend upon the comparatively limited length of the club-shaped expansions. The spores of such ears are, besides, more conspicuous and pressed together, so as to form on both sides a sort of zone or violet aureola, at a little distance from the stem ; and, between this and the periphery, runs a clearer intermediate zone, where the viscid substance is less coloured, but yet capable of disguising the sterigmata. The light proceeding from the condenser must, in fact, cross first the deep violet zone, which is next the slide, then the intermediate clearer substance (the index of refraction of which is identical perhaps with that of the peduncles), and finally the opaque violet zone, which overlooks the cover-glass. At any rate, the result of this optical combination is to hide the peduncles. When in the solution of iodine we come across such ears, the zones become mixed up, and we perceive, on the whole, a triplex series of coloured granules, as is shown in Fig. 13 of the previous Memoir, incompletely represented, which would lead to the supposition that in these ears the secretion of glair is more abundant and thick.*

It appears, besides, that such ears are even more compact and resisting to the mechanical agents of disintegration ; also, their

---

* After presenting this Memoir, we have made further researches (especially on the presence of such ears in the sputum of pneumonia, and on the manner of detecting the peduncles), which we will soon make known.

fragments always exhibit a cohesion of the single particles, and their brilliant colouring becomes more conspicuous with aniline. We remember having often found similar fragments in sputa ; but, not then knowing their nature, we overlooked them. We also remember that some sound ears, or fragments of the same, were found mixed with many minute ears in fructifications upon small flakes of urethral mucus, as we mentioned in the previous Memoir. We may, however, state that, in the preparations of the dental patina, ears of this sort are scanty in comparison with those of the preceding form.

In considering now those more robust and conspicuous forms of fructification, the mind tries, through analogy, to connect them with the process of fecundation, and finds, although indirectly, its existence is confirmed. In the preceding Memoir, we have described some pseudo-inflorescences *in tufts*, having points varying in shape and size, which we held to be future spindle-like bacilli (*Bacillus tremulus* of Rappin) and future comma, or serpentine bacilli, destined, after being dissevered from the stem and becoming free, to perform the functions of spermatia or antherozoids. We gave the reason for such hypothesis, as we also pointed out the likeness of those elements (supposed male organs), with the spermatia of certain well-known fungi, like *Sphærella sentina, Fumago salicina, Apiosporium citri*, etc.

The *antherozoids* or *spermatozoids* in sea weeds, and the *spermatia* in fungi were considered as elements of fertilisation. The first (*migratory filaments, spiral filaments, seminal corpuscles*), now cylindrical, now ribbon-shaped, furnished with cilia and endowed with spiral movement in various directions, are originally contained in a cellule or male organ (*antheridium*). The spermatia corpuscles, oval or in rods, straight or curved, also very motile, like the analogous forms of the mouth, were, nevertheless, held to have no cilia. They are originally sometimes contained in an appropriate cellule or male organ (*spermogonium*), sometimes they grow freely on the apex of the filaments, and get dissevered simply through disjunction. In our parasite we thought, at first, that the fertilising elements belonged to this last type, and were formed *in a free state* on the stems; but we shall see, by and by, that perhaps even they originate within apposite sheaths, and therefore may be

referred to the first type.   In general, the spermatia, unable to
multiply through fission, have been seen, at times, to germinate
on their own account;   one common example of this kind is
exhibited in the ergot.

Now, we repeat, the existence of fructifications more conspicuous and distinct from the others (through their large club-like
stem and their two zones of colouring, etc.), in the normal parasite
of the mouth, would be quite explained, admitting them to possess
fertilising elements constituted by spindle-like, comma, and serpentine bacilli, already described by us, and holding the other spores
as agamous and temporary.   Perhaps the persistent spores in
this parasite are destined to go through the intestinal tube uninjured, withstanding the dissolving action of the gastric juices, and
emerge into the external world, maintaining in the faeces their
vitality for the future diffusion of the species.

As regards the function of conjugation, it may be performed
on the already formed filaments, as we see in many other cryptogams, where sometimes the act takes place between two contiguous
filaments, the male organ of the one penetrating the female organ
of the other; but nothing prevents us from believing that a fertilisation of another kind may have taken place between the male
element (spindle-like, comma, or serpentine bacillus) and the
mother spore, before the germination of the fertile filament.

Against these views of ours, a quite opposite hypothesis might
be produced, namely, the hypothesis of a commensalism or symbiosis.   In such hypothesis the small sporules, and especially the
productions by *points*, would not be proper phases of *Leptothrix*,
but *parasites* of the *parasite*, or new micro-organisms of another
species, come to implant themselves and thrive at the expense of
the original parasite in the same way as *Leptothrix parasitica*,
Kützing, which with its slender filaments lodges itself on the larger
filaments of *Zygnema* and *Cladothrix dichotoma*, as we see in
Fig. 21 (stained with vesuvine and methyl violet, then with solution
of iodine, magnified to 600 diameters).   But, in this way, one
might object, for argument's sake, that grapes are so many parasites of the vine on which they fructify.   In fact, consulting the figure
in question, anyone can see that the secondary parasitical shoots,
c, c, are less thick than our points, which engraft themselves round

the stem, like the hairs of a bottle brush. They are not, besides, methodically arranged, but stretch out very much, like stems destined to vegetate on their own account, rather than to complete the organism bearing them. The filaments of *Leptothrix parasitica* implant themselves also upon the stem of *Zygnema*, *a*, *a*, or of *Cladothrix*, *b*, *b*, which feeds them, by means of bulbs, or spores originated in them, *s*, *s*, as the Fig. 22 shows still better (same staining, magnified to 1,600 diameters): spores which are not seen at all at the insertion of our points. And still less the filaments of *Leptothrix parasitica* are seen to drop at last from the central stem, and swim in the medium with the same briskness of our spindle-like, comma, or serpentine bacilli.

VARIOUS ASPECTS AND FORMS.

In the preceding Memoir we have spoken of various forms and appurtenances of the parasite in question; we specially point out the bifurcations and trifurcations towards the seat of certain filaments, with tiny radical swellings, like haustoria; and then the more pronounced ones, some at the knots, some at the apex: the latter like small heads. We have already mentioned the other apical swellings (fertile filaments), club-shaped.

We now go on to describe a third form of apical expansions, very scarce, which, provisionally, we shall denominate *sheath expansion*.

Such an expansion, represented in Fig. 28 (stained with gentian violet, magnified to 3,100 diameters), is very pale, has streaks in its contour, not detected with inferior objectives. Its paleness cannot be attributed to insufficient colouring, because the examined form rose on the top of a filament (likewise pale) in the midst of a thick and very pretty tuft of ears brilliantly stained (of the kind shown at Fig. 12), and nearly surpassed them with its point. The external contour is very slender from the base of the expansion up to its point, and between the internal stem and the exterior contour are seen numerous slender, tiny, transverse threads, which are attached to the external sheath by means of more prominent small dots.

At first we could not understand the probable meaning of this structure. That it was not to be taken for the club-like expansions of Fig. 26 was evident from its paleness, being the antipodes of the bright colouring of these, as well as the presence of the internal fine beams. On the contrary, after deeper reflection, we thought we might refer it to other forms already described in the previous Memoirs. One of these forms would be that of the pseudo inflorescences *in tufts* (preceding Memoir, Fig. 14). Comparing the two figures, it will not appear unlikely that the slender threads of Fig. 28 in the present Plate, distended like fine beams between the internal stem and the sheath, growing more and more in a transverse direction, may end by breaking the external envelope and become free *points*, perfect fertilising elements (spindle-like, comma, or serpentine bacilli), at first only free from the surrounding sheath, in order to constitute the tufts of the preceding Plate, and at last becoming disjoined from the central stem, so that they may fulfil their function through that stirring motion, mentioned in the previous work.

The other form, possibly analogous, would be that delineated in the first Memoir (Fig. 2, *n*, lower down), similar to a large bacillus, singularly veined throughout, and for a certain tract having traces of lineal bacilli. Even this is a rather rare form, and we have never found it on the top of any filament, but quite by itself. We have already hinted in the previous Memoir that that veined bacillus might be a sort of receptacle for the future fertilising elements, as an antherid or a spermogone.

Now, supposing that interpretation true, the transmutation from one to the other of the three forms would become clear enough. We should have, in the first form (or *sheath expansion*) of Fig. 28 (present Plate) an antherid or spermogone, hardly shown, and in the second form (the tufts of Fig. 14, previous Memoir) a male organ in full development. The third form (the veined bacillus of Fig. 2, *n*, below, first Memoir) would be an intermediate form to the other two (an arrested form), or an antherid or spermogone, prematurely fallen from its stem, having been unable to attain to the adult form of tuft; and then strayed from its destination and remained, as it were, unripe, or even returned to a neutral condition, which is, we think, common to the severed or truncated filaments, as, having been unable to attain the fructification, they limit themselves to a reproductive function of a lower degree *(fissiparous* multiplication) through the increase of the granules or the lineal minute elements, contained in the interior of their sheaths. If this were confirmed by farther researches, it would lead us to rectify the first supposition about the formation of the *fertilising elements.* They would not form themselves freely on the top of the respective filaments, but within a receptacle or male organ properly so called, (sheath expansion).

These are only simple conjectures, aiming at connecting the various forms hitherto described, reconducting the appurtenances of our parasite to the general laws of the cryptogamic flora, and far from pretending to give herewith a full and exact explanation of them. We shall be quite satisfied if the features of the facts we have endeavoured to describe can be proved by further researches.

But, even upon a simple descriptive ground, we should perhaps overstep the limits of a simple preliminary study if we were to dilate longer in the investigation of other particulars, before seeing confirmed and set up the points already demonstrated (such being the most important) by competent authorities. Neither is it our business to solve the question whether the discussed parasite is a fungus or an *alga.* We shall only say that it appears to us to partake of the characters of both families, to thrive as an alga, but to fructify similarly to certain fungi. We shall, therefore, limit our remarks about some apparent irregularities, in the aspect of the described ears, which, through inattention, might pass for true irregularities or anomalies of structure.

In the first place, we refer to some gibbosities or irregular projections, which are met sometimes by the side or upon some ears, which might lead to the belief that the ears themselves are perhaps constituted without any order, or that the series or longitudinal rows of the sporules are not always six in number, but at times more.

Now, one of the more frequent causes of such irregular appearance is very simple. The breaking up of a contiguous ear, and the adherence of that extraneous fragment to the ear that we examine ; having frequently verified this occurrence, we considered it superfluous to draw a similar specimen on our plate. The other case, less frequent, is drawn in Fig. 29 (saturated with acidulated solution of iodine, magnified to 1,700 diameters).

Here it is not the superposition of an extraneous fragment, but the folding up of the ear itself. In the figure referred to, the third superior and the point *b* of the ear are turned up and pressed back upon the middle third, but in a direction somewhat oblique to its axis, so that, at first sight, or with inferior objectives, they simulate a gibbosity. But focussing with the fine adjustment, it is easily perceived that there is, in reality, a folding due to a mechanical cause, and accidentally rendered even more pronounced by the pressure of the cover-glass.

We do not speak of a third apparent irregularity, which might deceive us only when using inadequate optical means—we mean the accidental apposition of extraneous cocci, of bacteria and comma bacilli around some ears, as shown in Fig. 23, *b, b.* Under lower power objectives, such cumuli or groups, especially if more conspicuous, may, in fact, simulate real protuberances.

There is, however, a special apposition of cocci and bacteria, sometimes visible on certain ears ; perhaps, those remained longer with their points in contact with the labial mucous, and there became incrusted with the above bacterial elements, through a cement of viscid mucus. Such ears, in fact, are never augmented on the opposite side, but only on the top, like a very oblong pear ; and the increase, gradually narrowing itself, seldom goes beyond the half of their length (Fig. 30, stained with gentian violet, magnified to 1,170 diameters).

When they reach the first colourising stage with the gentian violet, we can easily perceive there is in reality a sort of cap (at first, more pale and granular) constituted by cocci and bacteria, only slightly coloured, in various layers, towards the point *a*, and sloping towards the half, *a′, a′* ; whilst the ear, *b, b*, with its sporules, is seen brightly coloured in the interior. In the second stage, even these adventitious cocci and bacteria become coloured, and the cap is no longer distinguished from the internal ear.

CONCLUSION.

We have seen how Billet describes the evolutionary cycle of the Bacteriaceæ (which would constitute for our parasite only the inferior cycle), and that his remarks mostly agree with those we have made on the same parasite. His interpretations of the various phases of that cycle do not differ from ours, excepting in what affects the production of bacteria included in filaments, which Billet considers are real endogenous spores ; and we cannot positively deny that such is the case in the four species studied by him ; whilst in our parasite they ought, in our opinion, to be held as simple gemmules of reserve.

But from the exposition of the facts in our previous Memoir, and confirmed in the present one, it clearly results that the evolutionary cycle, so nicely delineated by Billet, cannot include all the morphological phases of our *Leptothrix racemosa*, but only some of them. In this parasite, besides that first cycle which we call *inferior*, there is another—the *superior*, which comprises the organs of genuine reproduction and fructification. Finally, together with these two *normal* evolutionary cycles, there is another one—*accidental*, called *virulent*, in which (according to laboratory experiments) it seems that certain elements, derived from the parasite itself, may, as Pommay says, develop themselves in the sense of virulence.

The first two cycles would, therefore, constitute the *morphological series*, and the third, or virulent, cycle would, in modern language, constitute the *biological series* of our parasite.

MORPHOLOGICAL SERIES.—*Inferior Cycle.*—The inferior cycle embraces the following phases :—

I.—*Phase of Vegetation.*—The characters of this phase are those assigned by Billet to the filamentous state, only that he considers the intertwined state (*état enchevêtré*) as a distinct and later phase ; whilst we believe that it, in our parasite, accompanies the filamentous state, in the same manner as the mycelium or creeping vegetation in fungi. In other words, the intertwined state is even posterior to the isolated filaments which lead a wandering life in the liquid substrata, being unable to attain a more vigorous and stable one ; but it does not constitute a distinct phase, as it is quite natural that, when the passage to the superior phases is precluded *(aërial vegetation)* for want of a fit soil, the filaments intertwine and drop to the bottom, without being able to spread like the mycelium of fungi. However, under favourable conditions, a kind of mycelium (a more complete phase of the entangled state) may be formed, giving birth to an aërial vegetation, as may be specially observed

in the patina of the tongue and in the deep strata of the patina
dentaria.

II.—*Phase of internal gemmulation or budding and dissemination.* This partly corresponds to the dissociated state of Billet,
through the disarticulation of the knots of the single filaments, or
through setting free the included bacteria ; and it is clear that the
unstable condition of the nutrient medium continuing, and consequently the passage to the superior phases being prevented, no
other way of perpetuating the species is left to these micro-organisms than the inferior reproduction or simple multiplication by
shoots and gemmules. Looked at in this way, this phase would
even comprise that held by Billet as endogenous sporulation, but
which are, at least in our parasite, gemmules of reserve, properly
destined to the multiplication of the species, in a neuter state,
through simple fission of the elements, when the genuine reproduction, by means of seminules or spores, is not possible, or when
the fertile filaments have been dissevered or cut by mechanical
injuries.

III.—*Protective phase.* This phase fully corresponds to the
zoöglœic state and to certain conditions of the dissociated state of
Billet, as we have pointed out in the first paragraph. The presence
of such forms has been undoubtedly detected in the mouth, and
even within the relative epithelia. They would appear to be a
kind of reserved fund, preserved in case of any alteration of the
future conditions of pabulum and surroundings, as the author
properly says. We shall speak by-and-by of the relationship of
this phase with the diplococcus of pneumonia.

Here Billet would end the evolutionary cycle, for us, on the
contrary, these first phases would only constitute a cycle, at times
preliminary, at times succedaneous to the second or superior cycle.
The varied elements of the first cycle, being taken separately and
held as special beings or complete living individuals (filaments,
bacilli, bacteria, and cocci), are in reality only particles, trunks,
organs, series of cellules, or cellules endowed mostly with fissiparous multiplication, but destined to constitute a more complex
organism.

SUPERIOR CYCLE.—This comprises the phases of the properly
called life of reproduction, and these phases are three.

IV.—*Agamous Fructification*.—Referring to the general laws of multiplication of the phanerogams by means of bulbs, tubers, shoots, or buds, and of their genuine reproduction by means of seeds, it should be borne in mind that with transplantation are transmitted the accidental modifications brought about by domestication, grafting, etc., which, in the long run, may end in the degeneration of the plant; whilst with seeds, on the contrary, the species reverts to the natural vigour of the wild state.

Now, we incline to believe that something similar may happen even with cryptogams. Probably in those which have, besides a fissiparous multiplication, a true reproduction through spores, the sporulation will mean that the species resumes its native vigour, in spite of the attacking or enfeebling causes which, as in our parasite, may impair its vigour.

The seminule or spore may germinate without previous fertilisation, and (as we have already said) we believe this may happen more frequently in our parasite. In the species studied by Billet, this external sporification or fructification was wanting, and it is remarkable that the cocci, which are found abundantly in the buccal contents, were likewise wanting. It remains to be seen whether this want of fructification is really the rule, or simply a consequence of the nature of its pabulum and the material upon which Billet based his researches. Under other conditions, upon media not only stable, but favourable to the production of a mycelium-like growth, a true and proper fructification might, even in the above-named species, take place.

In our parasite, the conditions indispensable for the fructification are :—*a*, The solidity and nature of the soil; *b*, the protection against attrition; and *c*, moisture of the saliva.

V.—*Organs and Fertilising Elements*.—By the side of the agamous sporulation we have, in many cryptogams, that of conjugation. This admits of male organs and fertilising elements or spermatic threads, of which we have already spoken. In our parasite, likewise, we have male organs and fertilising elements. The male organs would first show themselves in the state of young spermogones or antheridia (or of spermogones or aborted antheridia, not developed or reverted to the neuter state), and afterwards in the state of tufts of ripe pseudo-inflorescences, proceeding from the first type. On the other hand, the fertilising elements would be constituted by the spindle-like, comma, or serpentine bacilli, already formed in the described organs, and finally set free, through the disarticulation of the tufts already mentioned.

VI.—*Conjugated Fructification*—*i.e.*, that with clavated stem, a dual zone of colourisation, clusters, or more conspicuous ears, more bulky and compact spores, destined, perhaps, to cross uninjured the alimentary canal, and to remain alive in the fæces, withstanding the dissolving action of the gastro-intestinal juices.

Nature, as we see, has been prodigal to this parasite, by its various manners of multiplication, adapting each of them to this or that condition of the nutrient substratum, in order to preserve and multiply its species, in the midst of numberless and very varied difficulties. As soon as a higher phase is precluded, or a nobler element is thrown back, by external injuries, to an inferior degree, it does not stop from disseminating everywhere particles apt to germinate and spread *extensively* the species when it cannot do it intensively. We are reminded that Nature has even wished to endow this tiny plant with such a tenacious life, that its elements on the human teeth cannot be destroyed for centuries, as exhibited by the tartar on the teeth of Egyptian mummies (*vide* the preceding Memoir).

If here ends, at least provisionally, the *Morphological Series* of our parasite, it only remains to us to say two words on the *Biological Series* in the modern sense, or, rather, on the pathological phenomena, assigned to some of its forms, or to specific bacteria similar to the latter.

BIOLOGICAL SERIES.—Under this title we comprise but three forms : the *Pneumococcus*, the *Bacillus* of Koch, and the *Gonococcus* of Neisser, already demonstrated, according to our views, in the two previous Memoirs.

I.—*Pneumococcus.* Few now hold that the pneumococcus is a specific bacterium, arising externally. It is generally considered to be either an habitual germ of the mouth (*Micrococcus of salivary septicæmia*), or, in common with Pommay, a saprophytic bacterium evolving itself in the sense of virulence. Perhaps it will in time be known as a simple dissemination of the zoöglœic phase of our parasite, following on the formation of the pulmonitic exudation, as its greater abundance in the last stages of pneumonia, its presence in traumatic pneumonia, and other evidences will prove it to be so. But even admitting that the evolution in the sense of virulence may take place in the mouth and not in the respiratory organs affected (as regards virulence, which may be inoculated); admitting that it has preceded and not followed pneumonia, we cannot necessarily infer that the disease proceeded from this coccus. But even the colonies of these diplococci, when repeatedly transplanted in other culture media, or even the media themselves, may become contaminating through inoculation, this circumstance may also occur with other salivary bacteria (for example, *Bacillus crassus sputigenus*), and might be interesting in experimental pathology; but, botanically speaking, it does not implicitly imply a separate specific entity. We repeat that we do not impugn, but rather try to conciliate the results of the inoculations with those of the morphological research.

II.—*Bacillus of Koch.* Our remarks concerning this bacillus are very similar, and we hope at a future time to devote a special Memoir to it. The reasons which induced us to believe the bacillus in rosaries to be a dissemination of the small spores of *Leptothrix* spread over the tubercular spots, and the rod-shaped bacilli as proceeding from other elements of *Leptothrix*, have been already given in the preceding Memoirs. Their greater dissemination in phthisis than under other conditions may depend upon different causes, and partly, perhaps, from the following simple reason :—Breathing through the nose, as we generally do, only a few germs of a purely buccal origin are inhaled ; but the position becomes altered when we breathe through the mouth as well, as happens in phthisis, because of panting or burning heat :—

> " .   .   open'd wide his lips,
> Gasping as in the hectic man for drought,
> One towards the chin, the other upward curl'd."
>
> —Dante, *Inferno*, xxx., Cary's Transl.

Now, the small sporules of *Leptothrix* exclusively originate from the patina dentaria, and without that particular form of breathing (*i.e.*, by the mouth) they cannot gather in any great number into the air-passages. Many also, in sleeping, breathe likewise with their mouth ; but in the normal conditions, however, the ciliary action continuing unhurt through the air-passages, the inhaled leptothrical elements cannot reach the pulmonary tissue, but are instead thrown out and expectorated with the sputa, as stated in the other Memoir.

III.—*Gonococcus of Neisser.*   Probably, says Pommay, a normal bacterium of the urethra is the progenitor of the gonococcus of Neisser; it is useless to repeat here that relationship.   Let it suffice to record the fact that we have detected in the urine numberless gonococci, in a case of carcinoma of the bladder, independently from any gonorrheal contagium whatever.*   According to the morphological characteristics, the gonococci in question do not differ from the ordinary arched diplococci, and therefore can be held, till proved otherwise, to be derivations from the state of sarcina (one form of the zoöglœic state).   As regards the manner by which the leptothrical elements penetrate from the external genitals into the urinary passages, as well as the presence of analogous forms (curved diplococci) in the contents of the mouth, in sputa or the middle ear ; and also as regards the discovery in the urine or sperm of capsulated forms (forms analogous to pneumococci), we refer to the previous Memoirs.

From the morphological sketch we have just given, nothing perfectly absolute and incontroverted is to be inferred, either for or against any previously acknowledged systematic view ; but perhaps few people will doubt that, in the field of modern bacteriology, much still remains to be done, and a great deal of discussion, combined with careful study, will be required before any definite and satisfactory results will be arrived at concerning the pathogenesis of infectious diseases (we are, of course, now alluding to those cases which are strictly clinical).   Many will, we are sure, agree with us that, with regard to bacteria, our knowledge is so limited that no method of investigation will appear superfluous in order to study aright their genesis and properties in order thoroughly to understand and appreciate their function in our economy.

F. VICENTINI, Corresponding Member.

Chieti; May, 1892.

* *Atti della R. Accademia Medico-Chirurgica di Napoli*, tomo XLIII., 1890.

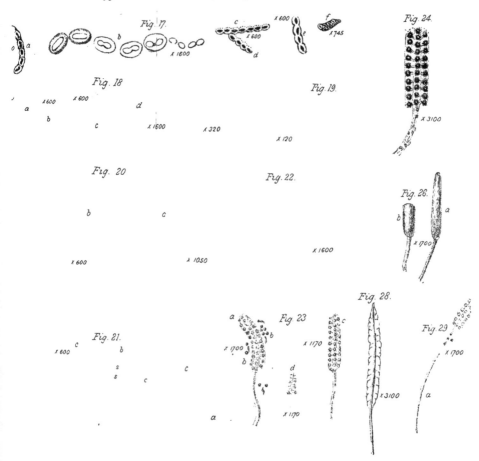

Fig. 17.

Fig. 18

Fig. 19.

Fig. 24.

Fig. 20

Fig. 22.

Fig. 26.

Fig. 28.

Fig. 21.

Fig. 23

Fig. 29

i' del.

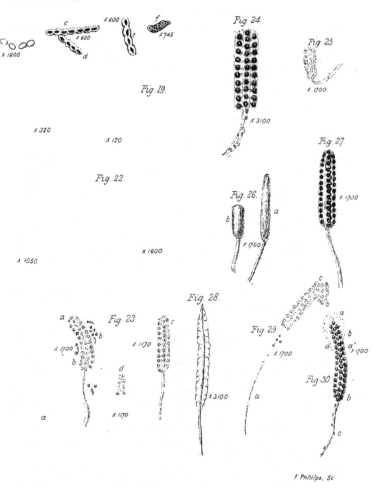

x 1600

c
x 600

x 600

d

x 600

f
x 745

e

Fig. 19.

Fig 24.

x 3100

Fig 25

x 1700

x 320

x 120

Fig. 22.

Fig. 27.

x 1700

Fig. 26.

b

a

x 1700

x 1600

λ 1050

Fig. 28.

a

Fig. 23

b

x 1700

c

x 1170

Fig. 29

c

b

a

x 1700

a

b

a'

x 1700

d

b

x 3100

d

x 1170

a

Fig 30.

b

a

x 1170

a

c

F Phillips. Sc

EXPLANATION OF PLATE XX.

FIGURES TAKEN FROM BILLET :

Fig. 17.—Articulations analogous to the pneumococci, belonging to various bacteriological species on the point of passing to the *dissociated* or to the *zoöglœic* state. *a*, A fragment of six articulations of *Cladothrix dichotoma* in disintegration, × 320. *b*, Articulations of the same *alga* in more advanced disintegration, × 1600. *c, d*, *Bacterium osteophilum*, ready to pass to the zoöglœic state, × 600. *e*, A fragment of four articulations of *Bacterium osteophilum* on the point of passing to the scorpioidal state, × 600. Small zoöglœic group of *Leptothrix parasitica*, Kützing, × 745.

" 18.—Other analogous articulations. *a*, Rectilineal forms of *Cladothrix dichotoma* in the act of passing to the zoöglœic state (stained first with solution of iodine, then with methyl violet or fuchsine), × 600. *b*, Beginning of the zoöglœic state of the *Bacterium osteophilum* (colouring with vesuvine, then with methyl-violet, and lastly with solution of iodine), × 600. *c*, Beginning of its scorpioidal state (same colouring), × 600. *d*, Same as *b*, enlarged, × 1600. *e*, Zoöglœic group of *Leptothrix parasitica*, Kützing (colouring as in *a*), × 320.

" 19.—Ramified zoöglœa of *Cladothrix dichotoma*, formed by 21 branches, upon a single peduncle (same colouring as Fig. 2, *a*), × 120.

" 20.—Superior endings, more enlarged, of two of the preceding branches (same colouring). *a, b*, Tops of branches, × 600. *c*, A portion of one of them, more enlarged, containing bacteria and bacilli of different shape and form, × 1050.

" 21.—*Leptothrix parasitica*, Kützing, vegetating on two crossed stems : the one of *zygnema*, the other of *Cladothrix* (colouring with vesuvine and methyl-violet, then with solution of iodine), × 600. *a, a*, Stem of *Zygnema*. *b, b*, Stem of *Cladothrix*. *c, c, c*, Filaments of *Leptothrix parasitica*. *s, s*, Their bulbs or engrafting spores.

" 22.—A part of the previous figure, more enlarged (same colouring), × 1600. *c, c, c*, Filaments of *Leptothrix parasitica*. *s, s*, Their bulbs or engrafting spores.

ORIGINAL FIGURES :

" 23.—Ears or clusters of *Leptothrix racemosa*, with acidulated solution of iodine. *a*, Young cluster, with small, thin spores upon their sterigms, × 1700. *b, b*, Groups of bacteria and comma bacilli. *c*, A matured cluster, with sterigms and thicker spores, × 1170. *d*, Fragment of ear, with sterigms without spores, excepting two spores, still adhering, × 1170.

" 24.—Cluster showing the peduncles or sterigms, funnel-shaped, and the central stem, containing minute gemmules of reserve, stained with weak gentian violet, × 3100.

Fig. 25.—Ear, foreshortened, seen upon a turned-up stem, with a terminal rosette of six sterigms and relative spores on the top, stained with acidulated solution of iodine, × 1700.

,, 26.—Clavated stems, stained with gentian violet, × 1700. *a*, Stem with a perfect clava, but not yet fructified. *b*, Stem with a short clava, in formation.

,, 27.—Ear with two colouring zones, with clavated stem (fructified) and more compact spores (teleutospores ?), stained with gentian violet, × 1700.

,, 28.—Sheath expansion (young antheridium or spermagonium ?), containing slender transversal fine beams (future fecundating elements ?), stained with strong gentian violet, × 3100.

,, 29.—Ear doubled up at the top, stained with acidulated solution of iodine, × 1700. *a*, Long stem. *b*, Point of the ear, doubled up obliquely and pressed down on the middle third. *c*, Elbow simulating a gibbosity.

,, 30.—Ear, apparently pyriform, through superposition of adventitious cocci and bacteria, stained with gentian violet, × 1170. *a*, Cap, at first stage of colourising, pale, formed by adventitious cocci and bacteria (through contact of lips ?) up to the middle, *à, à*, of the internal ear. *b, b*, Internal ear, brilliantly coloured. *c*, Stem.

Molluscs and Brachiopods.*

I T is with much pleasure that we are, through the kind courtesy of Messrs. Macmillan and Co., enabled to direct the attention of our readers to the first published volume of THE CAMBRIDGE NATURAL HISTORY.

The series, which is edited, and for the most part written, by Cambridge men, is to consist of ten volumes, each of which will contain about 500 pages and will be complete in itself. The volumes have been numbered on a definite plan, but will be published in the order in which they are ready for the press. The one before us is Vol. III. of the series.

The Cambridge Natural History is intended, in the first

---

* MOLLUSCS. By the Rev. A. H. Cooke, M.A., Fellow and Tutor of King's College, Cambridge.——BRACHIOPODS (RECENT). By A. E. Shirley, M.A., Fellow of Christ's College, Cambridge.——BRACHIOPODS (FOSSIL). By F. R. C. Reed, M.A., Trinity College, Cambridge. Royal 8vo, pp. xiii.—535. (London: Macmillan and Co. 1895.) Price 17/- net.

instance, for those who have not had any special training, and who are not necessarily acquainted with scientific language. At the same time, an attempt is made, not only to combine popular treatment with the latest results of scientific research, but to make the volumes useful to those who may be regarded as serious students in the various subjects.

The general plan of classification adopted in the work before us is not that of any single authority. It has been thought better to adopt the views of recognised leading specialists in the various groups, and thus place before the reader the combined results of recent investigation.

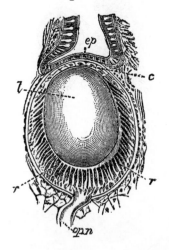

Fig. I.—Eye of *Helix pomatia*, L., retracted within the tentacle; *c.*, cornea; *ep.*, epithelial layer; *l.*, lens; *op. n.*, optic nerve; *r.*, retina. (After Simroth).

The volume opens with a Scheme of the Classification of the Mollusca adopted in this book; followed by chapters on the Position of Mollusca in the Animal Kingdom; Origin of Land and Fresh-water Mollusca; Their Habits and General Economy; Their Enemies; Means of Defence, etc. etc.

As the publishers have very kindly placed some of the illustrations at our disposal, we will now pass on to Chapter VII., which treats of the Organs of Sense: Touch, Sight, Smell, Hearing; The Foot and the Nervous System, and quote briefly—

The Organisation of the Molluscan Eye (p. 181).—The eye in Mollusca exhibits almost every imaginable form, from the extremely simple to the elaborately complex. It may be, as in certain bivalves, no more than a pigmented spot on the mantle, or it may consist, as in some of the Cephalopoda, of a cornea, a sclerotic, a choroid, an iris, a lens, an aqueous and vitreous humour, a retina, and an optic nerve, or of some of these parts only.

In most land and fresh-water Mollusca the eye may be regarded, roughly speaking, as a ball connected by an exceedingly fine thread (the optic nerve) with a nerve-centre (the cerebral ganglion). . . . . In *Helix* (Fig. 1) there is a structureless

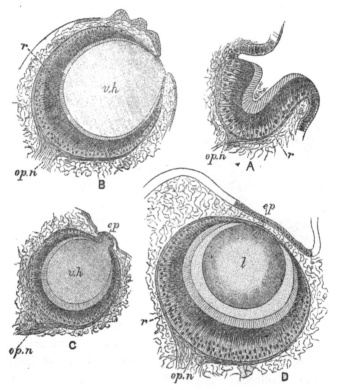

Fig. 2.—Eyes of Gasteropoda, showing arrest of development of successive stages : *A, Patella ; B, Trochus ; C. Turbo ; D, Murex ; ep.,* epidermis ; *l.,* lens ; *op. n.,* optic nerve ; *r.,* retina ; *v.h.,* vitreous humour (after Hilger).

membrane, surrounding the whole eye, a lens, and a retina, the latter consisting of a nervous layer, a cellular layer, and a layer of rods containing pigment, this innermost layer (that nearest the lens) being of the thickness of half the whole retina.

Comparing together the eyes of different Gasteropoda, we find that they represent stages in a general course of development.

Thus, in *Patella* the eye is scarcely more than an invagination or depression in the integument, which is lined with pigment and retinal cells. The next upward stage occurs in *Trochus*, where the depression becomes deeper and bladder-shaped, and is filled with a gelatinous or crystalline mass, but still is open at the top, and therefore permits the eye to be bathed in water. Then, as in *Turbo*, the bladder becomes closed by a thin epithelial layer, which finally, as in some *Murex*, become much thicker; while the 'eye-ball' encloses a lens (Fig. 2), which probably corresponds with the 'vitreous humour' of other types.

In Chapter VIII. is described The Digestive Organs, Jaw, and Radula, and Excretory Organs. As the mouth-organs are always specially interesting to microscopists, we make a few short extracts.

The *mouth* is generally, as in the common snail and periwinkle, placed on the lower part of the head, and may be either a mere aperture, circular or semi-circular, in the head-mass, or, as is more usual, may be carried on a blunt snout, which is capable of varying degrees of protrusion. From the retractile snout has doubtless been derived the long proboscis. which is so prominent a feature of many genera. . . . As a rule, Mollusca provided with a proboscis are carnivorous, while those whose mouth is on the surface of the head are vegetable feeders ; but this rule is by no means invariable.

The Pharynx, Jaws, and Radula.—Immediately behind the lips the mouth opens into a muscular throat, pharynx, or buccal mass. The pharynx of the Glossophora—*i.e.*, of the Gasteropoda, Scaphopoda, and Cephalopoda—is distinguished from that of the Pelecypoda by the possession of two very characteristic organs for the rasping or trituration of food before it reaches the œsophagus or stomach. There are (*a*) the *jaw or jaws*, and (*b*) the *radula*,* *odontophore*, or *lingual ribbon*. The jaws bite the food, the radula tears it up small before it passes into the stomach to undergo digestion. The jaws are not set with teeth like our own ; roughly speaking, the best idea of the relations of the molluscan jaw and radula may be obtained by imagining our own teeth removed from our jaws and set in parallel rows along a greatly prolonged tongue.

---

* *Radere*, to scrape; ὀδούς, tooth ; φέρειν, to carry.

The radula itself is a band or ribbon of varying length and breadth, formed of chitin, generally almost transparent, sometimes beautifully coloured, especially at the front end, with red or yellow. It lies enveloped in a kind of membrane, in the floor of the mouth or throat, being quite flat in the forward part, but usually curving up so as to line the sides of the throat farther back, and in some cases eventually forming almost a tube. The upper

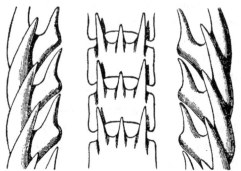

Fig. 3.—Portion of the radula of *Melongena verpertilio*, Lam., Ceylon, × 30.

surface—*i.e.*, the surface over which the food passes—is covered with teeth of the most varied shape, size, number, and disposition, which are almost invariably arranged in symmetrical rows. These teeth are attached to the cartilage, on which they work by muscles, which serve to erect or depress them ; probably also the radula,

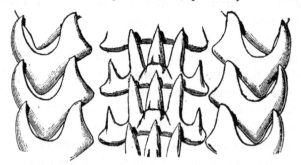

Fig. 4.—Portion of the radula of *Eburna japonica*, Sowb., China, × 30.

as a whole can be given a forward or backward motion, so as to rasp or card the substances which pass over it.

The extreme importance of a study of the radula depends upon the fact that in each species, and, *a fortiori*, in each genus and family, the radula is characteristic; thus, in *Melongena vesper-tilio* (Fig. 3), the central tooth is tricuspid, the central cusp being the smallest, while the laterals are bicuspid; in *Eburna japonica* (Fig. 4) the central tooth is 5-cusped. the two outer cusps being much the smallest. The teeth, on the whole, are sharp and hooked, with a broad base and formidable cutting edge.

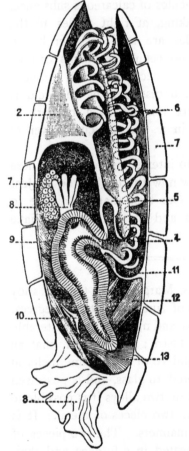

Fig. 5.—View of the left half of *Cistella (Argiope) neapolitana*, which has been cut in two by a median longitudinal incision to show the disposition of the organs. Partly diagrammatic. The inorganic part of the shell only is shown. The tubular extensions of the mantle and the organic outer layer are not included, and hence the pores appear open.

1.—The ventral valve.

2.—The dorsal valve.

3.—The stalk.

4.—The mouth.

5.—Lip, which overhangs the mouth and runs all round the tentacular arms.

6.—Tentacles.

7.—Ovary in dorsal valve.

8.—Liver diverticula.

9.—Occlusor muscle; its double origin is shown.

10.—Internal opening of left nephridium.

11.—External opening of left nephridium.

12.—Ventral adjustor. The line from 10 crosses the dorsal adjustor.

13.—Divaricator muscle.

We will now turn to Chapter XVII., which treats of Recent Brachiopoda. The body of a Brachiopod is enclosed within a bivalve shell, but the two halves are not, as they are in the Pelecypoda, one on each side of the body, but occupy a different position with regard to the main axis of the body (Fig. 5).

The shell of a Brachiopod is secreted partly by the general

surface of the body, which is situated at the hinder end of the shell and partly by the two leaf-like extensions of the body, which are termed the dorsal and ventral mantles.   These are, in fact, folds in the body-wall, and into them the body cavity and certain of its contents, such as the liver and generative glands, etc., extend.

Microscopic examination of thin sections of the shell shows that it consists of small prisms or spicules of calcareous substance, whose long axis lies, roughly speaking, at right angles to the surface of the shell.   These spicules are held together by an organic matrix, in which, however, no cellular elements can be detected.   In sections made through a decalcified shell the position of the spicules which have been dissolved by the acid is indicated by spaces, and the matrix remains as a network of fibrils, which end on the outside in a thin cuticular layer of organic matter.

We trust we have said sufficient to interest our readers in these very interesting classes, and to show how thoroughly the subject has been treated by the authors.   Our best thanks are due to publishers for the use of the electros, and for permission to make the above extracts.

SOLDERS FOR GLASS.—Mr. Chas. Margot finds that an alloy composed of 95 parts of tin and 5 of zinc melts at 200 degrees, and becomes firmly adherent to glass, and moreover is unalterable and possesses a beautiful metallic lustre ; and, further, that an alloy composed of 90 parts of tin and 10 of aluminium melts at 390 degrees, becomes strongly soldered to glass, and is possessed of a very stable brilliancy.   With these two alloys it is possible to solder glass as easily as it is to solder two pieces of metal.   It is possible to operate in two different manners.   The two pieces of glass to be soldered can either be heated in a furnace and their surfaces be rubbed with a rod of the solder, when the alloy as it flows can be evenly distributed with a tampon of paper or a strip of aluminium, or an ordinary soldering iron can be used for melting the solder.   In either case, it only remains to unite the two pieces of glass and press them strongly against each other, and allow them to cool slowly.—*Sci. American.*

[445]

Reviews.

OPEN-AIR STUDIES: An Introduction to Geology Out-of-doors.
By Grenville A. J. Cole, M.R.I.A., F.G.S., etc. Crown 8vo, pp. xii.—322.
(London : Chas. Griffin & Co. 1895.) Price 8/6.
The aim of the author of this interesting book has been to keep in view
the fact that Geology, like true zoology and true botany, is a study for the open
air. The twelve chapters into which the book is divided treat of The Material
of the Earth ; A Mountain Hollow ; Down the Valley ; Along the Shore ;
Across the Plain ; Dead Volcanoes ; A Granite Highland ; The Annals of the
Earth ; The Surrey Hills ; The Folds of the Mountains. There are 11 plates
and 33 illustrations in the text. ———

STUDIES IN THE EVOLUTION OF ANIMALS. By E. Bonavia,
M D. Cr. 4to, pp. xxxiv.—362. (Westminster : Archibald Constable & Co.
1895.) Price 21/- net.
In the preface to this handsomely got-up book the author tells us that,
" Thinking over the rosettes of the Leopards, and more especially those of the
Jaguar, and seeing spotted Horses constantly in the streets of London, some
new ideas flashed across my mind regarding the origin of all this spotting and
rosetting in mammals." The subjects of the curious callosities on the legs of
the Horse, its solitary leg digit, its possible close relationship to the pair of
digits in the ruminants, and various monstrosities, also came under his notice ;
and he was led to the conclusion that they must have a deeper meaning than may
have hitherto been attributed to them by evolutionists. We feel sure our readers will be deeply interested in a careful perusal of this work. There are 128
illustrations. ———

A LABORATORY GUIDE for the Dissection of the Cat: An
Introduction to the Study of Anatomy. By Fredric P. Gorham, A.M., and
Ralph W. Tower, A.M. 8vo, pp. ix.—87. (New York : Charles Scribner's
Sons. 1895.)
This book will prove a valuable laboratory guide for elementary classes in
anatomy. The instructions are very concisely given. There are 7 capital plates
showing Skeleton ; Superficial Muscles of Right sight ; Deeper Muscles on
Right side ; Superficial and Deeper Muscles of Ventral side; Arterial System;
Venous System ; and Nervous System. ———

YEAR BOOK OF THE SCIENTIFIC AND LEARNED SOCIETIES of
Great Britain and Ireland. 8vo, pp. v.—254. (London : Charles Griffin &
Co. 1895.) Price 7/6.
The 12th annual issue of this exceedingly useful work is before us. It
gives us a chronicle of the work done during the year 1894 by all the various
societies, together with information as to official changes. By referring to this
list we find full particulars of the various societies, date of formation, name of
the President, name and address of the Secretary, date and time of the Meet-
ings, and list of Papers read ; also where these Papers are published. We
find the Year-Book a valuable work for reference. ———

SCIENCE READERS. By Vincent T. Murche. Book IV. Cr.
8vo, pp. 216. (London : Macmillan & Co. 1895.)
This book follows the three noticed in our last issue ; the whole forming a
valuable and very instructive series of school reading books. The illustrations
are very effective.

CURIOUS and Instructive Stories about WILD ANIMALS and BIRDS. Cr. 8vo, pp. xii.—340. (Edinburgh: W. P. Nimmo & Co.) Price 2/6. We have here nine instructive and amusing stories, with some good illustrations.

A CHAPTER ON BIRDS, Rare British Visitors. By R. Bowdler Sharpe, LL.D., F.L.S., etc. Cr. 8vo, pp. x.—124. (London : Society for Promoting Christian Knowledge. 1895.) Price 3/6. Mr. Bowdler Sharpe, of the Zoological Department, British Museum, gives here very nice descriptions of eighteen of our very beautiful visitors, each description being accompanied by a fine coloured plate of the bird, drawn to scale, and the egg, natural size.

AN INTRODUCTION TO THE STUDY OF ZOOLOGY. By B. Lindsay, C.S., of Girton College, Cambridge. Cr. 8vo, pp. xix.—356. (London: Swan Sonnenschein & Co. 1895.) Price 6/-. The aim of the author of the volume before us has been to supply a simple outline sketch of the animal kingdom. Part I. treats of General Principles of Zoology ; Part II. of Systematic Zoology ; and Part III., Advice to Students on the Use of Books and on Practical Work. 'Animals as Fellow Creatures' is the title of the 3rd chapter of this part. There are 124 illustrations, and a glossary. The volume goes very thoroughly into the subject of which it treats, and the general student will doubtless gain much information from a careful study of it.

CONSIDER THE HEAVENS : A popular introduction to Astronomy. By Mrs. William Steadman Aldis. Cr. 8vo, pp. 224. (London : The Religious Tract Society. 1895.) Price 2/6. This book has been written for those who are at present quite ignorant of astronomy, and especially for such as have not much time for study. The book is written in a thoroughly interesting manner and is nicely illustrated.

HIDDEN BEAUTIES OF NATURE. By Richard Kerr, F.G.S. Foolscap 4to, pp. 256. (London : The Religious Tract Society. 1895.) 3/6. The chapters in this book contain, in simple language, the main points of lectures delivered to scientific societies, colleges and upper-class schools, and to large audiences in various parts of England. A few of the subjects are : On the Study of Nature ; How to Begin ; The Sea-Urchin ; Nature's Fireworks ; The Euplectella; Atlantic Ooze; Diatoms; Eggs of Insects; &c. &c. There are 59 beautiful illustrations.

LIGHT FROM PLANT LIFE : Truths derived from and illustrated by the Life History of Plants. By H. Girling. Cr. 8vo, pp. xiv.—178. (London : T. Fisher Unwin. 1895.) Price 3/6. This work is chiefly designed for those who desire to exercise their powers of thought, and sets forth the spiritual life as illustrated every day by plants and trees.

ANGLING AND HOW TO ANGLE : A practical guide to Bait-fishing, Trolling, Spinning, and Fly-fishing. By J. T. Burgess. Revised and brought down to date by R. B. Marston ; with a special article on Pike-fishing by A. J. Jardine. Cr. 8vo, pp. x.—212. (London: F. Warne and Co. 1895.) Price 1/- A comprehensive, practical, and handy manual, which is neither too large for the pocket, nor too brief to be useful. Besides some 70 illustrations of tackle, flies, etc., it contains a number of practical hints on the making and mending of fishing gear, fly-dressing, and other memoranda which will be duly appreciated.

THE PLANTS OF THE BIBLE. By the Rev. George Henslow, M.A., F.L.S., &c. Foolscap 8vo, pp. 128. (London : The Religious Tract Society.) Price 1/-

One of an interesting series of books published by the Society and known as "Present Day Primers." It gives some of the most interesting features relating to the 120 plants mentioned in the Bible; there are several illustrations.

THE STORY OF THE PLANTS. By Grant Allen. 12mo, pp. 232. (London : George Newnes. 1895.) Price 1/-

This is a most interesting and instructive little book, in which the author gives a short and succinct account of the principal phenomena in plant life, in language suited to the comprehension of unscientific readers.

DIE NATÜRLICHEN PFLANZENFAMILIEN. By A. Engler. Parts 120, 121, 122. (London: Williams & Norgate. Leipzig: Wilhelm Engelmann.)

These parts contain the completion of the Loganiaceæ, by H. Solereder ; Gentianaceæ, by E. Gilg ; Apocynaceæ. by K. Schumann, and the commencement of the Asclepiadaceæ, by the same author. There are 34 illustrations, consisting of 411 figures. The price of these numbers is 3 marks, or by subscription, 1·50 m. each.

THE ART OF MASSAGE : Its Physiological Effects and Therapeutic Applications. By J. H. Kellogg, M.D. 8vo, pp. xvi.—282. (Battle Creek, Mich., U.S.A.: Modern Medicine Pub. Co. 1895.) Price $3 = 12/6

The work before us goes thorougly into the subject on which it treats and is profusely illustrated with a great number of well-executed photo-mechanical plates. The author directs special attention to the classification of the different procedures of massage, and by a careful study of those described by the best authorities, and employed by expert manipulators, it was found possible to include all in seven different general classes with sub-divisions, each of which has been described with a very considerable degree of painstaking care. In this work there are 45 excellent plates, each showing on an average 4 distinct figures.

CHOLERA : Its Protean Aspects and its Management. By Dr. G. Archie Stockwell, F.Z.S.

WHOOPING COUGH. Vols. I. and II. By Dr. H. Richardière, Paris. Translated by Joseph Heleman.

ANTISEPTIC THERAPEUTICS Vols. I. and II. By Dr. E. L. Troussardt, Paris. Translated by E. P. Hurd, M.D.

MODERN CLIMATIC TREATMENT of Invalids with Pulmonary Consumption in Southern California. By P. C. Romondino, M.D.

CEREBRAL MENINGITIS : Its History, Diagnosis, Prognosis, and Treatment. By Martin W. Barr, M.D.

INTESTINAL DISEASES of Infancy and Childhood : Physiology, Hygiene, Pathology, and Therapeutics. By A. Jacobs, M.D.

A TREATISE ON DIPHTHERIA. By Dr. H. Bourges. Translated by E. P. Hurd, M.D.

PERNICIOUS FEVER : A clinical study of the Fevers of Rio de Janeiro. By Dr. Joãs Vincente Torres Homem. Translated by Surgeon George P. Bradley, U.S. Navy.

All the above are volumes of the PHYSICIAN'S LEISURE LIBRARY. They contain short practical treatises, prepared by well-known authors, and give the gist of what they have to say regarding the treatment of diseases commonly met with and of which they have made a special study. Many of the volumes are illustrated, and are published by George S. Davis, Detroit, Mich., U.S.A. Price 25c. each, in paper covers, or 50c., bound in cloth.

MICROBES AND DISEASE DEMONS. The Truth about the Anti-Toxin treatment of Diphtheria. By Edward Burdoe, L.R.C.P. Ed., M.R.C.S. Eng., etc. Cr. 8vo, pp. 93. (London: Swan Sonnenschein and Co. 1895.) Price 1/-

The author commences by saying "The most ancient and wide-spread theory of disease is the demon theory." On page 12 he says : " The disease demon has now reappeared as a germ. Some 36 diseases, many of which are the most terrible, which afflict men and animals are attributed by bacteriologists to micro-organisms." And goes on to say : " I cannot think it can ever be true scientific medicine to pour poison into the blood current to counteract other poison ; we may in one sense convey an antidote, but we may work untold mischief by our ignorant meddling, which we have no means of combating."

DREAMY MENTAL STATES. By Sir James Crichton-Browne, M.D., LL.D., F.R.S., etc. 8vo, pp. 32. (London : Baillière, Tindall, & Cox. 1895.) Price 1/-.

This is the Cavendish Lecture delivered before the West London Medico-Chirurgical Society, on Thursday, June 20th, 1895.

HERBAL SIMPLES approved for Modern Use or Cure. By W. T. Fernie, M.D. Cr. 8vo, pp. xvi.—432. (Bristol : John Wright & Co. London : Simpkin, Marshall, & Co. ; and Herschfield Bros. 1895.) Price 5/-

" Various British Herbalists," the author tells us, "have produced works, more or less learned and voluminous, about our native plants ; but no author has hitherto radically explained the why and the wherefore of their ultimate curative action. . . . Chemically assured of the sterling curative powers which our Herbal Simples possess, and anxious to expound them with a competent pen, the present author approaches the task with a zealous purpose." Some 300 plants are named and their properties described.

EYESIGHT AND SCHOOL LIFE. By Simeon Snell, F.R.C.S. Ed., etc. 8vo, pp. xii.—70. (Bristol : John Wright & Co. London : Simpkin, Marshall, & Co. ; and Herschfield Bros. 1895.) Price 2/6.

A large amount of valuable information is here given. The author states that there is abundant and convincing evidence that the vision of children, which should under normal conditions have remained good, is constantly deteriorating during the school period. He points out many of the causes of this deterioration, and suggests remedies. There are 15 illustrations.

A NEW ENGLISH DICTIONARY on Historical Principles. Ed. by Dr. James A. H. Murray. Deject—Depravation. (Oxford : The Clarendon Press. London : Henry Frowde. July, 1895.) Price 2/6.

This section (a portion of Vol. III.), which covers the words from DEJECT to DEPRAVATION, contains 1269 main words, 37 combinations explained under these, and 138 subordinate words ; to which may be added 125 obvious combinations recorded and illustrated without definition, bringing up the total to 1569 ; and of these 1365 are illustrated by quotations.

ARITHMETIC PRIZE PAPERS. By W. P. Workman. Foolscap
8vo, pp. 60. (London: Joseph Hughes & Co. 1895.) Price 2/-
For many years past a silver medal has been awarded annually at Kings-
wood School (of which the author is Head Master) 'to the best Arithmetician.'
The endowment for this purpose was left to the School at a time when Arithmetic was the only equipment of an educated gentleman. The author has
no doubt the donor meant 'the best Mathematician,' but he did not say so, and
consequently a few of the most difficult questions capable of being answered
by Arithmetic have been prepared. Four papers, each containing 12 to 15
questions, cover 8 pages of the book, the remaining 48 being required for their
working out. We think it would take a very clever boy to answer most of
them.

AN ELEMENTARY TEXT-BOOK OF MECHANICS. By William
Briggs, L.L.B., F.R.A.S., etc., and G. H. Bryan, M.A., F.R.S., etc. Cr.
8vo, pp. vii.—336. (London: W. B. Clive. 1895.) Price 3/6.
This is one of "The University Tutorial Series," in preparing which the
aim of the authors has been to afford beginners a through grounding in those
parts of Dynamics and Statics, which can be treated without assuming a previous knowledge of Trigonometry. The section devoted to Dynamics is divided into three parts :—I., Velocity and Acceleration ; II., Mass and Force ;
and III., The Parallelogram Law. And Statics :—I., Equilibrium of Forces
at one point; II., Moments and Parallel Forces; and III., Centres of Gravity.
The answers are at the end of the book.

ELEMENTARY TRIGONOMETRY. By Charles Pendlebury, M.A.,
F.R.A.S., etc. Cr. 8vo, pp. xvi.—336. (London : George Bell & Son.
1895.) Price 4/6.
The examples in this book are numerous and varied, and have been carefully graduated ; and at suitable places sets of oral examples have been inserted, similar to those in the "Arithmetic for Schools," by the same author ; and
towards the end of the work there will be found a set of questions on book-
work, based upon the text. Answers are at the end of the book.

MATHEMATICAL QUESTIONS and Solutions. Edited by W. J.
C. Miller, B.A. Vol. LXIII. 8vo, pp. 128. (London: F. Hodgson. 1895.)
This volume consists of Mathematical Questions with their Solutions, taken
from the *Educational Times*, with many others which were not published in
that Journal. There will be found in it contributions in all branches of Mathematics from many of the leading Mathematicians at home and abroad.

MATRICULATION DIRECTORY. June, 1895. Cr. 8vo, pp. 64
+ 132. (London Office : 32 Red Lion Square, Holborn.)
This volume of "The University Tutorial Series" contains articles on the
special subjects for January and June, 1896, with the Calendar for 1894—5.
Those students who contemplate going in for their 'Matric.,' will do well to
study this book. The Calendar for 1894—5 occupies the first part of the book.
The subjects treated in the Directory proper are : I.—Matriculation Regu-
lations ; II.—Text Books ; III.—Special Subjects for January and June, 1896 ;
IV. and V., which we advise all would-be candidates to study, give the papers
set on the various subjects in June, 1895, with their solutions.

STENOPAIC OR PIN-HOLE PHOTOGRAPHY. By F. W. Mills,
F.R.M.S., and A. C. Ponton. 8vo, pp. 27. (London : Dawbarn and Ward.
1895.) Price 1/-.

Good results may be obtained from pin-hole photography, as will be seen from the frontispiece of this little work. The authors state that a hole 1/100th of an inch in diameter, accurately drilled in a very thin plate, can be used with perfect success for ordinary out-of-door work, from 1-in. focus up to 5-in. focus, over almost any angle, and in studio work for enlarging the same sized hole can be advantageously used from 5-in. focus up to 30-in. focus.

TEXTE ET TABLES de la Collection des DIATOMEES DU MONDE ENTIER. By J. Tempere and H. Peragallo. 8vo, pp. 304+62. Price 18 francs.

Mons. Tempere appears to have given special attention to the various diatomaceous deposits, the first portion of this work being taken up with an enumeration of genera and species found in each deposit. On the last 50 pages is given a classified alphabetical list of genera and species, and where found. The diatomist will find this a valuable work of reference. We are sorry to notice there are no illustrations

SCIENCE PROGRESS : A Monthly Review of Current Scientific Investigation. (London : The Scientific Press.) Price 2/6 monthly. Subscription price, 25/- per annum post free.

The third volume of Science Progress, conducted by Henry C. Burdett, edited by J. Bretland Farmer, M.A., was completed with the August part. The volume contains articles by well-known scientists on the following subjects:—*Pathology:* Antitoxin ; Mountain Sickness ; The Antitoxins of Diphtheria ; and Pathological Results of the Royal Commission on Tuberculosis. *Geology, Mineralogy, and Palæontology:* Foreign Work among the Older Rocks ; Recent Contributions to the Geology of the Western Alps ; Methods of Petrographical Research ; Pithecanthropus Erectus ; Geology of the Sahara ; A Type of Palæolithic Plants ; and Views on Mineral Species. *Botany :* Insular Floras (Parts 4 & 5) ; Reserve Material of Plants ; Metamorphism in Plants ; What is a Tendency ? ; Notes on the Reproductive Organs of Olive-brown Seaweeds ; and New Aspects of an old Agricultural Question. *Physiology:* Peptone ; Coagulation of Blood ; Two Fundamental "Laws" of Nerve Action in relation to the Modern Nerve-Cell; and Formation of Lymph. *Animal Morphology:* Budding in Tunicata. *Anthropology :* Spanish Anthropology. *Chemistry and Physics :* The Arrangement of Atoms in a Crystal ; Light and Electrification ; Progress in Physical Chemistry during 1894 ; Ratio of the Specific Heats of Gases ; Recent Values of the Magnetic Elements at the Principal Magnetic Observatories of the World ; Chemical Affinity ; and Space Relation of Atoms. *Cytology :* On the Protoplastid Body and the Metaplastid Cell ; Notices of Books and Titles of Chemical Papers.

THE BURIED CITIES OF VESUVIUS: Herculaneum and Pompeii. By John Fletcher, M.D., D.Sc., &c. 8vo, pp. viii.—115. (London: Hazell, Watson, and Viney. 1895.) Price 3/6.

In this interesting work, illustrated with six plates, the author gives an account of Vesuvius and its Eruptions ; Rediscovery of Pompeii and Herculaneum ; Handicraft, Literature, and Art ; Social Life of the Pompeians ; Marriage, Divorce, Death, and Burial, etc. The book is nicely got up and photoprocess plates good.

FRIENDS IN FABLE : A Book of Animal Stories. By various Authors.

STORIES IN A SHELL. By various Authors. Cr. 4to, pp. 80 and 64. (London : Raphael Tuck and Sons. 1895.) Prices 3/6 and 2/6.

Two handsome and beautifully illustrated books for young people ; the stories are sure to delight the little folks, and the coloured plates are, we think, the best of the kind we have seen.

NOTABLE ANSWERS TO ONE THOUSAND QUESTIONS. Cr. 8vo, pp. xvi.—463. (London: George Newnes. 1895.) Price 2/6.

These are a reprint of the Sixth Thousand Questions in the Inquiry Columns of *Tit-Bits*, with the Replies. These questions appear to embrace almost every conceivable subject, and their answers afford a large amount of valuable information. ———

THE CHESS OPENINGS. By J. Gunsberg. Post 8vo, pp. xv.—104. (London: Geo. Bell and Son. 1895.) Price 1/-

Those who are not proficient at this most interesting game will derive much valuable information from the study of Gunsberg's openings. Here he will gain an idea of the best methods of opening a game, besides learning the methods and tactics of the best players. ———

CHESS PROBLEMS. By Philip H. Williams. 16mo, pp. viii.—58. (New Barnet: W. W. Morgan. London: Simpkin, Marshall, & Co. 1895.) Price 1/-

This little book contains 56 and two special problems, making 58 in all. Many have appeared in various newspapers, but all have been composed within the last four years. ———

THE BOOK OF THE FAIR. Parts XI. and XII. (Chicago: The Bancroft Publishing Co.) Price $1 each.

In these parts is concluded Chapter 14, which gives a full description of the Electricity section; Chapter 15 describes the Horticulture and Forestry section; and Chapter 16 is commenced, showing exhibits of Mines, Mining, Metallurgy. Too much cannot be said in praise of the very excellent illustrations in these and earlier parts; those relating to Horticulture and Forestry are especially fine. ———

· JARROLD'S GUIDE TO GREAT YARMOUTH and Neighbourhood.

JARROLD'S GUIDE TO CROMER and Neighbourhood. Cr. 8vo, pp. 116 and 133. (London: Jarrold & Son. 1895.) Price 6d. each.

Two handy and nicely illustrated little books which will be found useful to the tourist. ———

The Great Eastern Railway Company's TOURIST GUIDE to the Continent. Edited by Percy Lindley. Cr. 8vo, pp. 158. (London: 30 Fleet Street. 1895.) Price 1/-

This will be found a handy book for the continental tourist. Among its fresh features are a series of Continental Maps; a special chapter on Holland and its Exhibition, and Excursion round Amsterdam; and a chapter " Dull Useful Information," giving particulars as to the cost of Continental travel.

Index Vol. V.

Reviews.

C. SEERS & SON, PRINTERS, ARGYLE STREET, BATH.

Lightning Source UK Ltd.
Milton Keynes UK
UKHW012204051218
333536UK00014B/1368/P